PHOTODISINTEGRATION OF NUCLEI IN THE GIANT RESONANCE REGION

FOTORASSHCHPLENIE YADRA V OBLASTI GIGANTSKOGO REZONANSA

ФОТОРАСЩЕПЛЕНИЕ ЯДРА В ОБЛАСТИ ГИГАНТСКОГО РЕЗОНАНСА

The Lebedev Physics Institute Series

Editor: Academican D. V. Skobel'tsyn
Director, P. N. Lebedev Physics Institute, Academy of Sciences of the USSR

Volume 25	Optical Methods of Investigating Solid Bodies
Volume 26	Cosmic Rays
Volume 27	Research in Molecular Spectroscopy
Volume 28	Radio Telescopes
Volume 29	Quantum Field Theory and Hydrodynamics
Volume 30	Physical Optics
Volume 33	Studies of Nuclear Reactions
Volume 34	Photomesic and Photonuclear Processes
Volume 36	Photodisintegration of Nuclei in the Giant Resonance Region

In preparation:

Volume 31	Quantum Radiophysics
Volume 32	Plasma Physics
Volume 35	Electronic and Vibrational Spectra of Molecules
Volume 37	Electrical and Optical Properties of Semiconductors

Proceedings (Trudy) of the P. N. Lebedev Physics Institute

Volume 36

Photodisintegration of Nuclei in the Giant Resonance Region

Edited by
Academician D.V. Skobel'tsyn
Director, P.N. Lebedev Physics Institute
Academy of Sciences of the USSR, Moscow

Translated from Russian

Springer Science+Business Media, LLC 1967

ISBN 978-1-4899-2716-3 ISBN 978-1-4899-2714-9 (eBook)
DOI 10.1007/978-1-4899-2714-9

The Russian text was published by Nauka Press in Moscow in 1966 for the Academy of Sciences of the USSR as Volume XXXVI of the Proceedings (Trudy) of the P. N. Lebedev Physics Institute.

Фоторасщепление ядер в области гигантского резонанса
Труды Физического института им. П. Н. Лебедева
Том XXXVI

Library of Congress Catalog Card Number 67-27905

CONTENTS

SOME PROBLEMS IN MANY-BODY FIELD THEORY

G. M. Vagradov

In recent years, the methods of field theory have been widely used for the quantum-mechanical description of particle systems. In a number of cases, successful results not based on ordinary perturbation theory have been obtained with the help of these methods. This is particuarly important in the case of strongly interacting particles. We consider a system of fermions having that kind of interaction at zero temperature; it is assumed that the forces acting between particles are paired. The use of field-theoretical methods in this problem makes it possible to establish general relationships between various characteristic functions (vertex parts, effective potentials, etc.). In particular, a study is made of the problem of the ground-state stability of a system of fermions between which there exists an interaction potential containing two parts: a short-range repulsive portion, and a long-range attractive portion. In doing this, one should first keep in mind the use of results obtained in nuclear theory where the empirical interaction potential between free nucleons can be represented fairly accurately by the sum of two such parts.

1. We consider an infinite system of identical fermions having a paired interaction of the following simple form:

$$v(x_1, x_2) = v(|\mathbf{r}_1 - \mathbf{r}_2|), \tag{1}$$

where the x_i indicates the totality of space ($\mathbf{r}_i$) and spin coordinates of the particles.

We introduce a time-dependent effective potential V, defining it by the relation

$$V(1,2) = v(1,2) + iW(1)W(2) - i\int d1'\,d2'\,v(1,1')\,v(2,2')\,\langle|\,T(\rho(1')\,\rho(2'))\,|\rangle, \tag{2}$$

where the numbers indicate four-dimensional coordinates (for example, $1 = (x_1, t_1)$); the integral sign indicates integration over space and time, and summation over spins; $v(1, 2) = \delta(t_1 - t_2)v(|\mathbf{r}_1 - \mathbf{r}_2|)$; $|>$ is the ground-state vector for a system of particles in the second quantization representation ($T(\ldots)$ signifies t-product operators); $W(1) = -i\int d1' v(1, 1')G(1', 1'^{+})$, $G(1, 2) = -i < |\,T(\psi(1) \times \overline{\psi}(2))\,| >$ is the one-particle Green's function ($\overline{\psi}$ and ψ are fermion creation and destruction operators in the Heisenberg representation), where $G(1, 1^{+}) = G(x_1, t_1; x_1, t_1 + \varepsilon)_{\varepsilon \to +0}$; $\rho(1) = \overline{\psi}(1)\psi(1)$ is the particle density operator.

From the definition of the two-particle Green's function $G(1, 2, 3, 4) = <|\,T(\psi(1)\psi(2)\overline{\psi}(3)\overline{\psi}(4))| >$, there follows the relation:

$$\langle|\,T(\rho(1)\rho(2))\,|\rangle = -G(1, 2, 1^{+}, 2^{+}). \tag{3}$$

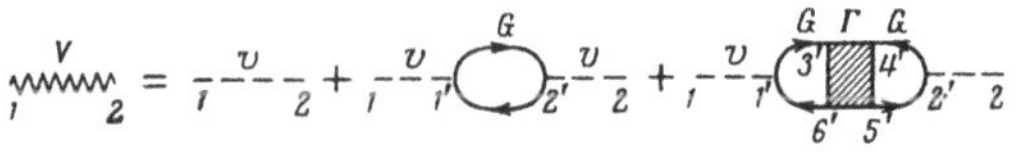

Fig. 1

Representing the function G(1, 2, 3, 4) in the form [1]

$$G(1, 2, 3, 4) = G(1, 3)\,G(2, 4) - G(1, 4)\,G(2, 3) + \\ + i\int d1' \ldots d4' G(1, 1')\,G(2, 2')\,\Gamma(1', 2', 3', 4')\,G(3' 3)\,G(4', 4),$$

we obtain from (2) and (3) a relation between the potential V and the vertex function Γ:

$$V(1, 2) = v(1, 2) - i\int d1' d2' v(1, 1')\,v(2, 2')\,G(1', 2')\,G(2', 1') - \\ - \int d1' \ldots d6' v(1, 1')\,v(2, 2')\,G(1', 3')\,G(2', 4')\,\Gamma(3', 4', 5', 6')\,G(5', 2')\,G(6', 1') \tag{4}$$

(this relation is shown diagrammatically in Fig. 1). In the momentum representation, (4) takes the form*:

$$V(q) = \nu_q + \nu_q^2 \Pi_0(q) - \frac{\nu_q^2}{(2\pi)^8}\int dp\,dp' G(p)\,G(p')\,\Gamma(p, p', q)\,G(p-q)\,G(p'+q),$$
$$q = (\mathbf{q}, \omega),\; p = (\mathbf{p}, \varepsilon),\; p' = (\mathbf{p}', \varepsilon'),\; \int dp = \int d\mathbf{p}\,d\varepsilon,\; \nu_q = \int d\mathbf{r}\,v(r)\,e^{-i\mathbf{q}\mathbf{r}}, \tag{4'}$$

where $q=(\mathbf{q}, \omega)$, $p=(\mathbf{p}, \varepsilon)$, $p'=(\mathbf{p}', \varepsilon')$, $\int dp = \int d\mathbf{p}\,d\varepsilon$, $\nu_q = \int d\mathbf{r}\,v(r)e^{-i\mathbf{q}\mathbf{r}}$, $\Pi_0(q)$ is the polarization operator of lowest order

$$\Pi_0(q) = -\frac{i}{(2\pi)^4}\int dp\,G(p)\,G(p-q). \tag{5}$$

The function Γ(p, p', q) is determined from the relation

$$(2\pi)^4\,\Gamma(p, p', q)\,\delta(p'+q-p'') = \Gamma(p, p', p-q, p'') = \int d1 \ldots d4\,\Gamma(1, 2, 3, 4)\,e^{-ip1-ip'2+i(p-q)3+ip''4}.$$

From the formal expansion of the function Γ in a series in the potential v (Fig. 2a), it is easy to obtain an infinite sum of diagrams which represent another expansion of Γ in the effective potential V (Fig. 2b). Furthermore, the new series will not contain interaction lines with internal loops of the type shown in Fig. 2c.

*For simplicity, we do not consider the spin variables from here on.

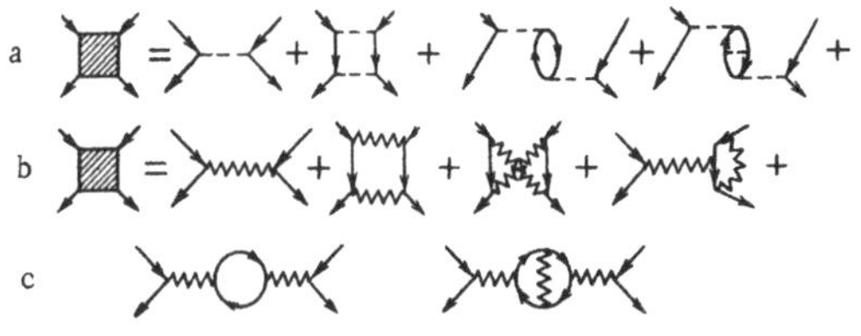

Fig. 2

Fig. 3

Fig. 4

We consider several general properties of the function V. To do this, we rewrite (1) in the following manner:

$$V(1, 2) = v(1, 2) + iW(1)W(2)$$
$$-i\sum_n \int d1'd2'v(1, 1')v(2, 2')$$
$$\{\theta(t_1' - t_2')\langle|\rho(1')|n\rangle\langle n|\rho(2')|\rangle + \theta(t_2' - t_1')\langle|\rho(2')|n\rangle\langle n|\rho(1')|\rangle\},$$

where $H|n\rangle = E_n|n\rangle$ ($n = 0$ corresponds to the ground state of the system, H is the total Hamiltonian). In the momentum representation, this equation is rewritten in the form

$$V(q) = \nu_q + \nu_q^2 \sum_n \frac{2\varepsilon_n |\langle|\rho_q|n\rangle|^2}{\omega^2 - \varepsilon_n^2 + i\alpha}, \tag{6}$$

where $\rho_q = \sum_p a_p^+ a_{p+q}$, $\varepsilon_n = E_n - E_0$. From that, one arrives at the dispersion relation:

$$V(q\omega) = \nu_q + \frac{1}{\pi}\int_0^\infty \frac{d\omega'^2 \operatorname{Im} V(q\omega')}{\omega'^2 - \omega^2 - i\alpha}. \tag{7}$$

From what has been said, one can conclude that the effective potential V has a meaning like that of an exact boson propagator in field theory. Continuing the analogy, one can also introduce the triangular vertex γ describing the interaction of a particle with the field:

$$\gamma(1,2,3) = \int \frac{dq\,dp\,dp'}{(2\pi)^{12}} \gamma(q, p, p') e^{-iq1+ip2-ip'3};$$
$$\gamma(q, p, p') = (2\pi)^4 \delta(q - p + p')\gamma(qp). \tag{8}$$

The function γ can be represented in the form of a formal series in v (Fig. 3a). However, one should keep in mind that this expansion must not contain diagrams which can be broken up into two parts joined by only a single interaction line, i.e., diagrams of the type shown in Fig. 3b are excluded. Just as in the case of the function Γ, the series for the vertex γ can be transformed into an infinite sum of diagrams in V (Fig. 3c).

It is easy to see that the functions γ, Γ, V, and G are related by the equation (Fig. 4):

$$V(q)\gamma(q, p) = \nu_q - \frac{i\nu_q}{(2\pi)^4}\int dp' G(p')\Gamma(p', p, q)G(p' - q). \tag{9}$$

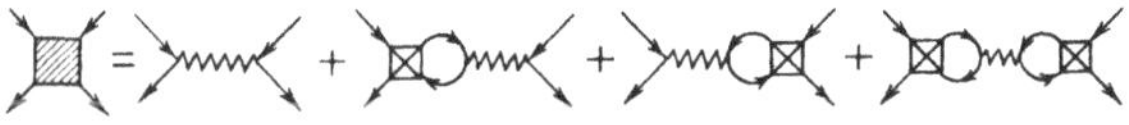

Fig. 5

Fig. 6

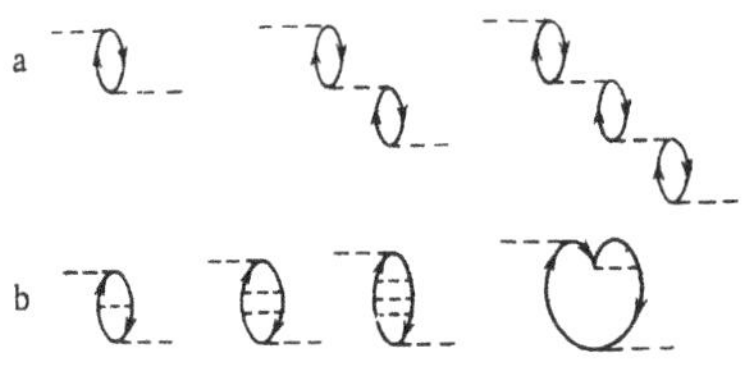

Fig. 7

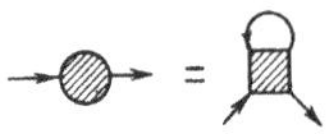

Fig. 8

Some general relationships can be derived from the following discussion. We introduce arbitrarily a certain "meson field" $\varphi(1)$ satisfying the equation

$$\hat{L}(1)\,\varphi(1) = \rho(1),$$

where the operator $\hat{L}(1)$ is determined by the equation

$$\hat{L}^{-1}(1)\,\delta(1,1') = v(1,1')$$

$$(\delta(1,1') = \delta(t_1 - t_{1'})\,\delta(x_1 - x_{1'})).$$

Hence

$$\varphi(1) = \int d1'\,v(1,1')\,\rho(1').$$

Now the effective potential $V(1,2)$ can be represented in the form

$$V(1,2) = v(1,2) + \langle\,|\,T(\varphi(1)\varphi^{+}(2))\,|\,\rangle.$$

We introduce the function

$$\beta(1,2,3) = \langle\,|\,T(\psi(2)\bar{\psi}(3)\varphi(1))\,|\,\rangle.$$

Considering the equation for $\varphi(1)$, we obtain for β

$$\beta(q,p,p') = -(2\pi)^4\,\nu_q\,\delta(q-p+p')\,G(p)\,G(p-q)\left\{1 - \frac{i}{(2\pi)^4}\int dp_1 G(p_1)\,\Gamma(p,p_1,q)\,G(p_1+q)\right\}.$$

From this equality, and from (9), there follows:

$$\beta(q,p,p') = -(2\pi)^4\delta(q-p+p')\,G(p)\,G(p-q)\,V(q)\,\gamma(qp). \tag{9'}$$

Considering that

$$i\partial_{t_1}\rho(1) = \left(\frac{\hat{p}_1^2 - \hat{p}_2^2}{2m}\,\bar{\psi}(1)\,\psi(2)\right)_{r\to 1} \quad (\hat{p}^2 = -\nabla^2),$$

and considering the derivative $i\partial_{t_1}\beta(1,2,3)$, we obtain from (9') the Ward difference identity:

$$\omega V(q)\,\gamma(q,p) = \nu_q\Big\{G^{-1}(p) - G^{-1}(p-q) + \omega_q^{(0)}(p-q) - $$
$$- \frac{i}{(2\pi)^4}\int dp'\,\omega_q^{(0)}(p')\,G(p')\,\Gamma(p,p',q)\,G(p'+q)\Big\}, \tag{9''}$$

where $\omega_q^{(0)}(\mathbf{p}) = [(\mathbf{p}+\mathbf{q})^2 - \mathbf{p}^2]/2m$. The Ward differential identity [1] follows if we pass to the limit $\omega/|\mathbf{q}| \to 0$ or $|\mathbf{q}|/\omega \to 0$ in (9").

2. Now we represent the vertex function Γ in the following form (Fig. 5):

$$\Gamma(1, 2, 3, 4) = \delta(1, 3)\delta(2, 4)V(1, 2) + f(1, 2, 3, 4) -$$

$$- i\delta(2, 4)\int d2'd4'd5' f(1, 2', 3, 4') G(5', 2') G(4', 5') V(5', 2) -$$

$$- i\delta(1, 3)\int d1'd3'd5' f(1', 2, 3', 4) G(5', 1') G(3', 5') V(5', 1) -$$

$$- \int d1' \ldots d6' f(1, 2', 3, 4') G(5', 2') G(4', 5') V(5', 6') G(6', 1') G(3', 6') | f(1', 2, 3', 4), \quad (10)$$

where the function f cannot be decomposed into parts connected by only a single line V. It is easy to see that that the vertex γ is connected with function f by the following relation (Fig. 6):

$$\gamma(q, p) = 1 - \frac{i}{(2\pi)^4}\int dp' G(p') f(p', p, q) G(p' - q).$$

Transforming (10) to the momentum representation, and considering expression (4'), we obtain

$$V(q) = \frac{v_q}{1 - v_q \Pi_0(q) - v_q R_q}; \quad (11)$$

$$R_q = -\int \frac{dp\,dp'}{(2\pi)^8} G(p) G(p') f(p, p', q) G(p - q) G(p' + q). \quad (12)$$

We now consider the case where the interaction radius r_0 of the potential v is considerably greater than the mean distance between particles, i.e., $k_F r_0 \gg 1$ (k_F is the limiting Fermi momentum). There will then exist diagrams with small momentum transfer ($|\mathbf{q}|/k_F \ll 1$) (Fig. 7a), and one can show [2] that the effective potential in this case is written in the form*:

$$V(q) = \frac{v_q}{1 - v_q \Pi_0(q)}. \quad (13)$$

Indeed, when $k_F r_0 \gg 1$, the function f can be neglected because it contains only unimportant diagrams (Fig. 7b). As is well known, the potential (13), as a function of ω, can have poles corresponding to plasma oscillations in the case of a repulsive interaction ($\nu_q > 0$).

In the case of attraction ($\nu_q < 0$), imaginary poles in ω arise in (13) because of which it is impossible to consider the system stable. From the relation (Fig. 8)

$$M(p) = -i\int \frac{dp'}{(2\pi)^4} G(p') \Gamma(p, p', 0) e^{i\varepsilon'\alpha}_{\alpha \to +0},$$

we find that the mass operator is simply a constant independent of p when $k_F r_0 \gg 1$ because then $f(p, p', q) \simeq 0$, therefore, $\Gamma(p, p', q) \simeq V(q)$. This means that consideration of the leading terms in the series does not lead to rearrangement (change in occupation numbers and the appearance of gaps in the energy spectrum). Apparently, an imaginary pole in (13) is associated with the fact that a system of particles with only attractive interaction must collapse to a point.

---

*We assume here that all the supplementary conditions, discussed in detail by D. A. Kirzhnits [2], are fulfilled. These conditions are the following:

$$\left|\frac{v_q}{\varepsilon_F}(k_F r_0)^3\right|_{q \to 0} \gg 1, \qquad \left|(k_F r_0)\frac{v_q}{\varepsilon_F}\right|_{q \to 0} < 1.$$

The situation can be changed in cases where the potential v contains in addition to the attractive part already mentioned, a repulsive part with an interaction radius r_c less than the average spacing between particles ($r_c k_F \ll 1$). Furthermore, it is impossible to neglect the R_q term in expression (11). For an evaluation, we make use of the following: the product $G(p)G(p+q)$, for small q, can be written in the form [1]

$$G(p)G(p+q) \simeq 2\pi i \frac{qp/k_F}{\omega - qp/m} \delta(\varepsilon)\delta(|p| - k_F) + \varphi(p)$$

[$\varphi(p)$ is the regular part of the product]. Considering only the polar terms of this expression, we obtain the approximation

$$R_q \cong \frac{k_F^4}{(2\pi)^6} \int dn dn' f(n, n') \frac{qn \cdot qn'}{\left(\omega - \frac{qnk_F}{m}\right)\left(\omega - \frac{qn'k_F}{m}\right)}, \qquad (14)$$

where we set

$$f(n, n') = f(p, p', q)\Big|_{\substack{\varepsilon, \varepsilon' \to 0 \\ |p|, |p'| \to k_F \\ q \to 0}}. \qquad (14')$$

We now show that the necessary condition for the absence of imaginary poles in (11) for all values $|\mathbf{q}|$ can be written

$$|v_q(\Pi_0(q) + R_q)|_{\omega \to 0} < 1. \qquad (15)$$

To do this, we first consider a polarization operator $\Pi_0(q)$, which has the following form:

$$\Pi_0(q) = \frac{1}{(2\pi)^3} \int\limits_{\substack{|p+q| > k_F \\ |p| < k_F}} \frac{dp 2\omega_q^{(0)}(p)}{\omega^2 - (\omega_q^{(0)}(p))^r + i\alpha}.$$

If we limit ourselves to only imaginary values of ω, $|\Pi_0(q)|$ attains its maximum value when $\omega = 0$. Consequently, if inequality (15) is fulfilled when $\omega = 0$, the left side of (15) will be less than unity for all imaginary values of ω, and there will be no imaginary poles in (11). Substituting the values

$$\Pi_0(q, 0) = -\frac{k_F m}{2\pi^2}, \quad R_{q,0} = \left(\frac{k_F m}{2\pi^2}\right)^2 f_0^k,$$

$$f_0^k = \frac{1}{(4\pi)^2} \int dn\, dn' f^k(n, n')$$

in equality (15), we obtain

$$\left|\frac{v_q m k_F}{2\pi^2}\left(1 - \frac{mk_F}{2\pi^2} f_0^k\right)\right| < 1$$

or

$$\frac{mk_F}{2\pi^2} f_0^k > 1 - \frac{2\pi^2}{mk_F |v_q|}. \qquad (16)$$

Thus, imaginary poles will be absent if condition (16) is fulfilled. It is easy to see that $f^k(\mathbf{n}, \mathbf{n}')$ agrees with the amplitude in Fermi-fluid theory [1] in the case of a short-range potential.

Finally, we turn briefly to the following. An approximate method for describing the properties of nuclear matter was proposed [3] which used an empirical inter-nucleon interaction potential having an infinite, repulsive core at small distances and a long-range attractive portion. It is easy to show that an imaginary pole in (11) will not be present for the numerical values of the parameters used in that paper. It need only be pointed out that it is necessary to take account of terms in higher orders of the parameter $r_c k_F$ when computing f (terms up to the third order in $r_c k_F$ were considered in [3]). This is because the quantity f is proportional to the second derivative of the total energy with respect to particle density, leading to a more important role for the terms in higher order of $r_c k_F$. However, this situation has little effect on the results in the paper mentioned.

In conclusion, the author expresses his deep gratitude to D. A. Kirzhnits for valuable advice and for fruitful discussions.

LITERATURE CITED

1. A. A. Abrikosov, L. P. Gor'kov, and I. E. Dzyaloshinskii, Quantum Field Theoretical Methods in Statistical Physics. Fizmatgiz (1962).
2. D. A. Kirzhnits, Field-Theoretical Methods in Many-Body Theory. Atomizdat (1963).
3. G. M. Vagradov and D. A. Kirzhnits, Zh. Eks. i Theor. Fiz., 38:1499 (1960); 43:1301 (1962).

EXCITED STATES IN F^{19}

D. A. Zaikin

INTRODUCTION

Experimental data on the low-lying levels in F^{19} and on the probabilities for electromagnetic transitions between them [1-3] indicate the existence of a strong asphericity in this nucleus. From the point of view of the collective model, which treats an odd nucleus as an aspherical, even core in whose field an external nucleon moves, the low-lying levels in F^{19} can be described as two rotational bands [4, 5] with $K = 1/2$ located in two states of neighboring energies for the odd portion in the field of a deformed core. The projection of the proton momentum on the symmetry axis is $1/2$ for these states and the parities are different. In the Nilsson scheme [6], levels 4 and 6 are such states. The experimental values for the decoupling parameters of the rotational bands in F^{19} with positive and negative parity are 3.2 and 1.1, respectively.

It should be pointed out that calculations of the low-lying excited states in F^{19} carried out in terms of the shell model and using wave functions in the SU_3 representation (for example, see [7-9]) lead to results which are essentially similar to those given by the collective model (rotational spectrum levels, structure of wave functions, significant quadrupole asphericity, etc.). These calculations satisfactorily describe the experimental data for the location of levels of E2 transitions and for their probabilities. However, none of the existing theoretical descriptions has managed to explain the large value for the probability of the E3 transition between the ground state ($1/2^+$) and the $5/2^-$ state at 1.35 MeV. The experimentally derived probability for this transition [2] is 12 ± 4 single-particle (Weisskopf) units, while the theoretical estimates based on the collective model with Nilsson wave functions, and on shell-model computations, give an identical result for this quantity of the order of one Weisskopf unit.

The existence of such a strong E3 transition leads to assumptions about the existence of a large octupole deformation (static or dynamic) in the F^{19} nucleus. However, the simplest assumption about the existence of static octupole deformation leads to a contradiction with experimental data for the location of the levels (see below), namely: the decoupling parameters for the rotational bands, in this case, are found to be equal in absolute value and opposite in sign.* On the other hand, a computation of the decoupling parameters using Nilsson wave functions without considering octupole deformation gives results that are qualitatively correct although somewhat low in comparison with experimental results. This makes it clear that the magnitude of the decoupling factors is extremely sensitive to the nature of the coupling of the motion of the odd nucleon with the octupole deformation (or octupole oscillations).

Therefore there is interest in an attempt to describe in a consistent manner the experimental data both for the location of the levels in F^{19} and for the E3 transition, using the collective model and without being confined to the simplest case – static octupole deformation.

We turn now to a formulation of the problem where one takes into account the coupling between an octupole deformation of the nuclear surface and the motion of an odd nucleon located in an ellipsoidal axial field.

*This result was pointed out in [10].

1. FORMULATION OF THE PROBLEM

The collective model Hamiltonian for an odd nucleus has the form [11, 12]

$$H = \sum_{i=1}^{3} \frac{(I_i - j_i)^2}{2F_i} + \mathcal{H},$$

where I_i and j_i are the projection operators onto the principal axes of the nucleus for the total momentum of the nucleus and of the odd nucleon; F_i are the nuclear moments of inertia with respect to these axes; and $\mathcal{H}$ is the Hamiltonian for the internal motion. In the case of interest to us, $K \equiv I_3 = j_3 = ^1/_2$ so that the wave functions of the rotational states take the form

$$\Psi^{I\pm}_{M^1/_2\alpha} = \sqrt{\frac{2I+1}{16\pi^2}} \{D^I_{M^1/_2}\chi^{\pm}_{^1/_2\alpha} + (-)^{I+^1/_2} D^I_{M-^1/_2\alpha}\chi^{\pm}_{-^1/_2\bar{\alpha}}\}, \tag{1}$$

where D^I_{MK} is the Wigner D-function; χ are the wave functions for the internal motion described by the Hamiltonian $\mathcal{H}$; the subscript α indicates the set of all quantum numbers of the internal state with the exception of j_3; generally speaking, the state α does not coincide with the state α, and is obtained from the latter by a rotational transformation of π around axis 2:

$$\chi^{\pm}_{-^1/_2\bar{\alpha}} = R_2(\pi)\,\chi^{\pm}_{^1/_2\alpha}.$$

The internal Hamiltonian $\mathcal{H}$ in our case can be represented in the form

$$\mathcal{H} = H_n + H_0 + H_I, \tag{2}$$

where H_n is the Hamiltonian for the odd nucleon in an axial field of the Nilsson type; H_0 is the Hamiltonian for the collective octupole oscillations; and, finally, H_I is an operator for the interaction of the odd nucleon with an octupole deformation:

$$H_I = -k(r) \sum_{\mu=-3}^{3} \alpha_{3\mu} Y_{3\mu}(\Omega). \tag{3}$$

Here, $\mathbf{r} = (r, \Omega)$ is the position vector of the nucleon; and $\alpha_{3\mu}$ are. the collective variables which describe the octupole deformation of the nucleus. In the case of a Nilsson potential, $k(r) = \hbar\omega_0 r^2$, where ω_0 is the frequency of the Nilsson oscillator.

Our problem is to determine, or, at least to study, the properties of the internal wave functions χ in connection with the possible occurrence of an octupole deformation. As indicated in the introduction, the values of the decoupling parameters for the rotational bands are very sensitive to the nature of this deformation. For that reason, it is expedient to investigate the effect of octupole deformation on the decoupling parameters first.

2. EFFECT OF OCTUPOLE DEFORMATION OF DECOUPLING PARAMETERS

We discuss qualitatively what consideration of the interaction between an external particle and octupole degrees of freedom in the nuclear surface might lead to. For simplicity, we limit ourselves to an axial deformation $\alpha_{30} \equiv \eta$, setting all the remaining $\alpha_{3\mu}$ to zero. Then H_0 and H_I, which appear in the total internal Hamiltonian $\mathcal{H}$, take the form:

$$H_0 = -\frac{\hbar^2}{2B}\frac{\partial^2}{\partial\eta^2} + \frac{1}{2}C\eta^2; \tag{4}$$

$$H_I = -k(r)\,\eta Y_{30}(\Omega), \tag{5}$$

where B and C are the mass parameter and the rigidity parameter for octupole surface oscillations, respectively.

Following the usual adiabatic procedure, we find single-particle wave functions including interaction with octupole oscillations. Furthermore, we limit ourselves to the two lowest single-particle states, in which the projection of the total momentum of the nucleon on the symmetry axis is ½ and the parities are opposite.* Since, as was indicated above, these states are close in energy, we can consider them as approximately degenerate for the purposes of the present discussion. We indicate the wave functions for these states by $\psi^+_{1/2}$ and $\psi^-_{1/2}$. Then the problem of finding new single-particle functions reduces to the solution of a simple secular equation

$$\begin{vmatrix} -\varepsilon & -p\eta \\ -p\eta & -\varepsilon \end{vmatrix} = 0,$$

where $p = \langle \psi^+_{1/2} | k(r) Y_{30}(\Omega) | \psi^-_{1/2} \rangle$. The new single-particle functions and the corresponding energies are

$$\psi_1 = \frac{1}{\sqrt{2}}(\psi^+_{1/2} + \psi^-_{1/2}), \quad \varepsilon_1 = -p\eta;$$

$$\psi_2 = \frac{1}{\sqrt{2}}(\psi^+_{1/2} - \psi^-_{1/2}), \quad \varepsilon_2 = p\eta.$$

As a result, the potential energy of the octupole oscillations acquires the form $\frac{1}{2}C(\eta \pm \eta_0)^2$, where $\eta_0 = p/C$. The corresponding wave functions will be ordinary oscillator functions which describe oscillations around the equilibrium values $\pm\eta_0$ and which transform into one another by transformations of inversion and rotation by π around any axis perpendicular to the symmetry axis. We designate these functions as $\varphi(\eta - \eta_0) \equiv \varphi_\uparrow$ and $\varphi(\eta + \eta_0) \equiv \varphi_\downarrow$.

Thus the eigenfunctions of the operator $\mathcal{H}$ in the adiabatic approximation are the products

$$f_1 = \frac{1}{\sqrt{2}}(\psi^+_{1/2} + \psi^-_{1/2})\,\varphi(\eta - \eta_0) \equiv \psi_1\varphi_\uparrow;$$

$$f_2 = \frac{1}{\sqrt{2}}(\psi^+_{1/2} - \psi^-_{1/2})\,\varphi(\eta + \eta_0) \equiv \psi_2\varphi_\downarrow.$$

As is easily seen, the functions f_1 and f_2 belong to the same energy value. Note also that these functions do not have a definite parity and transform into one another by inversion because the pairs ψ_1, ψ_2 and $\varphi_\downarrow$, $\varphi_\uparrow$ possess that property. Therefore, one ought to choose as wave functions having definite parity the combinations

$$\chi^+_{1/2} = \frac{1}{\sqrt{2}}(\psi_1\varphi_\uparrow + \psi_2\varphi_\downarrow);$$

$$\chi^-_{1/2} = \frac{1}{\sqrt{2}}(\psi_1\psi_\uparrow - \psi_2\varphi_\downarrow). \tag{6}$$

---

*These states are degenerate with respect to the sign of the projection of nucleon angular momentum K on axis 3. Generally speaking, therefore, states with $K = -\frac{1}{2}$ should also be taken into consideration in the discussion. Since the operator H_I was chosen in the form (5), it couples only states having the same sign for K. Therefore, we can confine ourselves to states with only positive K in the present discussion, and obtain the corresponding results for negative K on the basis of the symmetry properties of the problem.

To obtain expressions for the internal wave functions which describe states with the negative value for the projection of the particle momentum on the symmetry axis, it is necessary to rotate the functions (6) around axis 2 by an angle π:

$$\chi^{+}_{-1/2} = R_2(\pi)\,\chi^{+}_{1/2} = \frac{1}{\sqrt{2}}(\bar{\psi}_1\varphi_{\downarrow} + \bar{\psi}_2\varphi_{\uparrow});$$

$$\chi^{-}_{-1/2} = R_2(\pi)\,\chi^{-}_{1/2} = \frac{1}{\sqrt{2}}(\bar{\psi}_1\varphi_{\downarrow} - \bar{\psi}_2\varphi_{\uparrow}), \tag{7}$$

where

$$\bar{\psi}_1 = \frac{1}{\sqrt{2}}(\psi^{+}_{-1/2} + \psi^{-}_{-1/2});$$

$$\bar{\psi}_2 = \frac{1}{\sqrt{2}}(\psi^{+}_{-1/2} - \psi^{-}_{-1/2}).$$

Thus the total wave function for the nucleus has the form (1) where (6) and (7) are taken as the internal wave functions χ.

Now we compute the decoupling parameters for both rotational bands

$$a^{\pm} = -\langle \chi^{\pm}_{1/2} | j_+ | \chi^{\pm}_{-1/2} \rangle.$$

It is easy to show that

$$a^{+} = \frac{1}{2}[(a_0^{+} + a_0^{-})\,w + (a_0^{+} - a_0^{-})];$$

$$a^{-} = \frac{1}{2}[(a_0^{+} + a_0^{-})\,w - (a_0^{+} - a_0^{-})], \tag{8}$$

where $w = (\varphi_{\uparrow}, \varphi_{\downarrow})$ is the covering integral for vibrational functions which correspond to mirror values of the effective equilibrium octupole deformation ($\pm\,\eta_0$), and $a_0^{\pm} = -\langle \psi^{\pm}_{1/2} | j_+ | \psi^{\pm}_{-1/2} \rangle$ are the decoupling parameters with no octupole oscillations. When $w = 1$, which corresponds to $\eta_0 = 0$, i.e., to octupole oscillations around an equilibrium shape having only quadrupole deformation, one obtains the natural result: $a^{\pm} = a_0^{\pm}$. As w decreases (which corresponds to an increase in η_0), the decoupling parameters decrease. Further, their difference remains constant at $a_0^{+} - a_0^{-}$. In the limiting case $w = 0$, which would correspond to the existence of a clearly expressed octupole deformation, we obtain the result mentioned in the introduction

$$a^{\pm} = \pm\frac{1}{2}(a_0^{+} - a_0^{-}),$$

which disagrees with experimental data. Considering the dependence of $a^{\pm}$ on w, and the fact that estimates of $a_0^{\pm}$ by means of Nilsson wave functions give results correct in order of magnitude but always low when compared with experimental data, it seems reasonable to consider octupole oscillations of F^{19} around an equilibrium shape having only quadrupole deformation ($\eta_0 = 0$). Such consideration can be carried out simply by means of perturbation theory.

3. CALCULATION OF INTERNAL WAVE FUNCTIONS BY PERTURBATION THEORY

We now find the characteristic wave functions for the Hamiltonian (2) considering H_I as a perturbation. As before, we shall consider that only the two single-particle states $K = 1/2^{+}$ and $K = 1/2^{-}$, which correspond to levels 6 and 4, play an essential role in the interaction of single-particle and octupole degrees of freedom. Each of these states is doubly degenerate with respect to the sign of the momentum projection of the symmetry axis. Then, in first approximation perturbation theory, only octupole degrees of freedom corresponding to

$\mu = 0, \pm 1$ will be excited as the result of interaction between nucleon and surface, i.e., only three terms($\mu = 0, \pm 1$) will play a part in the operator H_I. In solving the problem, it is convenient to transform from the variables α_{30}, α_{31}, and α_{3-1} to the variables η, ξ, and ψ which are defined by the relations

$$\alpha_{30} = \eta, \quad \alpha_{31} = \xi \exp(-i\psi), \quad \alpha_{3-1} = -\xi \exp(i\psi),$$

where

$$-\infty < \eta < \infty, \quad 0 \leqslant \xi < \infty, \quad 0 \leqslant \psi < 2\pi.$$

The Hamiltonians for the octupole oscillations and the interaction operator in the variables η, ξ, and ψ take the forms

$$H_0 = -\frac{\hbar^2}{2B}\frac{\partial^2}{\partial\eta^2} + \frac{1}{2}C\eta^2 - \frac{\hbar^2}{4B\xi}\frac{\partial}{\partial\xi}\xi\frac{\partial}{\partial\xi} - \frac{\hbar^2}{4B\xi^2}\frac{\partial^2}{\partial\psi^2} + C\xi^2; \tag{9}$$

$$H_I = -\hbar\omega_0 r^2\{\eta Y_{30}(\Omega) + \xi e^{-i\psi}Y_{31}(\Omega) - \xi e^{i\psi}Y_{3-1}(\Omega)\}. \tag{10}$$

If H_I is considered to be a perturbation, the unperturbed wave functions and the corresponding energies can be represented in the form

$$|K^\pi n_0 n_1 m\rangle = \psi_K^\pi(\boldsymbol{r})\,\varphi_{n_0}(\eta)\,f_{n_1 m}(\xi)\,F_m(\psi);$$

$$E^\pi_{Kn_0n_1m} = \varepsilon_K^\pi + \hbar\omega(n_0 + 2n_1 + |m| + {}^3/_2);$$

$$\varphi_{n_0}(\eta) = \left(\frac{BC}{\hbar^2}\right)^{1/8}(2^{n_0}n_0!\sqrt{\pi})^{-1/2}\exp\left(-\sqrt{\frac{BC}{4\hbar^2}}\eta^2\right)H_{n_0}\left(\sqrt[4]{\frac{BC}{\hbar^2}}\eta\right);$$

$$f_{nm}(\xi) = 2\sqrt[4]{\frac{BC}{\hbar^2}}\frac{\sqrt{(n_1+|m|)!}}{|m|!\sqrt{n_1!}}\exp\left(-\sqrt{\frac{BC}{\hbar^2}}\xi^2\right)\Phi\left(-n_1, |m|+1.2\sqrt{\frac{BC}{\hbar^2}}\xi^2\right);$$

$$F_m(\psi) = \frac{1}{\sqrt{2\pi}}\exp(im\psi). \tag{11}$$

Here, $\psi_K^\pi(\mathbf{r})$ and ε_K^π are the single-particle Nilsson functions for the states under consideration and the corresponding energies; $H_{n_0}(x)$ and $\Phi(-n_1, |m|+1, z)$ are Hermite polynomials and confluent hypergeometric polynomials, respectively; ω is the frequency of the octupole oscillations, equal to $\sqrt{C/B}$. The quantum numbers n_0, n_1, and m take on the following values: $n_0 = 0, 1, 2, \ldots$; $n_1 = 0, 1, 2, \ldots$; $m = 0, \pm 1, \pm 2, \ldots$

Using (11), it is not difficult to calculate the matrix elements for the interaction operator (10) and to obtain expressions for the internal wave functions in first approximation perturbation theory. The internal wave functions for the lowest states of interest to us take the following form in this approximation.

$$\chi^+_{1/2} \equiv \chi^+_{1/2\,000} = |{}^1/_2{}^+000\rangle + \frac{\hbar\omega_0 p_0\alpha}{\hbar\omega+\Delta}|{}^1/_2{}^-100\rangle - \frac{\hbar\omega_0 p_1\alpha}{\hbar\omega+\Delta}|-{}^1/_2{}^-001\rangle;$$

$$\chi^+_{-1/2} \equiv \chi^+_{-1/2\,000} = |-{}^1/_2{}^+000\rangle - \frac{\hbar\omega_0 p_0\alpha}{\hbar\omega+\Delta}|-{}^1/_2{}^-100\rangle - \frac{\hbar\omega_0 p_1\alpha}{\hbar\omega+\Delta}|{}^1/_2{}^-00-1\rangle;$$

$$\chi^-_{1/2} \equiv \chi^-_{1/2\,000} = |{}^1/_2{}^-000\rangle + \frac{\hbar\omega_0 p_0\alpha}{\hbar\omega-\Delta}|{}^1/_2{}^+100\rangle - \frac{\hbar\omega_0 p_1\alpha}{\hbar\omega-\Delta}|-{}^1/_2{}^+001\rangle;$$

$$\chi^-_{-1/2} \equiv \chi^-_{-1/2\,000} = |-{}^1/_2{}^-000\rangle - \frac{\hbar\omega_0 p_0\alpha}{\hbar\omega-\Delta}|-{}^1/_2{}^+100\rangle - \frac{\hbar\omega_0 p_1\alpha}{\hbar\omega-\Delta}|{}^1/_2{}^+00-1\rangle, \tag{12}$$

where

$$p_0 = \langle\psi^-_{1/2}|r^2 Y_{30}(\Omega)|\psi^+_{1/2}\rangle; \quad p_1 = \langle\psi^-_{1/2}|r^2 Y_{31}(\Omega)|\psi^+_{-1/2}\rangle;$$

$$\alpha^2 = \left(\frac{\hbar^2}{4BC}\right)^{1/2} = \overline{\eta^2} = \overline{\xi^2}; \quad \Delta = \varepsilon^-_{1/2} - \varepsilon^+_{1/2}.$$

The total wave functions for the rotational states has the form (1) where the functions in (12) are used for χ.

4. E3 TRANSITION PROBABILITY

Using the expressions obtained in the previous section for the wave functions, one can estimate the reduced E3 transition probability from the ground state of F^{19} to the $5/2^-$ state. Neglecting the single-particle part in the E3 transition operator in comparison with the collective part, we obtain

$$B\,(E3,\ {}^1/_2{}^+ \to {}^5/_2{}^-) = \frac{27}{28\pi^2} Z^2 e^2 R_0^6 \frac{(\hbar\omega_0)^2 (\hbar\omega)^2 \alpha^4}{[(\hbar\omega)^2 - \Delta^2]^2} \left(p_0 - \frac{2}{\sqrt{3}} p_1 \right)^2, \tag{13}$$

where R_0 is the nuclear radius. The quantities p_0 and p_1 are easily evaluated by using Nilsson wave functions:

$$p_0 = \frac{12\sqrt{2}}{5\pi\sqrt{7}} (\sqrt{3} a_{10} a_{20} - a_{11} a_{21});$$

$$p_1 = \frac{12\sqrt{2}}{5\pi\sqrt{7}} \left(\sqrt{2} a_{11} a_{20} - 2\sqrt{\frac{2}{3}}\, a_{10} a_{21} \right).$$

Here, a are the coefficients of the single-particle Nilsson functions, published in tabular form [6].

Neglecting Δ^2 in comparison with $(\hbar\omega)^2$ in (13) and transforming to Weisskopf units $B_{s.p}$, we have

$$B\,(E3,\ {}^1/_2{}^+ \to {}^5/_2{}^-) \approx 10 \left(\frac{\hbar\omega_0}{\hbar\omega} \right)^2 \alpha^4 N B_{s.p}, \tag{14}$$

where $N = [\sqrt{3} a_{10} a_{20} - a_{10} a_{21} - 2\sqrt{(2/3)}\, a_{11} a_{20} + 4(\sqrt{2/3})\, a_{10} a_{21}]^2$ is a quantity which varies from 6.1 to 3.7 when the equilibrium deformation parameter β_0 varies from 0.1 to 0.3. Therefore, it is easy to estimate the order of magnitude of the quantity $B(E3,\ 1/2^+ \to 5/2^-)$. It is reasonable to assume $\hbar\omega_0 \approx 15$ MeV for the F^{19} nucleus. Then, when $\hbar\omega = 4$ MeV and $\alpha^2 = 0.05$

$$B\,(E3,\ {}^1/_2{}^+ \to {}^5/_2{}^-) \approx (1.3 - 2.1)\, B_{s.p},$$

and when $\hbar\omega = 3$ MeV and $\alpha^2 = 0.1$

$$B\,(E3,\ {}^1/_2{}^+ \to {}^5/_2{}^-) \approx (9 - 15)\, B_{s.p}.$$

It is clear from the examples presented that it is necessary the amplitude of the zero-order oscillations be rather large ($\alpha \approx 0.3$), in order to explain the high probability of E3 transitions in the F^{19} nucleus through the interaction of the odd proton with octupole oscillations, i.e., that the nucleus be rather "soft" with respect to octupole oscillations.

CONCLUSION

In evaluating the results obtained in this paper, one should keep in mind that the application of the collective model to a light nucleus such as F^{19} carries a very approximate connotation if only because it makes little sense, strictly speaking, to talk of the surface of this nucleus and even less to talk of such details as octupole deformation. In addition, even in the confines of the collective model, which treats an odd nucleus as a combination of a deformed even core and an external nucleon, the representation of the wave functions of low-lying states in the form (1) is a crude approximation because such a representation is based on the assumption of the adiabaticity of rotation with respect to internal motion (in particular, with respect to motion of the nucleon in the field of the core). This assumption is hardly justified for odd nuclei. Therefore, the conclusion about the "softness" of the F^{19} nucleus with respect to octupole deformation is, of course, qualitative.

Nevertheless, there is definite interest in comparing the octupole rigidity parameter C for the F^{19} nucleus which must be chosen in order to obtain quantitative agreement between the results of this paper and experiment with the corresponding quantity for neighboring nuclei. The nucleus closest to F^{19} for which the characteristics of the octupole oscillations are well known is O^{16}. The reduced probability $B(E3,\ 0^+ \rightarrow 3^-)$ for the excitation of the 3^- level in O^{16}, which is at 6.13 MeV, is $0.31\ e^2 \cdot 10^{-74}\ cm^6$, or 29 $B_{s.p}$ [13]. If this level is considered as a single-phonon vibration, it is easy to determine that the rigidity parameter of the O^{16} nucleus with respect to octupole oscillations is about 19 MeV. On the other hand, if we replace ω and α, expressed by the parameters B and C, in (14), there results

$$B(E3,\ {}^1\!/_2{}^+ \rightarrow {}^5\!/_2{}^-) \approx 10 \left(\frac{\hbar\omega_0}{C}\right)^2 \frac{N}{4} B_{s.p}.$$

This clearly indicates that if $B(E3,\ {}^1\!/_2{}^+ \rightarrow {}^5\!/_2{}^-) \approx 10\ B_{s.p}$, the parameter C must be approximately equal to $\hbar\omega_0$, for F^{19}, i.e., 15 MeV. In other words, the rigidity of the F^{19} nucleus with respect to octupole oscillations is approximately the same as that for the neighboring magic nucleus O^{16}. This comparison demonstrates that the result for the "softness" of the F^{19} nucleus with respect to octupole oscillations is completely reasonable.

In conclusion, the author expresses his deep gratitude to Professor Aage Bohr for stimulating this investigation and for numerous discussions. The author also thanks V. I. Belyak for valuable discussions.

LITERATURE CITED

1. J. D. Prentice, N. W. Gebbie, and N. S. Caplan, Phys. Letters, 3:201 (1963).
2. A. E. Litherland, M. A. Clark, and C. Broude, Phys. Letters, 3:204 (1963).
3. D. Beder, Phys. Letters, 3:206 (1963).
4. E. B. Paul, Phil. Mag., 15:311 (1957).
5. G. Rakavy, Nucl. Phys., 4:375 (1957).
6. S. G. Nilsson, Mat. Fys. Medd. Dan. Vid. Selsk., Vol. 29, No. 16 (1955).
7. M. Harvey, Phys. Letters, 3:209 (1962).
8. M. Harvey, Preprint (1963).
9. R. M. Dreizler, Phys. Rev., 136B:231 (1964).
10. M. Harvy, AECL-1594 (1963).
11. A. Bohr, Mat. Fys. Medd. Dan Vid. Selsk., Vol. 26, No. 14 (1952).
12. A. Bohr and B. R. Mottelson, Mat. Fys. Medd. Dan. Vid. Selsk., Vol. 27, No. 16 (1953).
13. G. R. Satchler, R. H. Bassel, and R. M. Drisko, Phys. Letters, 5:256 (1963).

A STUDY OF THE STRUCTURE OF THE O^{16} AND C^{12} γ-RAY ABSORPTION CROSS SECTIONS IN THE GIANT DIPOLE RESONANCE REGION BY THE ABSORPTION METHOD*

B. S. Dolbilkin

Introduction

Photonuclear reactions have decided advantages for nuclear studies when compared with reactions involving nucleon interactions because the properties of the electromagnetic field are well known. In this connection, the study of the nuclear photoeffect in the giant dipole resonance region (10-30 MeV nuclear excitation region) is of considerable interest.

One of the most important quantities characterizing the interaction of photons with nuclei is the γ-ray absorption cross section. A theoretical description of the absorption cross section requires no additional assumptions about the dynamics involved in the decay of the excited nucleus. Therefore, it is simpler and more reliable than computations of individual photodisintegration processes. A comparison of experimental absorption cross sections with theoretical predictions offers, in principle, a possibility for checking ideas about the photon absorption mechanism. Detailed studies for the γ-ray absorption cross section may give valuable information about highly-excited states of the nucleus. From this point of view, the study of light nuclei is particularly promising. According to theoretical concepts, the number of levels excited by photons in the giant resonance region is relatively small in such nuclei (particularly in those with closed shells), and thus there is a practical possibility of resolving these levels, or closely-spaced groups of levels, and of studying their characteristics. The refinement of giant resonance parameters, particularly integrated γ-ray absorption cross sections, is also of great importance for light nuclei.

The detection of breaks (changes in slope) in the yield curves for the (γ, n) reaction in O^{16} and C^{12}, which, it was supposed, were associated with isolated resonances in the absorption cross section, intensified interest in data on the detailed energy dependence of the absorption cross section.

At the start of the studies which are described in this paper (1957), there was no quantitative information on the γ-ray absorption cross section in light nuclei. Data for the total nuclear cross section did not exist. The basic means for obtaining information about the absorption cross section was summation of the cross sections for individual reactions. The relative contributions of the latter depend on the mass number A and the nuclear excitation energy. For nuclei with $A > 100$, the total nuclear cross section can be replaced with satisfactory accuracy by the cross section for the production of photoneutrons, whose main component is the (γ, n) reaction. With decrease in the mass number, reactions involving the emission of protons gradually play an increasing part. In light nuclei ($A < 40$), the cross sections for the (γ, n) and (γ, p) reactions become com-

*Dissertation offered in satisfaction of the requirements for the academic degree of candidate in the physical and mathematical sciences. Defended June 14, 1965, at the P. N. Lebedev Institute of Physics, Academy of Sciences of the USSR (abridged version).

parable in magnitude. The (γ, p) cross section is determined mainly from a proton spectrum, and is subject to large error because, not knowing the decay scheme, one usually assumes that the resulting nucleus remains in the ground state after proton emission. Therefore, the data for total nuclear cross sections in light nuclei was considerably less reliable than that for heavy nuclei, and it was of a purely qualitative nature. Another indirect method for the determination of absorption cross sections was the study of elastic scattering, which was connected with them by dispersion relations. Significant information was not obtained by this method because of experimental difficulties.

This paper gives the results of a study of the γ-ray absorption cross sections in O^{16} and C^{12}. The selection of these nuclei was natural for the first stage of this work: in them were first observed indirect indications of a fine structure in the giant resonance (breaks), and peaks were found in proton spectra somewhat later. O^{16} is a doubly-magic nucleus with a closed 1p shell and the $1p_{3/2}$ subshell is filled in C^{12}; as a result, the level density in them should be minimal. From the point of view of the shell model, O^{16} and C^{12} are two of the simplest nuclei for detailed theoretical calculations of absorption cross sections. Consequently, such calculations have been performed, which undoubtedly has increased the value of experimental information about these nuclei.

CHAPTER I

First Data on the Structure of the Giant Resonance in Light Nuclei

1. Breaks in the (γ, n) Reaction Yield Curves. Interpretation

The fine structure of the giant resonance in light nuclei was first observed by Katz's group [1, 2] during calibration of the energy scale of a betatron. The most detailed studies were carried out for O^{16} and C^{12}, therefore, only data for these nuclei are given below. During this work, the yield curves for the O^{16} (γ, n) and C^{12} (γ, n) reactions were measured with considerably more care than had been the case previously; the energy instability of the upper limit of the betatron bremsstrahlung spectrum was reduced to ± 5 keV, the energy change per step was 30-50 keV, and the statistical accuracy of the measurements was increased. As a result, sharp changes in slope, or breaks, were discovered in the yield curve

$$Y(E_0) = \int_0^{E_0} \Gamma(E_0, k)\, \sigma_{\gamma, n}(k)\, dk,$$

where $\Gamma(E_0, k)$ is the bremsstrahlung spectrum; $\sigma_{\gamma,n}(k)$ is the cross section for the (γ, n) reaction; and E_0 is the upper energy limit of the bremsstrahlung spectrum.

The breaks were detected at γ-ray energies from the (γ, n) reaction threshold to approximately 22-23 MeV (an energy range of 4-6 MeV) in yield curve measurements by the induced activity method [1] and by direct detection of neutrons [2]. Their existence was also established for some other light nuclei: Li^7 and F^{19} [2]. A series of experiments showed that the presence of breaks in the yield curves was not associated with instrumental or other experimental effects, but was a feature of the process under investigation.

Somewhat later, more detailed measurements of the yield curve for the O^{16} (γ, n) reaction were made by Penfold and Spicer [3]. The instability of the upper energy limit was reduced to ± 3 keV, and the change in E_0 per step to 10-20 keV, which made it possible to resolve a considerably larger number of breaks.

The presence of breaks was interpreted as an indication that photon absorption occurred in well-defined levels of the target nucleus [1, 2]. This was based on the following considerations.

The cross section for the (γ, n) reaction can be written in the form of a product of two quantities:

$$\sigma_{\gamma, n}(k) = \sigma(k)\, G_n(k),$$

where $\sigma(k)$ is the absorption cross section; and $G_n(k)$ is the probability for the emission of a neutron by the excited nucleus.

Substituting this expression in the yield curve, we obtain

$$Y(E_0) = \int_0^{E_0} \Gamma(E_0, k)\, \sigma_\gamma(k)\, G_n(k)\, dk.$$

Breaks in $Y(E_0)$ may be associated with irregularities in the photon spectrum, in the absorption cross section, or in the probability of neutron emission. Since irregularities in the yield curves were observed at different energies for different nuclei, the spectra can be excluded from consideration. A change in the probability G_n must be connected with the appearance of new, competing processes as the energy increases. In that situation, the slope of the curve should be reduced. Since this is not so, one must conclude that the observed irregularities are connected with a fine structure in the absorption cross section. Further, the location of a break is shifted with respect to the energy at a maximum in the absorption cross section by approximately the width of a level Γ_i, i.e.,

$$E_i = E_b + \Gamma_i.$$

The relative location of breaks was determined with an accuracy of ± 40 keV in the work of Penfold and Spicer [3]. The error in the absolute calibration of the energy scale was considerably larger. It was assumed to be ± 30 keV in the 18 MeV region and it increased to 300 keV at 23 MeV.

The integrated cross section for a maximum in the (γ, n) reaction, and if the probability for neutron emission G_n is known, for a resonance in the absorption cross section corresponding to a break in the yield curve, was determined from the expression

$$G_n \int \sigma_\gamma^i(k)\, dk = \text{const}\, \frac{\Delta Y}{\Gamma(E_0, E_i)},$$

where ΔY is the difference between the yields $Y(E_0^{i+1}) - Y(E_0^i)$.

It is natural that summation of the integrated cross sections corresponding to individual breaks lead to a quantity close to that obtained by treatment of the smooth curve. Among other things, this agreement is interpreted as additional verification of the existence of fine structure in the absorption cross section.

In order to verify the indirect reasons for the association of breaks and fine structure, Penfold and Spicer made a calculation of the cross section for the (γ, n) reaction in the range of excitation energies associated with the break at 16.03 MeV [3]. (The break is now placed at 16.17 MeV as the result of a recalibration of the energy scale [4].) The portion of the yield resulting from breaks at lower energies was subtracted. The remaining portion of the curve was analyzed in a range of the order of 100 keV by the method of photon differences in 10 keV steps and subsequent introduction of corrections for thick target effects. As a result, it was established that the region of yield curve considered led to a symmetric maximum in the cross section for the (γ, n) reaction with a width $\Gamma = 18 \pm 5$ keV. Noting that the shape of the curve was the same both near the threshold and in the neighborhood of the maximum of the giant resonance, Penfold and Spicer suggested that all the breaks were produced by levels of comparable widths, and assumed the width to be 25 keV. By the reverse computation – from a cross section curve of given shape to a yield curve – it was shown that with a level widths $\Gamma > 70$ keV, the contributions did not lead to abrupt changes in the slope of the yield curve, and that to obtain a yield curve with a shape like the experimental one, it was necessary to assume level widths $\Gamma \leq 40$ keV.

The reliability in detection of breaks is greatest close to the threshold for the (γ, n) reaction. In this region, changes in slope of the yield curve are comparable in magnitude with the slope itself. Change in slope is proportional to the ratio

$$\frac{\Gamma(E_0, E_i)\int\limits_{E_i}^{E_i+\Delta}\sigma^i_{\gamma, n}(k)\,dk}{\int\limits_0^{E_i}\Gamma(E_0, k)\,\sigma_{\gamma, n}(k)\,dk}.$$

As one moves away from the threshold, the denominator increases rapidly, and only "strong" breaks will be detected. In this connection, it is apparent that the existence of breaks was established only up to the maximum of the giant resonance.

The stability of accelerator energy determines the separation of breaks and the resolving power. With an instability of the order of 50 keV, most of the breaks shown in the papers of Katz, et al. [1], and of Penfold and Spicer [3], disappeared and the yield curve became smooth.

The conclusions reached in these papers with respect to the existence of fine structure in the O^{16} and C^{12} absorption cross sections attracted a great deal of attention. However, as can be seen from the discussion, these conclusions were founded on indirect considerations. The only direct demonstration that breaks correspond to isolated resonances, though not in the absorption cross section but in the cross section for the (γ, n) reaction, was the treatment of the yield curve near 16 MeV. The extrapolation of this conclusion to all observed breaks was done purely qualitatively. While Penfold and Spicer [3] asserted that the widths of levels corresponding to breaks should not exceed 40 keV, Tzara [5] found that $\Gamma \geq 400$ keV. The estimate of Tzara actually placed in doubt the idea of discrete levels because an average frequency of five levels per MeV interval was found in the work of Penfold and Spicer [3].

Shortly after, a series of experiments were performed which were aimed at checking hypotheses on the absorption of photons in the giant resonance region by isolated levels. Their purpose was to obtain information about fine structure in partial reactions by methods more direct than breaks in yield curves.

2. Structure in Photoproton Spectra and Partial Reaction Cross Sections

Proton spectra were measured in the first attempts at such experiments. Johansson and Forkman [6, 7] measured the proton spectra from O^{16} by the photoemulsion method for several values of maximum bremsstrahlung energy: 20.5, 23, and 26 MeV. By means of difference spectra, the authors estimated the contribution from protons leaving N^{15} in an excited state, and found indications of structure in the cross section for the (γ, p) reaction. In the opinion of the authors, this cross section coincided with the cross section for the (γ, n) reaction in the 16-19.5 MeV energy range if one assumed that it consisted of narrow peaks corresponding to breaks [3] in this region. However, if one considers the magnitude of the resolution of this method for detecting photoprotons and the low statistical accuracy of the measurements, it is evident that one can consider that there is in this region a broad maximum (~ 1 MeV) in the cross section for the (γ, p) reaction at 17.3 MeV. In addition, resonances were observed in the cross section at γ-ray energies of 14.5, 20.7, and 21.9 MeV, the first of them having been found somewhat earlier by Spicer [8] also. The general conclusion of [6, 7] was that the cross section obtained for the (γ, p) reaction was in accord with the idea of the existence of narrow isolated peaks in the (γ, n) reaction cross section. As will be shown later on (Chapter V), this conclusion proved to be incorrect.

Structure in the photoproton spectra from O^{16} and C^{12} was also found by Cohen et al. [9] when those nuclei were irradiated by a bremsstrahlung spectrum with a maximum energy of 25 MeV. According to the authors' estimate, the resolution of their measurements was no better than 0.2 MeV. Two small peaks on the low-energy side of the giant resonance at 19.6 and 20.6 MeV were found in O^{16}. Several maxima were also resolved in the C^{12} proton spectrum. Their identification was complicated by the difficulty of distinguishing transitions to the excited states of B^{11} (the first excited state of B^{11} is at 2.14 MeV). Using data for the cross section for the (γ, p_0) reaction which was obtained from the cross section for the $B^{11}(p, \gamma_0)C^{12}$ reaction by the principle of detailed balance [10], Cohen et al. [9] suggested that all the peaks in the spectrum were produced by transitions to the ground state of B^{11}. For both nuclei, there was agreement between the structure found and the fine structure in the yield curve for the (γ, n) reaction [1].

The purpose of Livesey's work [11] was to obtain information about structure from proton spectra at higher O^{16} excitation energies. He did not succeed in obtaining clear data from spectra produced by irradiating targets with γ-rays having a maximum energy of 70 MeV. In spectra measured with γ-ray maxima of 30 and 35 MeV, peaks were observed at 20.7, 21.9, and 24.0 MeV. The accuracy in the determination of the energy of the maximum was ± 0.2 MeV. The main maximum was located at 21.9 MeV. Its coincidence with the small peak at this energy found in [7], cannot be considered as agreement between the two.

Maxima were observed by Spicer [12] in the cross section for the (γ, n) reaction on the low-energy side of the giant resonance. In this work, the yield curve for the O^{16} (γ, n) reaction was measured by means of the positron activity in O^{15}. The measurements were made with a statistical accuracy of 1%. The spectral maximum varied by 0.2 MeV. The cross section for the (γ, n) reaction was calculated from the yield curve by the Penfold-Leiss method [13] in 1 MeV steps. Two small peaks were found at excitation energies of 17.5 and 19.5 MeV. The probable error of the individual points in the cross section curve was about 10%. It is significant that the usual procedure for smoothing the yield curve was not used in transforming from yield to cross section.

It is apparent that one can only note irregularities in the cross section curve of the order of 1 MeV by "differentiation" of the yield curve in steps of that size. Thus a comparison with fine structure is very problematical. Nevertheless, the 17.5 MeV maximum was explained [12] as the result of a grouping of levels discovered by study of the fine structure. The level corresponding to the break at 19.2 MeV [3] was associated with the maximum at 19.5 MeV. However, smooth cross section curves were obtained [14, 15] as the result of measurements carried out at almost the same time and with comparable accuracy. It is clear that this indicates the unsatisfactory reliability of Spicer's results [12].

A common feature of all the work discussed in which proton spectra were measured was the low statistical accuracy; 10-30 recorded events occurred in each interval of a histogram. The value of such information was also reduced because of difficulties in identification of resonances in spectra with maxima in the cross sections. Thus the reliability of the data obtained required verification as to agreement with the fine structure revealed by the breaks in the yield curves.

The signs of the existence of structure in partial reaction cross sections, such as breaks, which were observed in these investigations, were indirect indications of the presence of resonances in the γ-ray absorption cross sections. There was no direct data on this point.

CHAPTER II

Formulation of Problem and Method

The obtaining of information about γ-ray absorption cross sections in light nuclei is a complex problem which involves: (1) a solution of the question of the existence of structure in the absorption cross section; (2) a determination of the quantitative characteristics of the cross section – integrated cross sections, etc.

In order to solve these problems, L. E. Lazareva, in 1956, proposed using a direct method for measuring the γ-ray absorption cross sections in light nuclei – the absorption method with high-resolution detectors.

The essence of the absorption method is the measurement of the attenuation of a γ-ray flux after passage through an absorber consisting of the material under investigation. Under good geometric conditions, the number of photons of a given energy recorded by a detector is

$$N = N_0 e^{-n\sigma},$$

where N_0 is the number of γ-rays incident upon the absorber; n is the number of nuclei per cm^2 in the absorber and σ is the cross section for all processes which remove γ-rays from the primary beam during penetration of the sample. Hence, one can determine the total γ-ray absorption cross section in the material under study. The total cross section consists of several components, of which one is the nuclear absorption cross section. If the contribution from all other processes is known, it is possible to determine the nuclear absorption cross section.

The technique of determining the total interaction cross section of particles by passing a flux of the particles through matter has been used fruitfully in other branches of nuclear physics, for example, in neutron spectroscopy.

The special characteristics of the use of the absorption method for studying nuclear γ-ray absorption cross sections are associated with the small magnitude of the nuclear photoeffect in comparison with other processes for interaction of γ-rays with matter and with the need for using γ-ray bremsstrahlung spectra.

1. A high "background" level places rigid requirements on the accuracy of the measurements. The magnitude of the total interaction cross section is mainly determined by the pair-production process and by the Compton scattering process. The relative magnitude of the nuclear absorption cross section is a few percent of the total cross section. It can be estimated roughly from existing data for the partial reactions. For example, the cross section for the (γ, n) reaction in C^{12} at the maximum is 8 mb according to the data of Barber et al. [16]. The cross section for the $C^{12}(\gamma, p)$ reaction is less accurately known because it is determined from spectra on the assumption that the residual nucleus remains in the ground state. We take the maximum value to be 14 mb [9]. Since the contribution from other photodisintegration processes and from nuclear γ-ray scattering in the region of the maximum of the giant resonance is small in comparison with that from the sum of the (γ, p) and (γ, n) reactions 22 mb is an estimate of the maximum nuclear absorption cross section. The sum of the pair-production and Compton scattering cross sections for 23 MeV γ-rays in carbon is about 300 mb. Consequently, the maximum contribution of the nuclear cross section is only about 7% of the total. In order to determine the nuclear absorption cross section with reasonable accuracy, say 10%, measurements of the total cross section with an accuracy of the order of 0.5% are necessary under conditions where its non-nuclear portion is presumed known.

The nuclear cross section increases in proportion to A with increasing mass number A, if one allows for variation in the width of the giant resonance. This follows from the fact that the integrated cross section $\sigma_{int} \simeq 0.015$ A. The pair-production cross section, which makes the main contribution to the total cross section for γ-ray energies greater than 15 MeV, and for mass number $A > 25$, increases in proportion to Z^2. Thus the determination of the nuclear absorption cross section for medium and heavy nuclei demands even more accurate determination of the total cross section and hinders the use of this method for nuclei with $A > 40$.

An additional difficulty in the determination of nuclear γ-ray absorption cross sections is associated with uncertainty in the values for the non-nuclear processes. Such uncertainty is estimated to be several percent for 10-30 MeV γ-rays. Because the nuclear cross section itself is 7-10% of the total cross section, this situation can lead to considerable error in the absolute value of the cross sections.

2. It was pointed out above that only a continuous spectrum of bremsstrahlung radiation can serve as a γ-ray source in these studies. Monochromatic γ-lines from nuclear reactions cannot be used for a systematic study of the nuclear photoeffect because of their scarcity and because they cover a small range of nuclear excitation energies.* The necessity for using a continuous γ-ray spectrum places considerably more rigid requirements on the detection system because the energy resolution of the method in this case is determined by the detector resolution.

Spectrometers suitable for the detection of γ-rays in the 10-30 MeV region have the following energy resolutions: scintillation spectrometers, ~10%, magnetic Compton and pair spectrometers, 1-2%. The problem faced – the establishment of structure in the absorption cross section – required the use of a detector with maximum possible high-energy resolution. From experimental data on structure in proton spectra available to use in 1957, before the initiation of this investigation (see Chapter I), it was assumed that the lower limit of the resolution required for such studies was approximately 1.5%.

*More recently, quasimonochromatic γ-ray beams have been successfully produced in connection with the startup of high-current, linear electron accelerators. The energy spread in such beams is about 3% at the present.

To estimate the minimum resolution, the following considerations based on data concerning breaks, can be used. According to the results of Penfold and Spicer [3], the average spacing between breaks was 0.2 MeV for O^{16} excitation energies above 18 MeV. In order to differentiate the peaks corresponding to them, a resolution of at least approximately 100 keV, or 0.5%, is required.

It is clear that a scintillation spectrometer does not satisfy the requirements that have been set. In a pair spectrometer, the resolution can be made 0.6-0.7% for γ-rays with energies of the order of 10 MeV [17]. The use of specially constructed slits limiting the entrance aperture of the counters later made it possible to obtain a maximum resolution of 40-50 keV [18]. However, such improvement in resolution is associated with considerable loss in efficiency. The efficiency of the two types of spectrometers are comparable at γ-ray energies of 10-15 MeV. With further increase in energy, the efficiency of the pair spectrometer increases, and the efficiency of the Compton spectrometer decreases. Therefore, the pair spectrometer possesses a slight advantage in resolution and efficiency, which can prove to be important. Taking all these considerations into account, a pair spectrometer was selected as detector. Because of the low transmission of such a detector, the choice made the measurements extremely time-consuming.

The selection of optimal absorber thickness is associated with the condition for obtaining the minimum statistical error dependent on the magnitude of the γ-ray absorption in the sample. It is easy to show (see [19], for example) that the square of the relative fluctuation in the measurement of the ratio x of two intensities is

$$\delta^2 = c\frac{1+\sqrt{x}}{\ln x}.$$

The minimum δ^2 is found when $x = 13$. The curve of δ^2 as a function of x has a broad, flat minimum roughly from $x = 5$ to $x = 30$ making it possible to vary the sample thickness within rather broad limits while insignificantly (10-20%) increasing the statistical error.

In the first experiment, carried out for C^{12} with a single-channel pair spectrometer (excitation energy region, 22.4-24.0 MeV), it was established that it was possible, in principle, to obtain data on nuclear γ-ray absorption cross sections by this method [20]. At the same time, this experiment demonstrated that the investigation of the giant resonance region for a single element required several years of continuous measurements at an accelerator like the 30-MeV synchrotron of the FIAN (Institute of Physics, Academy of Sciences of the USSR).

The use of the γ-ray beam from the FIAN-260-MeV synchrotron reduced the time for achieving acceptable accuracy by a factor of 15-20. As a result, sufficiently reliable data was obtained for the O^{16} nuclear absorption cross section in the 21-27 MeV excitation energy range [21] and for C^{12} in the 13-27 MeV range [22]. In the excitation energy range below 21 MeV, where the percentage of nuclear cross section drops to approximately 2% of the total interaction cross section and where the efficiency of the pair spectrometer falls at the same time, the accuracy of the O^{16} measurements [23] was insufficient for reaching definite conclusions about cross section structure. A subsequent increase in the transmission of the method was achieved by the use of a multichannel pair spectrometer whose efficiency was approximately an order of magnitude greater than the previous one. The O^{16} nuclear %-ray absorption cross section in the 13.5-22 MeV range was obtained with this detector [24]. Thus, the two possibilities mentioned above – an increase in the statistical reliability of the results and a reduction in measurement time – were realized, making it possible subsequently to obtain in an acceptable period of time data for F^{19}, Mg^{24}, Al^{27}, and Ca^{40} in the 10-30 MeV excitation energy range [25, 26].

The absorption method was used independently for studies of the γ-ray absorption cross section in the giant resonance region by Zeigler [27, 28], by the Yugoslav group at the Stefan Institute in Ljubljana [29-31], and by Tzara and associates [32]. Specific data for structure in the cross section was not obtained in these experiments; this was primarily connected with insufficient intensity of the γ-ray beam used. Apparently, valuable information about the nuclear cross section can be derived from a study of inelastic electron scattering. As yet, only qualitative data for structure in the region close to the maximum of the giant resonance has been obtained [33].

Obtaining the accuracy mentioned above requires lengthy measurements during the course of which both the γ-ray spectrum from the accelerator and the spectrometer efficiency may change. These instabilities are avoided by making relative measurements. The number of coincidences without an absorber in the γ-ray beam, referred to some monitor reading, is

$$N_0 = \int_{k_0}^{\infty} \Gamma(E_0, k)\,\varepsilon(k)\,\Delta(k - k_0)\,dk. \tag{2.1}$$

Here, $\Gamma(E_0, k)$ is the γ-ray spectrum; $\varepsilon(k)$ is the spectrometer efficiency; $\Delta(k - k_0)$ is the resolution function of the spectrometer, whose half-width is assumed to be the resolution; k_0 is the minimum γ-ray energy recorded by the spectrometer for a given value of the magnetic field H_0.

If the exponential law for γ-ray attenuation is valid for a given E_0 and experimental geometry, the number of coincidences when there is an absorber in the γ-ray beam, referred to the same monitor reading, is

$$N = \int_{k_0}^{\infty} \Gamma'(E_0, k)\,\varepsilon'(k)\,\Delta(k - k_0)\,e^{-n\sigma(k)}\,dk, \tag{2.2}$$

where Γ' and ε' are the spectrum and efficiency during measurement with the sample.

The γ-ray spectrum $\Gamma(E_0, k)$ and the spectrometer efficiency are smooth functions of the energy, which change slowly in the energy range where the resolution function is different from zero. $\Gamma(E_0, k) \sim 1/k$; the efficiency of a pair spectrometer roughly increases like k^n, where $n < 2$, with increase in energy above 10 MeV, and $\varepsilon(k) \sim k$ in the 15-25 MeV range. Consequently, the product $\Gamma(E_0, k)\varepsilon(k)$ can be considered a constant and taken outside the integral sign in expressions (2.1) and (2.2):

$$N_0 = [\Gamma(E_0, k^*)\,\varepsilon(k^*)] \int_{k_0}^{\infty} \Delta(k - k_0)\,dk, \tag{2.3}$$

$$N = [\Gamma'(E_0, k^*)\,\varepsilon'(k^*)] \int_{k_0}^{\infty} e^{-n\sigma}\Delta(k - k_0)\,dk, \tag{2.4}$$

where k^* is an energy in the range where the resolution function is different from zero.

The resolution function is normalized:

$$\int_{k_0}^{\infty} \Delta(k - k_0)\,dk = 1.$$

Dividing N by N_0, we obtain

$$\frac{N}{N_0} = \frac{[\Gamma'(E_0, k^*)\,\varepsilon'(k^*)]}{[\Gamma(E_0, k^*)\,\varepsilon(k^*)]} \int_{k_0}^{\infty} e^{-n\sigma}\Delta(k - k_0)\,dk. \tag{2.5}$$

In order to equate Γ' to Γ and ε' to ε with sufficient accuracy, it is necessary to measure the values of N and N_0 for each given energy one right after the other in a period of time during which the γ-ray spectrum and spectrometer efficiency can be considered constant. Under these conditions, the experimentally obtained ratio

$$\frac{N}{N_0} = \int_{k_0}^{\infty} e^{-n\sigma(k)} \Delta(k - k_0)\, dk \tag{2.6}$$

is independent of γ-ray spectrum and spectrometer efficiency.

The γ-ray cross section for interaction with matter can be divided into two parts:

$$\sigma(k) = \sigma_n(k) + \sigma_a(k),$$

where $\sigma_n(k)$ is the cross section for interaction with nuclei in the absorber; and $\sigma_a(k)$ is the cross section for all other interaction processes leading to attenuation of the γ-ray flux during passage through the absorber. This is a slowly-varying, smooth function of the γ-ray energy relative to the resolution. One can therefore write

$$\frac{N}{N_0} = e^{-n\sigma_a(k_m)} \int_{k_0}^{\infty} e^{-n\sigma_n(k)} \Delta(k - k_0)\, dk, \tag{2.7}$$

where k_m is an energy corresponding to the maximum of the instrumental function.

Let us consider how data for $\sigma_n(k)$ can be obtained from the experimental values $N/N_0\,(k_0)$.

The γ-ray attenuation in the absorber because of nuclear interactions is

$$A = 1 - \frac{N}{N_0} e^{n\sigma_a(k_m)} = \int_{k_0}^{E_0} \Phi(k)\, \Delta(k - k_0)\, dk, \tag{2.8}$$

where $\Phi(k) = 1 - e^{-n\sigma_n(k)}$. Expanding $e^{-n\sigma_n(k)}$ in a series, we obtain

$$A = \sum_{m=0}^{\infty} (-1)^m \frac{n^{m+1}}{(m+1)!} \overline{\sigma_n^{m+1}},$$

$$\overline{\sigma_n^{m+1}} = \int_{k_0}^{E_0} \sigma_n^{m+1} \Delta(k - k_0)\, dk. \tag{2.8a}$$

Integrating (2.8) over a measured energy range (k_1, k_2), we find.

$$B = \int_{k_1}^{k_2} A\, dk_0 = \int_{k_1}^{k_1+2\Delta k} \Phi(k)\, \xi(k)\, dk + \int_{k_1+2\Delta k}^{k_2} \Phi(k)\, dk + \int_{k_2}^{k_2+2\Delta k} \Phi(k)\, \xi(k)\, dk, \tag{2.9}$$

with $0 \le \xi(k) \le 1$, when $k_1 \le k \le k_1 + 2\Delta k$ and $k_2 < k < k_2 + 2\Delta k$.

It is assumed roughly here that

$$\int_{k_0}^{k_0+2\Delta k} \Delta(k - k_0)\, dk \simeq 1,$$

i.e., that the resolution function is significantly different from zero only in the range $(k_0, k_0 + 2\Delta k)$. Since $\Delta k \ll k_2 - k_1$, $\Delta k \ll k_1$, and the first and third integrals are small in comparison with the second, then

$$B \simeq \sum_{m=0}^{\infty} (-1)^m \frac{n}{(m+1)!} \tilde{\sigma}_n^{m+1}, \quad \tilde{\sigma}_n^{m+1} = \int_{k_1+2\Delta k}^{k_2} \sigma_n^{m+1} dk. \tag{2.9a}$$

It is clear from (2.8) and (2.9) that, in the general case, the experimental quantity is associated with a sum of the moments of $\sigma_n(k)$.

Relationships of practical interest between $\sigma_n(k)$ and N/N_0 are given below which were obtained from (2.8) and (2.9) for definite assumptions about the parameters of the nuclear cross section.

I. $\sigma_n(k)$ is constant, or changes slowly in the range $(k_0 + 2\Delta k)$:

$$\sigma_n(k) = -\frac{1}{n} \ln(1 - A). \tag{2.10}$$

II. $n\sigma_n \ll 1$. In this case,

$$\bar{\sigma}_n = \int_{k_0}^{E_0} \sigma_n(k) \Delta(k - k_0) dk = \frac{A}{n}. \tag{2.11}$$

The cross section is averaged over an energy range approximately equal to $(k_0 + 2\Delta k)$; $\bar{\sigma}_n$ is close to σ_n if the region where the cross section changes markedly is greater than the resolution Δk. From formula (2.9a)

$$B = n \int_{k_1+2\Delta k}^{k_2} \sigma_n dk = n\sigma_{n.\text{int}}. \tag{2.12}$$

III. $n\sigma \sim 1$. In this case, one can limit oneself to the first few terms of the series in formulas (2.8a) and (2.9a).

IV. $\Gamma/\Delta k \ll 1$. We assume that there are l isolated, nonoverlapping levels in the region where the resolution function is noticeably different from zero. The cross sections which correspond to each of them are described by the Breit-Wigner formula (resonance energy k_{pi}, cross section at the maximum σ_{0i}, width Γ_i). Under the condition that $|(k_0 - k_{pi})/(\Gamma/2)| \gg 1$, we have

$$A \simeq \sum_1^l \Delta(k_{p_i} - k_0) \cdot n \cdot \frac{\pi}{2} \sigma_{0i} \Gamma_i \cdot \alpha_i, \tag{2.13}$$

where

$$\alpha_i = e^{-\frac{1}{2} n\sigma_{0i}} \left[I_0\left(\frac{1}{2} n\sigma_{0i}\right) + I_1\left(\frac{1}{2} n\sigma_{0i}\right) \right] = 1 + \sum_{m=1}^{\infty} \frac{(-1)^m (n\sigma_{0i})^m (2m-1)!!}{(m+1)!\, 2m!!}. \tag{2.14}$$

Here, I_0 and I_1 are Bessel functions of imaginary argument.

We shall estimate the limits of variability of α (for simplicity, we shall consider a single isolated level):

$$\text{if } n\sigma_0 \ll 1, \quad \alpha \simeq 1 \text{ and } A \sim \sigma_0 \Gamma;$$

$$\text{if } n\sigma_0 \gg 1, \quad \alpha \simeq \frac{2}{\sqrt{\pi n\sigma_0}} \left(1 - \frac{1}{4n\sigma_0}\right) \text{ and } A^2 \sim \sigma_0 \Gamma^2.$$

We can limit ourselves to the first two terms in (2.14) with good accuracy for $n\sigma_0 < 0.15$. Then

$$A = \Delta (k_p - k_0) \frac{n\pi}{2} \sigma_0 \Gamma \left(1 - \frac{n\sigma_0}{4}\right). \tag{2.15}$$

A similar particular case of formula (21.3) has also been obtained elsewhere [28].

For $n\sigma_0 > 10$

$$A^2 = \Delta^2 (k_p - k_0)\, n\pi\sigma_0\Gamma^2 \left(1 - \frac{1}{2n\sigma_0}\right). \tag{2.16}$$

In the case where $|(k_0 - k_p)/(\Gamma/2)| = b$, we obtain

$$A = \Delta (k_p - k_0) \frac{n\sigma_0\Gamma}{2} \left\{\left(\frac{\pi}{2} - \operatorname{arctg} b\right) + \sum_{m=2}^{\infty} (-1)^{m-1} \frac{(n\sigma_0)^{m-1}}{m!} \left[\frac{(2m-1)!!}{2^{m-1}(m-1)!}\left(\frac{\pi}{2} - \operatorname{arctg} b\right) - \right.\right.$$

$$\left.\left. - \frac{b}{2m-1} \sum_{s=1}^{m-1} \frac{(2m-1)(2m-3)\ldots(2m-2s+1)}{2^s (m-1)(m-2)\ldots} \frac{1}{(1+b^2)^{m-s}}\right]\right\}. \tag{2.17}$$

Consequently, the coefficient α depends not only on $(n\sigma_0)$ but also on Γ and $(k_0 - k_p)$ when the resonance energy is close to the minimum recorded energy k_0. Since $\Delta(k_p - k_0) \to 0$ at the same time, this region will not be considered below. We note only that when $b = 0$ and $n\sigma_0 \ll 1$

$$A = \Delta (k_p - k_0) \frac{n\pi}{4} \sigma_0 \Gamma.$$

The fact that the γ-ray attenuation effect depends on resonance parameters and on the resolution function $\Delta(k_p - k_0)$ when $\Gamma/\Delta k \ll 1$ follows from formulas (2.13)-(2.16). The magnitude of the attenuation is most significant when the resonance energy is close to the maximum of the resolution function. In this case

$$\Delta (k_p - k_0) \sim \frac{1}{\Delta k}, \quad \frac{A}{n} = \frac{\pi}{2} \sigma_0 \frac{\Gamma}{\Delta k} \quad (n\sigma_0 \ll 1). \tag{2.18}$$

Consequently, σ_{int} can be obtained from a measurement at a single energy k_0 if the resolution function and the resonance energy are known.

If $n\sigma_0 \gg 1$

$$\frac{A}{n} = \bar{\sigma} \frac{2}{\sqrt{\pi n\sigma_0}} \left(1 - \frac{1}{4n\sigma_0}\right) \tag{2.19}$$

and it is necessary to know the magnitude of σ_0 in addition in order to determine σ and σ_{int}.

As is clear from (2.18), the measured effect with identical values of σ_0 is considerably less for resonances which are narrow in comparison with the detector resolution than it is for maxima in which $\Gamma \gtrsim \Delta k$. In addition, it is further reduced if $n\sigma_0 \gg 1$ (see (2.19)). This situation significantly raises the requirements for accuracy of measurement in energy ranges where such resonances exist.

Integration over the variable k_0 makes it possible to eliminate the spectrometer resolution function whose shape may be unknown. Using formula (2.13) with $l = 1$, we obtain

$$B = \int_{k_1}^{k_2} A\,dk_0 = n\frac{\pi}{2}\sigma_0\Gamma\alpha, \tag{2.20}$$

$$\sigma_{\text{int}} = \frac{B}{n\alpha}. \tag{2.21}$$

The relation (2.20) can also be obtained from formula (2.9).

The resonance parameters, and therefore the energy dependence of the cross section, can be determined separately by making measurements with samples of varying thickness or by accepting one of the parameters on the basis of other data.

Formula (2.20) is similar to the result which is obtained from a consideration of the relation between the coefficient for neutron penetration through a sample and the resonance parameters in neutron cross sections [34].

Thus it is necessary to evaluate the quantity $n\sigma_0$ if there are narrow peaks in the measured range of nuclear excitation energies in order to establish the correct relationship between the experimental quantities N_0/N and the cross section by measurements with a single absorber. It is reasonable to select an absorber thickness so that the condition $n\sigma << 1$ is satisfied. In practice, this means that $n\sigma_n$ must be limited to the values 0.1-0.2. With such a choice of sample thickness, the coefficient α in the relations considered above will be close to unity. This makes it possible to use the simple expression (2.11) for calculation of the cross section, or the closely related formula (2.10) (on condition that $n\sigma_n \ll 1$), regardless of the magnitude of $\Gamma/\Delta k$.

The value of the resolution of the pair spectrometer equal to the width of the instrumental function at half-height was used repeatedly above. The shape of the instrumental function can be determined by theoretical calculations or by measurements with monochromatic γ-rays. We shall consider the latter possibility.

Let $N'(k - k_p)$ be the photon distribution, and N'_0 the total number of γ-rays incident on the spectrometer radiator per unit time. The number of coincidences is

$$N' = \int_{k_0}^{\infty} \varepsilon(k)\,N'(k-k_p)\,\Delta(k-k_0)\,dk.$$

Coincidences occur when the functions $N'(k - k_p)$ and $\Delta(k - k_0)$ overlap. Since ε and $\Delta(k - k_0)$ are functions which are slowly changing relative to $N'(k - k_p)$, one can write

$$N' = \varepsilon\Delta(k_p - k_0)\int_{k_0}^{\infty} N'(k-k_p)\,dk = \varepsilon N'_0\Delta(k_p - k_0) = \text{const}\cdot\Delta(H_p - H_0),$$

where H_p is the maximum value of the field for which coincidences are still recorded for γ-rays of energy k_p, and $\Delta(H_p - H_0)$ is the resolution function for monochromatic γ-rays.

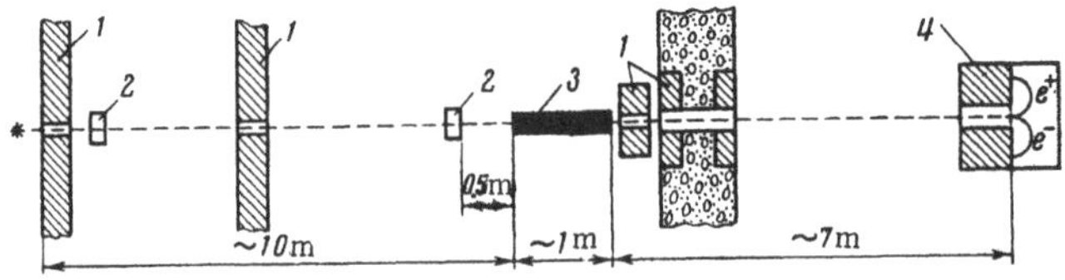

Fig. 1. Experimental arrangement: (1) collimators; (2) monitors; (3) absorber; (4) spectrometer.

Fig. 2. Chamber of the single-channel spectrometer.

CHAPTER III

Method of Measurement

Measurements were made with the FIAN 260-MeV synchrotron. The main portion of the data was obtained when the accelerator was operated at $E_0 = 260$ MeV and a γ-ray beam duration τ of 3000 μsec. The accelerator was also used with $E_0 = 230$ and 200 MeV and a beam duration of 2000 μsec.

The experimental arrangement, equipment, and measurement procedure will be described briefly below. Questions relating to the technical portions of the work carried out with a single-channel pair spectrometer have been discussed in detail [35].

Experimental Arrangement (Fig. 1). A collimated beam of photons from the accelerator, 50 cm in diameter, was incident on the absorber. The intensity of the beam was measured by a thin-walled monitor located 0.5 m in front of the absorber. A similar chamber, located in the γ-ray beam near the accelerator target, was continually used to check on the working monitor. By checking the γ-ray back scatter, it was determined that the readings of the working monitor did not depend on sample position. The transverse dimensions of the absorber were selected so that the absorber covered the beam with room to spare, assuring penetration of the beam through the entire thickness of the sample. The following absorbers were used:

1. The sample was distilled water for measurements of the γ-ray absorption cross section in oxygen. An absorber 100 g/cm^2 thick reduced a 20-MeV γ-ray flux by a factor of 6.

2. A graphite block 10 × 10 × 100 cm in size (170 g/cm^2 thick), which reduced a 20-MeV γ-ray flux by a factor of 14, was used as a sample for measurements of the γ-ray absorption cross section in carbon.

Behind the absorber, the γ-ray beam once again passed through a collimating system. The spectrometer was located about 7 m from the absorber, and was shielded from scattered γ-radiation by 1.3 m of lead and a concrete wall. (A deflecting magnet was used in the first series of measurements to clear the photon beam of charged particles. Experience showed that there were few such particles in the γ-ray beam, and the magnet was not used in subsequent measurements.)

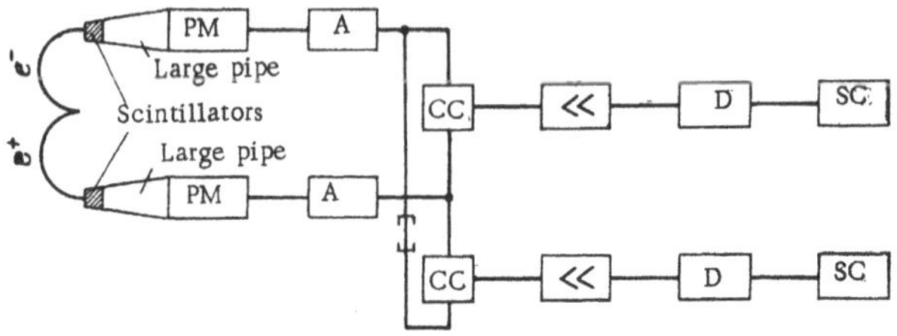

Fig. 3. Block diagram of the single-channel spectrometer: (CC) coincidence circuit; (D) discriminator; (SC) scaling circuit); (A) amplifier; (PM) photomultiplier.

TABLE 1. Spectrometer Resolution (keV)

Energy, MeV	Slit width	
	2 mm	4 mm
10	75	125
20	110	210
25	130	250

Equipment. The single-channel pair spectrometer was built in the Photonuclear Reactions Laboratory, FIAN, by B. S. Dolbilkin, L. E. Lazareva, and F. A. Nikolaev in collaboration with N. A. Burgov and G. V. Danilyan of the ITEF staff. The principles of operation of this kind of spectrometer and the experimental and theoretical aspects of its operation with γ-rays up to 10 MeV have been described in detail previously [17].

The magnetic field in the operating region (400 mm diam) was uniform within 0.05%. The time stability of the magnetic field was maintained by the nuclear resonance method with an accuracy of about 0.01%. The magnitude of the field was measured with this same system.

The vacuum chamber (pressure under operating conditions, 10^{-2} mm Hg) was rectangular in shape and $70 \times 70 \times 10$ cm in size (Fig. 2). The chamber was lined with 2 mm of polyethylene to reduce the background from pairs scattered by the internal surfaces. The entrance and exit windows of the chamber were made of thin aluminum ($80 \times 180 \times 0.3$ mm). The radiator was made of gold foil 5 μ thick (10 mg/cm^2), and had a working area of 60×50 mm. Background (counts with no radiator present) was 1%.

The pair components reached the counters through slits formed by tungsten plates $80 \times 15 \times 3$ mm in size. The width of the slits could be varied within the limits of 0 to 6 mm. Detection of electrons and positrons was accomplished with plastic scintillators located behind the slits. The scintillators were attached to light pipes of plastic 20 cm long which were in optical contact with FÉU-33 photomultipliers located outside the spectrometer chamber. The photomultipliers were protected against stray magnetic fields by four shields – the inner two of permalloy, and the outer two of iron.

A block diagram of the equipment is shown in Fig. 3. Pulses from the photomultipliers were fed through broad-band UP-1M amplifiers into a diode coincidence circuit [36] with a resolving time of $\sim 4 \times 10^{-9}$ sec. A second such coincidence circuit (with a time delay $\Delta t = 2 \times 10^{-8}$ sec in one of the channels), connected in parallel with the main coincidence circuit, simultaneously recorded chance coincidences. In order to obtain the stability needed for lengthy measurements, the portion of the counting channel including coincidence circuits, amplifiers, and discriminators was thermostatically controlled.

A portion of the data was obtained by measurements with a 9-channel pair spectrometer built in 1963 at the Photonuclear Reactions Laboratory, FIAN. The transmission of this spectrometer was an order of magnitude greater than the previous one. Sets of three counters were located symmetrically with respect to the center of the radiator. The counters were at equal distances from one another so that γ-rays of five energies were detected simultaneously for a given value of the magnetic field. Slit width during the measurements was 2 mm. Detailed descriptions of this spectrometer have been given by B. A. Zapevalov [37] and F. A. Nikolaev (p. 86, this volume).

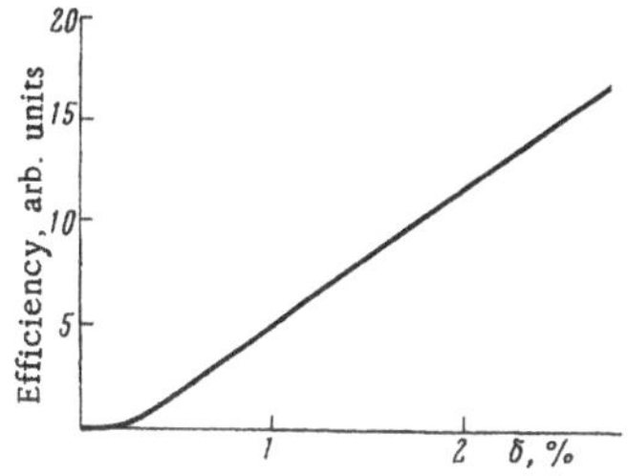

Fig. 4. Spectrometer efficiency as a function of relative slit width (k = 20 MeV).

Spectrometer resolution for several γ-ray energies is shown in Table 1.

Measurements. As mentioned above, it was necessary to carry out the measurements for an individual point in a rather short time in order to eliminate from the results the energy dependence of the γ-ray spectrum, which can change with changes in duration and also to eliminate the possibility of instrumental instability. Therefore, the ratio of the number of coincidences without absorber in the γ-ray beam (N_0) to the number of coincidences with an absorber in the beam (N) was determined in individual measurements which lasted for 10-20 min. Movement of the absorber was accomplished with a remotely-controlled motor. A ratio measurement was repeated if the accelerator shut down during the time of measurement or if pronounced changes in the intensity or the on-time of the γ-ray beam occurred.

The statistical accuracy of the determination of N_0/N in an individual measurement was 6-10%. Each measurement was repeated 10-15 times. Neighboring groups of 10-20 points in an excitation energy range of 0.5-1.5 MeV were measured successively in order to eliminate possible instabilities and systematic errors here also; then the measurements were repeated. Several sets of measurements were made in some energy ranges for more reliable establishment of structure. They were averaged during analysis. For example, measurements of the O^{16} absorption cross section in the 19-27 MeV energy range were carried out for 151 values of the magnetic field in the spectrometer, mainly for $\Delta f = 30$ kHz ($\Delta E = 50$ keV). A total of 250 points was obtained because measurements were repeated from 2 to 6 times for some values with subsequent sets of measurements usually being made after a considerable time lapse. The root-mean-square error for each point was about 1-1.5%. As a rule, results were averaged over the region of spectrometer resolution ($\Delta k \sim 100$ keV).

The O^{16} absorption cross section in the 13-19 MeV energy range, and the C^{12} absorption cross section in the 13-27 MeV energy range, were obtained with a 4 mm slit width, and the corresponding spectrometer resolution was approximately twice as poor as with a 2 mm slit width. This derived from the fact that the operating time in the accelerator beam was insufficient to produce results of given accuracy because of the efficiency of the spectrometer with 2 mm slit width. (Dependence of efficiency on slit width is shown in Fig. 4.) The magnetic field during measurements with 4 mm slits was varied through 125 kHz ($\Delta E \simeq 175$ keV).

On the average, random coincidences were about 15% with no absorber in the beam and about 1% with an absorber present. The accuracy of random coincidence counting in the parallel circuit was checked several times per day.

CHAPTER IV

Results

1. Attenuation Coefficient N_0/N

The immediate result obtained from absorption measurements is the dependence of the attenuation coefficient N_0/N on the NMR frequency f with which the magnetic field of the spectrometer was measured.

1. Values of N_0/N obtained with the single-channel pair spectrometer, using a water absorber, and in the frequency range from $f = 8{,}800$ MHz to $f = 18{,}730$ MHz, are shown in Fig 5. The rise in the curve at frequencies below 13.0 MHz is connected with the increase in the "non-nuclear" γ-ray interaction cross section for water at decreasing γ-ray energies. Above 13.0 MHz, the contribution from the absorption of γ-rays by oxygen nuclei is shown clearly.

The measurements were carried out with 2 mm slit width at frequencies above $f = 13{,}330$ MHz (solid circles). Particular attention was devoted to the region above 15 MHz in order to obtain more reliable data

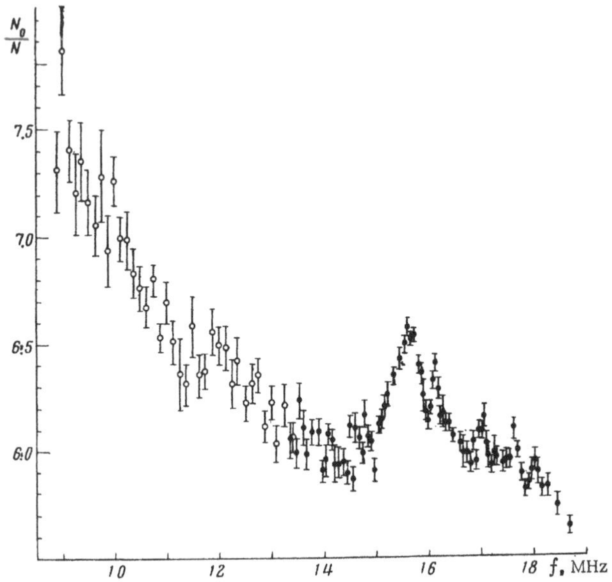

Fig. 5. Dependence of attenuation coefficient N_0/N on NMR frequency f (H_{20}).

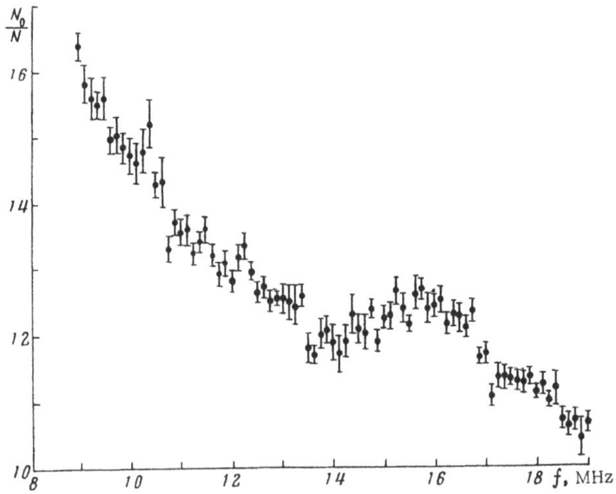

Fig. 6. Dependence of attenuation coefficient N_0/N on NMR frequency f (carbon, 4 mm slit width).

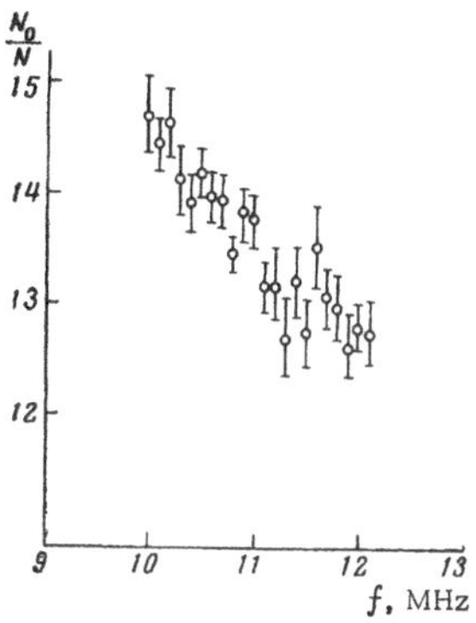

Fig. 7. Dependence of attenuation coefficient N_0/N on NMR frequency f (carbon 2 mm slit width).

on structure in the N_0/curve. As a result, the existence of a series of maxima in this region was established. Irregularities were also observed below 15 MHz. Conclusions about the possible existence of peaks there, and about their number, are not reliable because of the large errors. The measurements were made with 4 mm slit width below 13.3 MHz (open circles). However, the accuracy of the measurements, even with 4 mm slits, was insufficient for definite conclusions about structure in the N_0/N curve in this region because of the significantly lower value of the nuclear cross section and the reduction in spectrometer efficiency with decreasing energy.

2. Measurements using a water sample were made with the 9-channel pair spectrometer in the 10-15 MHz frequency range in order to obtain quantitative data in the frequency region below 15 MHz. Because of the considerably greater labor involved, the analysis was carried out on an electronic computer. The results will be presented below (see Section 6).

3. Results for the measurement of the quantity N_0/N with a carbon sample in the frequency range from 9 to 19 MHz are given in Fig. 6. The measurements were made with 4 mm slit width. The general behavior of the N_0/N curve is determined by the reduction in the "non-nuclear" γ-ray interaction cross section with increasing γ-ray energy. In contrast to the curve obtained with an H_2O sample, the N_0/N curve is considerably smoother in this case. A set of measurements was also made with 2 mm slits in order to refine the data in the 10-12 MHz range where considerable spread of the points was observed. The results are shown in Fig. 7.

The following sections will discuss the problems connected with the calculation of the nuclear γ-ray absorption cross section from the experimentally obtained attenuation coefficient N_0/N, namely:

1. Determination of the nuclear cross section.
2. Estimation of the deviation from exponential of the γ-ray attenuation law for the absorber (the so-called photon "transfer").
3. Conversion from the magnetic field scale, measured in NMR frequency units, to the usual energy scale (in MeV).
4. Evaluation of the accuracy of the theoretical calculations.

Following this, curves for the nuclear γ-ray absorption cross section will be given.

2. Determination of the Nuclear Cross Section

It was shown in Chapter II that the relations involving the nuclear cross section and the attenuation coefficient N_0/N as a function of the resonance parameters in the cross section under study – the ratio of resonance width to resolution $\Gamma/\Delta k$ and the product $n\sigma_{0n}$, where n is the number of nuclei per cm^2 in the absorber and σ_{0n} is the maximum value of the cross section.

For cross sections which change slowly in comparison with the resolution, the measured value agrees with the true value (formula (2.10)):

$$\sigma_n = -\frac{1}{n}\ln(1 - A) = \frac{1}{n}\ln\frac{N_0}{N} - \sigma_a, \tag{4.1}$$

where σ_a is the cross section for "non-nuclear" processes.

If the width of the resonance is comparable with the resolution, or is less than it, one measures a cross section which is averaged over the energy range in which the resolution function is different from zero. In the case $n\sigma_{0n} \ll 1$

$$\bar{\sigma} = \frac{1}{n}\left(1 - \frac{e^{n\sigma_a}}{N_0/N}\right). \tag{4.2}$$

Expression (4.2) is multiplied by the coefficient $1/\alpha$ if $\Gamma/\Delta k \ll 1$ and $n\sigma_{0n} \gg 1$. The value of α can be computed (see formula (2.14)) if the magnitude of $n\sigma_0$ is known for a given resonance.

We shall consider what data for the magnitude of the cross section and for the width of individual resonances can be obtained from the results of investigations of partial reactions in O^{16} and C^{12}. We shall make use of only the most recent experiments performed with higher resolution and accuracy.

Oxygen O^{16}. A study of the (p, γ_0) reaction has shown that the resonance widths in the cross section are greater than 0.35 MeV for excitation energies above 19 MeV [38]. This conclusion was also verified by the results of studies of photoneutron [39] and photoproton [40] spectra. However, the cross section for the (γ, n) reaction, which was obtained from the yield curve by the "second difference" method [41], contains a series of narrow peaks (with width $\Gamma \sim 20$-50 keV). It was established [38] that such resonances were either absent in the (γ, p_0) reaction, or, if they were present, $\Gamma_p < (1/30)\Gamma_n$. Furthermore, neither were narrow resonances found in the absorption cross section obtained by the absorption method with monochromatic γ-rays from the T(p, γ) reaction (resolution 40-50 keV) in the 20.5-22 MeV range [42].

We assume that there are narrow peaks in the cross section for the (γ, n) reaction despite the contradictory results of [41] and [39]. For the values $n = 3.3 \times 10^{24}$ cm^{-2} and $\sigma_0 \simeq 15$ mb used with measurements of the O^{16} nuclear absorption cross section [41], we obtain $n\sigma_0 \simeq 0.05$ and $\alpha \simeq 1$-0.01. Since the integrated cross section for the (γ, n) reaction associated with narrow resonances [41] is 1% of the total cross section, the use of $\alpha = 1$ in the analysis leads to a negligibly small error of the order of 0.01% in the measured integrated O^{16} absorption cross section.

Meanwhile, it was concluded [43] that the values of the integrated γ-ray absorption cross sections for O^{16} and C^{12} obtained in these measurements were apparently underestimated by 10-30%. The apparent difference is a consequence of the assumption made in [43] that the O^{16} γ-ray absorption cross section (as well as that for C^{12}) consists of narrow, isolated resonances with width $\Gamma = 40$ keV and $\sigma_0 = 150$ mb. Such a cross section was constructed from indirect data obtained from analysis of the breaks in the yield curves for the (γ, n) reaction [3]. More precise spectral measurements [39, 40], and measurements of the partial reaction cross sections [38, 41] did not verify this conclusion.

Carbon C^{12}. Narrow resonances were not observed in the cross section for the (p, γ_0) reaction [44] or in photoneutron spectra [45]. The width of levels corresponding to breaks in the photoneutron yield curves [47] was determined from comparison with peaks in the absorption cross section obtained with monochromatic γ-rays [47] ($\Gamma \sim 150$-300 keV). Treatment of the experimental data can be carried out in accordance with formula (4.2).

Measurements with the carbon sample were also carried out in the excitation energy region located below the giant resonance (the threshold for the (γ, p) reaction in C^{12} is about 16 MeV). A narrow resonance was observed in the γ-ray absorption cross section at $k_p = 15.11$ MeV [48] ($\Gamma = 64.5 \pm 10.4$ eV, $\sigma_0 = 29.7 \pm 1.1 \times 10^{-24}$ cm^2, $\sigma_{int} = 3.01 \pm 0.60$ MeV-mb). A sample with number of nuclei in $n \simeq 8.5 \times 10^{24}$ cm^{-2} was used in the carbon measurements. For this peak, $n\sigma_0 = 250$ and $\alpha = 0.0714$. According to formulas (2.19) and (2.20), we have in this case

$$A/n = 0.0714\,\bar{\sigma}; \tag{4.3}$$

$$B/n = 0.0714\,\sigma_{int}. \tag{4.4}$$

The spectrometer resolution for 15 MeV γ-rays during the C^{12} measurements was about 160 keV. Using the resonance parameters in [48], we have $A/n \simeq 1.33 \times 10^{-27}$ cm^2 and $B/n \simeq 0.21$ MeV-mb. As a check, the value of $A/n = \Phi(n, \sigma_0, \Gamma, \Delta k)$ for the 15.11 MeV resonance was determined by numerical integration (see Appendix). The resulting value $A/n = 1.30 \times 10^{-27}$ cm^2 is in good agreement with the results of general considerations. Therefore, the use of formulas (4.1) and (4.2) for the analysis of experimental data related to this excita-

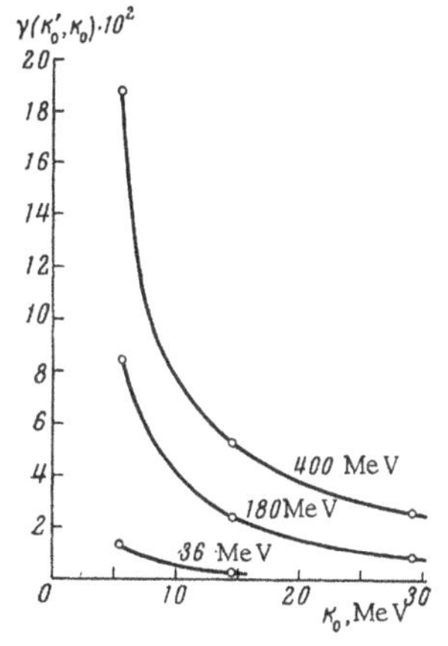

Fig. 8. "Transfer" of photons with energy k'_0 (H_2O).

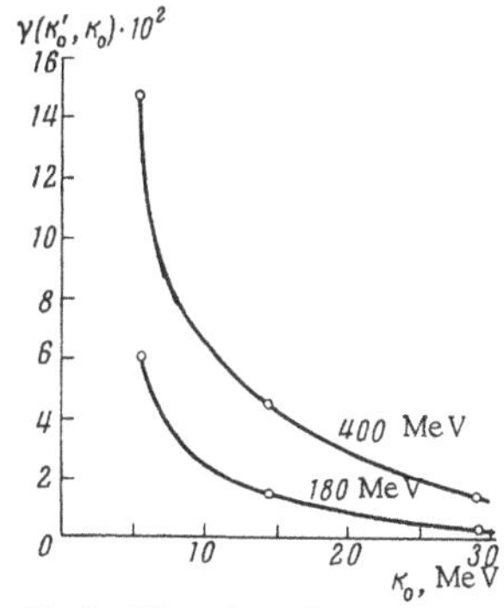

Fig. 9. "Transfer" of photons with energy k'_0 (C^{12}).

tion energy region in C^{12} led to the incorrect results $\sigma_0 = 1.33 \times 10^{-27}$ cm^2 and $\sigma_{int} = 0.21$ MeV-mb. A similar conclusion was reached in [43] where it was found from computer calculations that $\sigma_0 = 0.694 \times 10^{-27}$ cm^2 and $\int_0^\infty \sigma dE = 0.07143(\pi/2)\sigma_0\Gamma$. The quantity N_0/N must be measured with an accuracy of 0.02-0.3% in order that the level at 15.11 MeV be observed in such measurements. It was not observed in this experiment where the ratio N_0/N was measured with an error of approximately 1%. For this reason, the excitation energy region around 15 MeV in C^{12}, where the cross section was also calculated by formula (4.1), must contain an additional integrated cross section of 0.2 MeV-mb. It is considerably less than the experimental error, and will not be considered further.

3. Deviation from the Exponential Law

The well-known law for the attenuation of an γ-ray flux, $I = I_0e^{-\mu t}$, is valid for monochromatic γ-rays. It is necessary to estimate the magnitude of the correction to the exponential law associated with the use of a γ-ray bremsstrahlung spectrum. As mentioned, the measurements were made with a spectrum whose upper limit was approximately 260 MeV. Photons in the high-energy portion of the spectrum produce cascade showers when passing through matter. As a result, enhancement of the spectrum occurs in the medium- and low-energy portions.

TABLE 2. Photon "Transfer" Values for Water and Carbon Samples

k_0	Water			Carbon		
	ξ	$\theta \cdot 10^4$	$\eta \cdot 10^4$	ξ	$\theta \cdot 10^4$	$\eta \cdot 10^4$
10	0.475	1.8	0.8	0.247	1.8	0.5
15	0.412	3.0	1.25	0.195	3.0	0.6
20	0.318	4.3	1.4	0.158	4.3	0.7
25	0.228	5.3	1.2	0.100	5.3	0.5
30	0.475	1.8	0.8	0.247	1.8	0.5

We shall estimate the magnitude of photon "transfer" from cascade curves. The integral cascade curves for air, which give the number of photons with energy above k'_0 in a shower produced by a γ-ray with energy k'_0, are given in [49]. The differential spectra $\lambda(k'_0, k_0)$ for carbon ($t = 4.25\ t_0$) and water ($t = 3\ t_0$), where t is absorber thickness in radiation units t_0, were obtained from these curves by interpolation (Fig. 8 and 9). Consideration of the γ-ray bremsstrahlung spectrum leads to the following formula for the number of γ-rays transferred by cascade processes into unit energy range about energy k_0:

$$\xi(k_0) = k_0 \int_{k^*}^{E_0} \gamma(k_0', k_0)\, \Gamma(E_0, k_0')\, dE. \tag{4.5}$$

Here, k^* is the minimum primary photon energy which contributes to the transfer. The factor k_0 appears for normalization of the γ-ray spectrum in unit energy range. The calculation assumes that $\Gamma(E_0, k'_0) \sim 1/k'_0$. This approximation overestimates the quantity $\xi(k_0)$ because the actual spectrum (thick target, $t = 0.15\ t_0$, and time spread of the γ-ray beam) varies with energy more rapidly than $1/k'_0$.

In addition, it is necessary to consider γ-ray scattering. The solution of the shower equations for small angles of deviation (considering only multiple scattering [5]) gives a formula for the mean squared angle of deviation $\bar{\theta}^2$:

$$\sqrt{\bar{\theta^2}} = \frac{1.51\, E_k}{\beta} \sqrt{F(\Xi)}, \tag{4.6}$$

where $E_k = 21$ MeV, β is the critical energy, and $F(\Xi)$ is a function of k_0 given in [50].

For light materials, the formula is valid up to an angle $\theta \sim \pi$ if $\Xi \ll 1$ (Ξ is the energy in the units usually used in cascade shower theory). The spectrometer records γ-rays which deviate from the direction of the primary photons by no more than the solid angle Ω_0. Therefore, the number of γ-rays which appear at a given k_0 as the result of "transfer," and which are recorded by the spectrometer as well, is

$$\eta(k_0) = \xi(k_0)\, \Theta(\theta). \tag{4.7}$$

Values of $\xi(k_0)$, $\Theta(\theta)$, and $\eta(k_0)$ are given in Table 2. It is assumed that Θ is equal to the ratio of the solid angles Ω_0 and $\bar{\Omega}$, which is determined by the mean square angle of $\sqrt{\bar{\theta^2}}$.

Thus, the γ-ray spectrum in the 10-30 MeV energy range is greatly changed after penetration of the absorber. To the primary photon beam attenuated by the exponential law is added a comparable number of γ-rays produced as the result of a cascade shower. Only the experimental geometry reduces the "transfer" effect to a small correction.

The accuracy of the calculation of the quantity ξ, as far as it can be estimated from the accuracy of the approximations made in the computation, is about 30%. It is more difficult to estimate the error in formula (4.6) for determining the mean square angle of scattering. We will assume that the quantity η, which characterizes the deviation from the exponential law, was determined with an error of 100%. On this basis, one can consider that the correction is constant over the entire energy range and is 1×10^{-4} for measurements of γ-ray attenuation in water and carbon.

We now write formulas for N/N_0 and σ_n including the additions introduced by cascade showers:

$$\frac{N}{N_0}(k) = \frac{\int_{k_0}^{E_0} \varepsilon(k)\, \Gamma(E_0, k) \left[e^{-n\sigma(k)} + \Theta(\theta)\, k \int_{k^*}^{E_0} \gamma(k', k)\, \Gamma(E_0, k')\, dk \right] \Delta(k - k_0)\, dk}{\int_{k_0}^{E_0} \varepsilon(k)\, \Gamma(E_0, k)\, \Delta(k - k_0)\, dk} = \overline{e^{-n\sigma(k)}} + \eta(k_0). \tag{4.8}$$

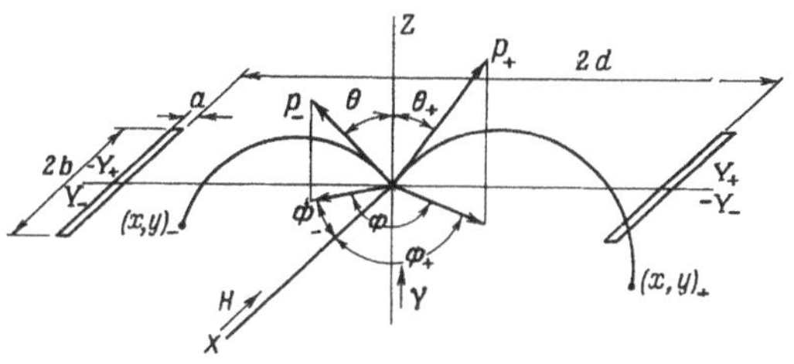

Fig. 10. Spectrometer geometry.

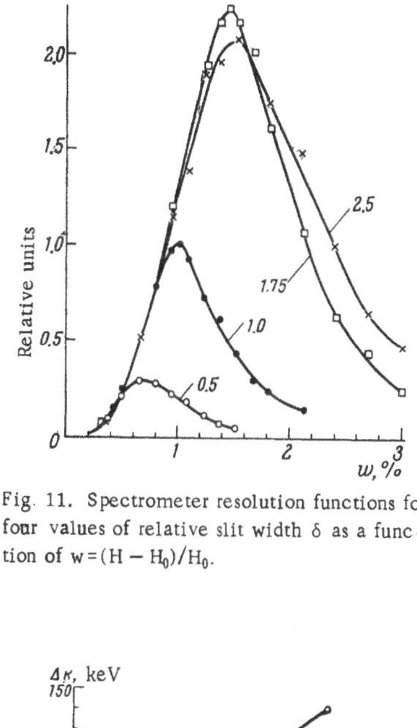

Fig. 11. Spectrometer resolution functions for four values of relative slit width δ as a function of $w = (H - H_0)/H_0$.

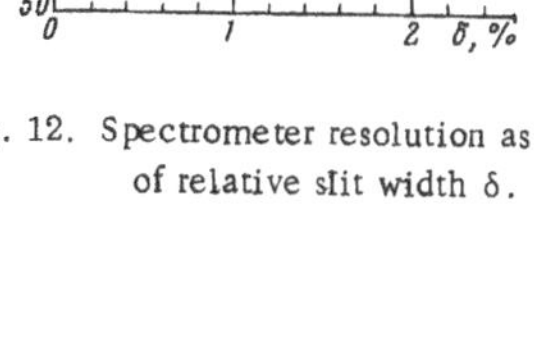

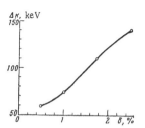

Fig. 12. Spectrometer resolution as a function of relative slit width δ.

Formulas (4.1) and (4.2), according to which one can compute the nuclear cross sections for these nuclei, as was shown in Section 2, are changed in the following fashion:

$$\sigma_n = \frac{1}{n} \ln \frac{N_0}{N} \left(1 + \eta \frac{N_0/N}{\ln N_0/N}\right) - \sigma_a; \tag{4.9}$$

$$\bar{\sigma}_n(k_0 + 2\Delta k) = \frac{1}{n}(A + \chi), \quad \chi = \eta e^{n\sigma_a(k_0+\Delta k)}. \tag{4.10}$$

The resulting correction is small in comparison with experimental error. A cross section calculated without consideration of the "transfer" effect turns out to be underestimated by 0.1%.

4. Determination of Nuclear Excitation Energy

As a result of the measurements, a set of $(N_0/N)_i$ was obtained corresponding to various values of the magnetic field H_{0i}. This section gives formulas for converting from the magnetic field scale to the usual energy scale (MeV). In this regard, it is necessary:

1. To determine the minimum γ-ray energy which is recorded by the spectrometer for a given magnetic field H_0;
2. to consider the dependence of spectrometer resolution Δk on the operating parameters of the spectrometer and on γ-ray energy. The resolution determines the energy range over which averaging of the measured cross section is performed; and
3. to determine the γ-ray energy k_i from the averaging range with which we agree to associate σ_i. The spectrometer resolution function has a shape that is close to triangular. The cross

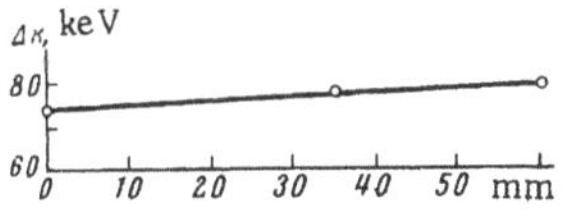

Fig. 13. Spectrometer resolution as a function of distance from the center of the radiator (2 mm slit width, 27 mm beam diameter).

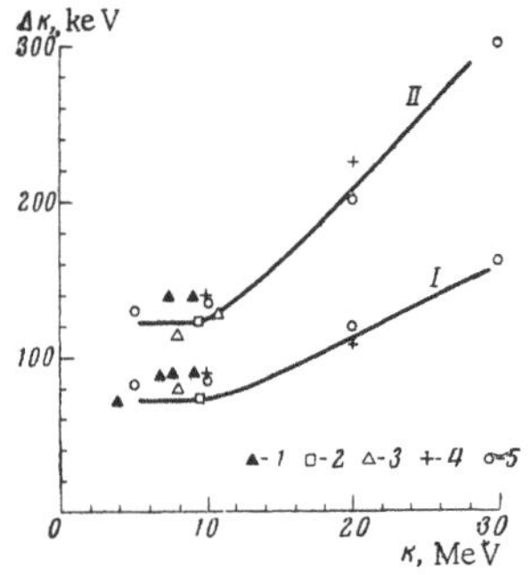

Fig. 14. Spectrometer resolution as a function of γ-ray energy. (I) $\delta = 0.01$; (II) $\delta = 0.02$. (1) Experimental data [17]; (2) experimental data; (3) theoretical data [17]; (4) theoretical data [51]; (5) data obtained by the graphical method.

section at γ-ray energy k_m, which is where the maximum is located, makes the greatest contribution to the averaged cross section. It is assumed that $k_i = k_m$ in the analysis of experimental data.

Determination of k_0. The coordinates of the pair components in the plane perpendicular to the direction of the γ-ray flux and the field H are (Fig. 10)

$$y_{\pm} = \frac{2p_{\pm}c}{eH} \cos\theta_{\pm}, \tag{4.11}$$

where $p_{\pm}$ are the positron and electron momenta; $\theta_{\pm}$ are the positron and electron angles of emission with respect to the normal to the plane of the radiator. The positron and electron intersect this plane (formed by the radiator and the slits which define the entrance apertures for the counters) at a distances

$$y_{+} + y_{-} = \frac{2c}{eH}(p_{+}\cos\theta_{+} + p_{-}\cos\theta_{-}), \tag{4.12}$$

and will be recorded, for a given field H_0 and infinitely long slits, if

$$2d \leqslant y_{+} + y_{-} \leqslant 2d + 2a,$$

where 2d is the distance between the inside edges of the slits; and a is the slit width. Detection of γ-rays with energy k_0 begins for the condition where

$$2d = y_{+} + y_{-} = \frac{2c}{eH_0}(p_{+}\cos\theta_{+} + p_{-}\cos\theta_{-})_{\max}. \tag{4.13}$$

The quantity in parentheses takes on a maximum value when $\theta_{\pm} = 0$ and $p_{+} = p_{-} = p_0$, i.e., the equality (4.13) holds for pairs produced half way between the inside edges of the slits by γ-rays with energy k_0 and emitted normally to the plane of the radiator. Taking this into account, one can write (4.13) in the form

$$2p_0c = edH_0. \tag{4.14}$$

Since $k_0 = 2p_0cS(\gamma)$, where $S(\gamma) = [1 - (2\gamma)^2]^{-1/2}$, and $\gamma = \mu/k_0$,

$$k_0 = edH_0S(\gamma). \tag{4.15}$$

Formula (4.15) was obtained in [17] for determining the resonance energy k_p of monochromatic γ lines.

In the experiments, H_0 was measured in NMR frequency units f, by means of which the magnetic field of the spectrometer was stabilized. H_0 is connected with f by the equality

$$2\pi f_0 = gH_0, \tag{4.16}$$

where $g = (2.67523 \pm 0.00006) \times 10^4 \sec^{-1} \cdot G^{-1}$ is the gyromagnetic ratio of the proton.

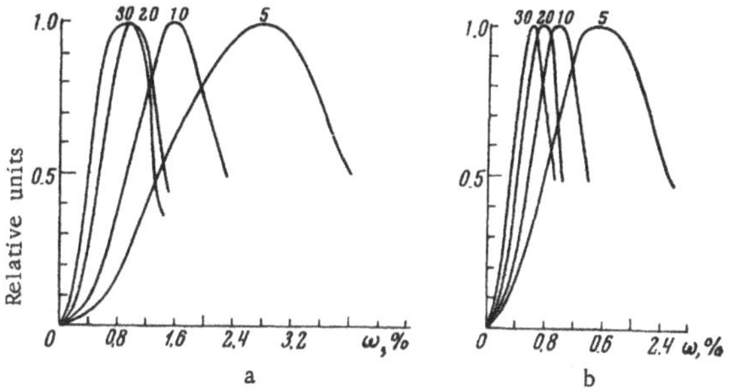

Fig. 15. Instrumental function of the spectrometer computed by a graphical method. (a) $\delta = 0.02$; (b) $\delta = 0.01$. Numbers on the curves are values of k in MeV.

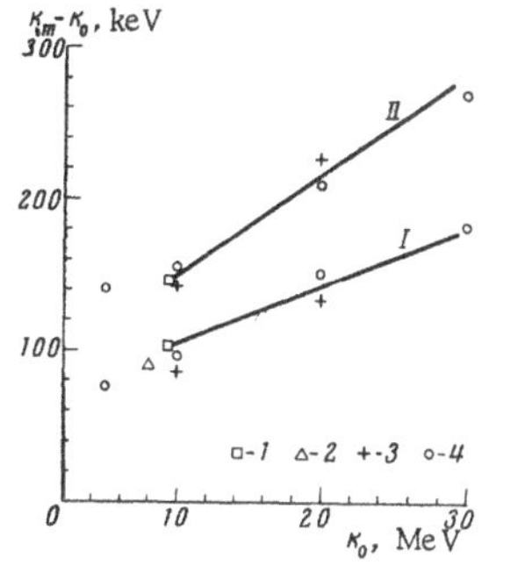

Fig. 16. Conversion values of $k_m - k_0$ depending on γ-ray energies. (I) $\delta = 0.01$; (II) $\delta = 0.02$. (1) experimental data; (2) theoretical data [17]; (3) theoretical data [51]; (4) data obtained by the graphical method.

For the single-channel spectrometer which was used in this work,

$$k_0 = 1.407502\, S(\gamma)\, f_0. \tag{4.17}$$

Here, k_0 is measured in MeV, f_0 in MHz, and $S(\gamma)$ is a slowly varying function of γ-ray energy which is close to unity. When $k_0 = 10$ MeV, $S(\gamma) = 1.005205$; $S(\gamma) = 1.000706$ when $k_0 = 27$ MeV.

Determination of the Resolution. The shape of the coincidence peak was measured at the ITEF reactor for monochromatic 9.716 MeV γ-rays from the Cr^{53} (n, γ) reaction. Results for four values of relative slit width δ are shown in Fig. 11. The spectrometer resolution Δk is shown in Fig. 12 as a function of δ. It was also found that the resolution was practically no poorer when an extended radiator of considerable dimensions was used (Fig. 13).

The resolution function of a pair spectrometer was computed theoretically for several values of γ-ray energy up to k = 10.8 MeV [17]. The calculations were checked experimentally. The shape of the coincidence peak for the pair spectrometer used was calculated by a method similar to that of Kinsey [17] for two γ-ray energies – 10 and 20 MeV [51]. In these papers, the resolution function was determined by integration of the pair production differential cross section along trajectories determined by spectrometer geometry. The Bethe-Heitler formula [52] was used for the pair production cross section without the inclusion of screening and Coulomb corrections. This made the computed resolution somewhat better than the true one. A similar effect on the resolution was produced by the assumptions made in the calculations: (1) an infinitely thin, point radiator midway between the slits; no scattering or electron energy loss occurs in the radiator; (2) there is no scattering by the edges of the slits or the walls of the chamber; (3) the magnetic field is uniform. No corrections were made for these effects.

At the same time, it was assumed that the slits were infinitely thin and long. The actual dimensions of the slits were taken into account by corrections which improved the resolution.

By comparing the calculated and experimental data for the resolution (Fig. 14), one can arrive at conclusions with respect to the importance of the factors 1-3, which have been enumerated above, and about the accuracy of the calculations. Theoretical values for the resolution Δk are given in Fig. 14 without corrections for the actual dimensions of the slits. Nevertheless, the experimental Δk is approximately 10-15% higher than the calculated value in Kinsey's paper [17]. Corrections for slit length and thickness are inversely proportional to γ-ray energy and the quantity δ. From the estimates of Kinsey, they amount to several percent at the energies under consideration. Evidently, one can consider that the accuracy of such calculations is of the order of 20%.

There is considerably less data on resolution in the 10-30 MeV γ-ray energy range of interest to us. Therefore, the shape of the coincidence peaks was calculated for four values of γ-ray energy k = 5, 10, 20, and 30 MeV by a qualitative, graphical method (Fig. 15). The spectrometer geometry determines the resonance shape of the resolution function (this was shown in [18] for σ_{pair} = const). The coincidence peak as a function of H can be represented by the difference of two coincidence curves determined by the inside and outside edges of the slits, respectively. The calculated curve in [17] for $\delta = 2.8\%$ was used as the first curve. The behavior of the other was found from consideration of spectrometer geometry.

The results of this calculation are shown in Fig. 14. They agree with the other data at 5, 10, and 20 MeV. Extrapolation to 30 MeV from the data in [51] leads to the conclusion that the results of the graphical method agree with the results of the theoretical method within the limits of error. The curves in Fig. 14 were drawn by eye for $\delta = 0.01$ and 0.02. It was assumed that Δk = const up to γ-ray energies of the order of 10 MeV in accordance with the results in [17, 18]; beyond that, the resolution increases linearly.

The quantity $2\Delta k$ determines within 10% the γ-ray energy range which is recorded by the spectrometer:

$$k_m - \Delta k \leqslant E \leqslant k_m + \Delta k,$$

where k_m is the energy corresponding to the maximum of the coincidence peak. Because of the asymmetry of the resolution function, the position of its center of gravity is shifted by 10-20% toward the high-energy side (see Fig. 11). It has been shown [18] that the magnitude of this shift is constant for $\delta = 0.009$ in the 5-11 MeV γ-ray range.

Determination of k_m. The energy of the maximum of the coincidence peak is determined from the same data as the spectrometer resolution. The accuracy of the determination of this quantity is somewhat greater than the accuracy of the calculation of the resolution because, for example, the corrections for finite slit thickness and length have practically no effect on the position of the maximum [51]. Values of the difference $k_m - k_0$ are given in Fig. 16 for a number of γ-ray energies. Line I and II are drawn through the data so that they agree with the experimental value for $k_m - k_0$ at the energy k_0=9.716. Equations for the lines are

$$\text{I. } k - k_0\,[\text{keV}] = 3.74\, k_0\,[\text{MeV}] + 66.6 \quad (\delta = 0.01); \tag{4.18}$$

$$\text{II. } k - k_0\,[\text{keV}] = 6.43\, k_0\,[\text{MeV}] + 85.7 \quad (\delta = 0.02). \tag{4.19}$$

From them, we obtain for k_m

$$k_m = 1.00374\, k_0 + 0.0666 \quad (\delta = 0.01); \tag{4.20}$$

$$k_m = 1.00643\, k_0 + 0.0857 \quad (\delta = 0.02). \tag{4.21}$$

Finally, substituting the value for k_0 in (4.17), we have

$$k_m = 1.41276\, S(\gamma) f_0 + 0.0666\ (\delta = 0.01); \tag{4.22}$$

$$k_m = 1.41655\, S(\gamma) f_0 + 0.0857\ (\delta = 0.02). \tag{4.23}$$

For the nine-channel spectrometer:

$$\left.\begin{aligned} k_m^{I} &= 1.33075\ S(\gamma) f_0 + 0.0666 \\ k_m^{II} &= 1.36596\ S(\gamma) f_0 + 0.0666 \\ k_m^{III} &= 1.40116\ S(\gamma) f_0 + 0.0666 \\ k_m^{IV} &= 1.43637\ S(\gamma) f_0 + 0.0666 \\ k_m^{V} &= 1.47158\ S(\gamma) f_0 + 0.0666 \end{aligned}\right\} (\delta = 0.01); \tag{4.24}$$

$$\left.\begin{aligned} k_m^{I} &= 1.32371\ S(\gamma) f_0 + 0.0857 \\ k_m^{II} &= 1.35892\ S(\gamma) f_0 + 0.0857 \\ k_m^{III} &= 1.39412\ S(\gamma) f_0 + 0.0857 \\ k_m^{IV} &= 1.42933\ S(\gamma) f_0 + 0.0857 \\ k_m^{V} &= 1.46453\ S(\gamma) f_0 + 0.0857 \end{aligned}\right\} (\delta = 0.02). \tag{4.25}$$

5. "Non-nuclear" Portion of γ-Ray Interactions with Materials Studied

The cross section for the interaction of γ-rays with nuclei is less than 10% of the total cross section for the interaction of γ-rays with matter even in light elements. It is obvious that if we do not have precise information about the behavior of 90% of the total cross section, we can only obtain information about the energy location of resonances in the cross section by this kind of measurement. In order to obtain quantitative data about the nuclear γ-ray absorption cross section in the materials being studied, it is necessary to know the "non-nuclear" part σ_a of the γ-ray interaction cross section. We cannot use the results of experimental work on the determination of σ_a because they contain considerable error and because the nuclear cross section was subtracted, which is exactly what is to be determined. Consequently, it is necessary to use theoretical values of σ_a. This section will give the results of theoretical calculations of the principal parts of the "non-nuclear" cross section – Compton scattering and pair production in the field of the nucleus and in the field of the electron.

Compton Scattering of γ-Rays. The fraction of Compton effect in the attenuation of γ-rays during penetration through carbon and water roughly varies from 75% at 10 MeV to 45% at 30 MeV. The cross section for Compton scattering by a free electron is given by the well-known Klein-Nishina formula [53]. The so-called radiative corrections (terms of higher order in the expansion of the Hamiltonian when perturbation theory is used) and double Compton scattering were not taken into account in the derivation of this formula. Both effects are small (~ 1%), and are of different sign [54, 55]. The accuracy of the expression for the Compton scattering cross section [53], which is estimated from the accuracy of the calculation of the corrections mentioned above, is about 0.3-0.5%.

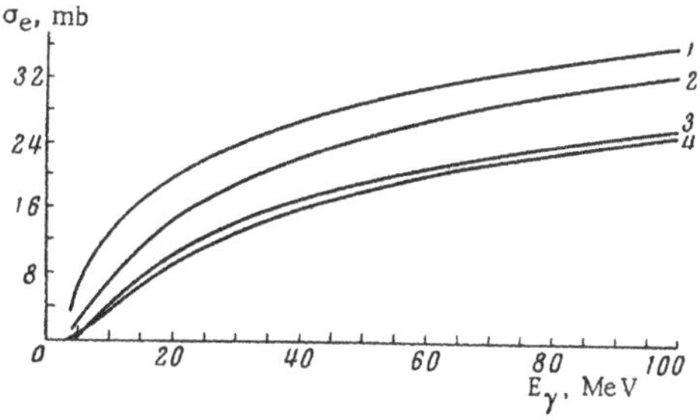

Fig. 17. Pair production cross section in the field of an electron [61]. (1) [57]; (2) [59]; (3) [58]; (4) [60].

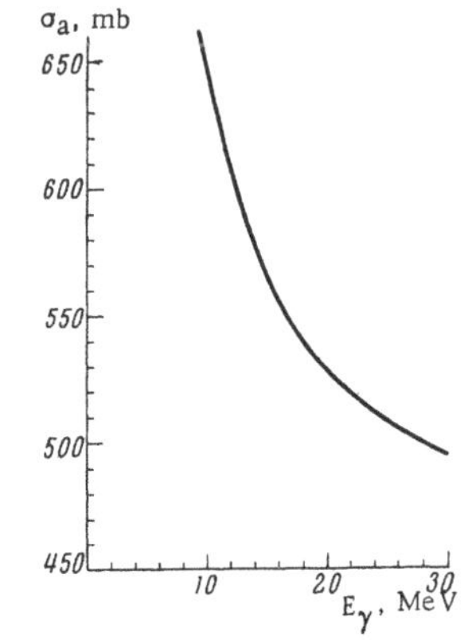

Fig. 18. "Non-nuclear" γ-ray interaction cross section for H_2O.

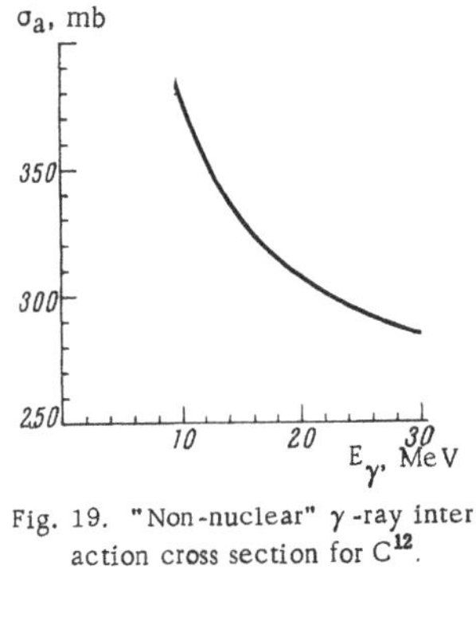

Fig. 19. "Non-nuclear" γ-ray interaction cross section for C^{12}.

Pair Production in the Field of the Nucleus. The pair production cross section for the nuclear field was obtained by Bethe and Heitler [52] in the Born approximation. Because of screening and the deviation of electron wave functions from plane waves (Coulomb correction) [56], the magnitude of the Bethe-Heitler cross section is reduced by approximately 2%. Since the screening correction is calculated insufficiently accurately, the uncertainty in the pair production cross section is evidently 1-2%.

Pair Production in the Field of an Electron. The probability of pair production in the field of an electron is approximately 1/Z of the Bethe-Heitler pair production cross section. However, the differences in the results of calculations by different authors are greater than the sum of the errors in the basic processes discussed in preceding sections. We will, therefore, consider these results in more detail.

Electrons in an atom can be considered free for the γ-ray energy range under consideration. We therefore exclude from consideration the results of Wheeler and Lamb [57] and of Joseph and Rohrlich [58], which were obtained for bound electrons. Calculations for free electrons were made by Borsellino [59] and Votruba [60].

Votruba calculated the differential pair production cross section for the free electron in a general form with an accuracy equivalent to that of the Bethe-Heitler formula [52]. However, corrections were required during integration which changed Votruba's results at the lower limit of the actual cross section. When $k \gg \mu$, the calculation of Votruba, with a small correction by Joseph and Rohrlich [58], gives

$$\sigma_e^V = \frac{r_0^2}{137}\left(\frac{28}{9}\ln\frac{2k}{\mu} - \frac{100}{9}\right). \tag{4.26}$$

σ_e^V differs from Φ_{pair}^{BH} (Bethe-Heitler pair production cross section for Z = 1) only by a constant equal to 82/27.

Borsellino obtained the total cross section σ_e^B in the form

$$\sigma_e^B = \Phi_{pair}^{BH} - \frac{r_0^2}{137}\frac{\mu}{k}\left[\frac{4}{3}\left(\ln\frac{2k}{\mu}\right)^3 - 3\left(\ln\frac{2k}{\mu}\right)^2 + 6.84\ln\frac{2k}{\mu} + 21.51\right]. \tag{4.27}$$

At very high energies, $\sigma_e^V \simeq \Phi_{pair}^{BH}$. The Borsellino calculation does not take into account volume effects, which reduce the cross section by an amount $\sim \mu/k$. Therefore σ_e^B is an upper limit of the pair production cross section for the free electron, practically coinciding with it in the energy range where volume effects can be neglected (γ-ray energies $k \gg 100$ MeV). The results of the above-mentioned calculations of the pair production cross section in the field of an electron are shown in Fig. 17.

Calculation of σ_e in the Weizsacker-Williams approximation (for $k \gg \mu$) leads to a value for the cross section lying approximately half way between σ_e^B and σ_e^V. Measurements of σ_e [62] also indicated that in the 40-100 MeV range the value of the cross section was located between the curves for σ_e^B and σ_e^V with the experimental points located closer to σ_e^V.

The calculation of Votruba is evidently more correct than the Borsellino calculation in the 10-30 MeV range because it is impossible to neglect volume effects there. Consequently, we used the Votruba results in the following. Their accuracy can be estimated to be within 50%.

Small Effects. Other processes have no noticeable effect on the attenuation of γ-rays during penetration through matter. We mention a few of them.

1. Photoeffect. The probability of this process in carbon is 2.5×10^{-31} cm^2 at a γ-ray energy $k = 15$ MeV.

2. Delbrück Scattering. For $k \gg \mu$, the total cross section is $\sigma^D \simeq 0.385\,(z/137)^4 r_0^2$ [63]. (For O^{16}, $\sigma^D = 1.2 \times 10^{-30}$ cm^2.)

Thus the attenuation of a γ-ray beam by "non-nuclear" processes during penetration of an absorber is determined by the quantity $\sigma_a = \sigma_c + \Phi_{pair} + \sigma_e$ which is known from calculations with an accuracy of 2-2.5%.

The total "non-nuclear" cross section, summed over the components discussed above, has been tabulated [64]. Values of σ_a for water and carbon are given in Figs. 18 and 19, respectively.

Better known, and more often used, are the cross sections σ_a derived in a more recent calculation by the National Bureau of Standards which are tabulated in the Siegbahn book, "Beta- and Gamma-Ray Spectroscopy" [65]. The values of σ_a obtained in this work differ from those derived in [64] chiefly because the results of the Borsellino calculations were used for the determination of the pair production cross section in the field of the electron. As indicated above, the approximations used in the Borsellino calculations are valid for higher-energy γ-rays ($k \sim 100$ MeV). In the 10-30 MeV γ-ray energy range, the Votruba calculation is more correct, as has evidently been verified experimentally [62].

6. Normalization of the "Non-nuclear" Cross Section. Nuclear γ-Ray Absorption Cross Section

In the preceding section, it was noted that the uncertainty in the magnitude of the "non-nuclear" cross section σ_a, obtained from theoretical calculations, was about 2-2.5%. Since the nuclear cross section is itself 5-10% of the total, its absolute magnitude may be determined with considerable error ($\sim$ 50%). In addition, some underestimation of the value of N_0/N is possible because of the effect of the transfer of photons into the measured portion of the spectrum when there is an absorber in the beam (see Section 3). Consequently, in order to obtain a sufficiently reliable absolute magnitude for the nuclear cross section, it is necessary to normalize the theoretical curve to experimental data by using additional information about the nuclear cross section.

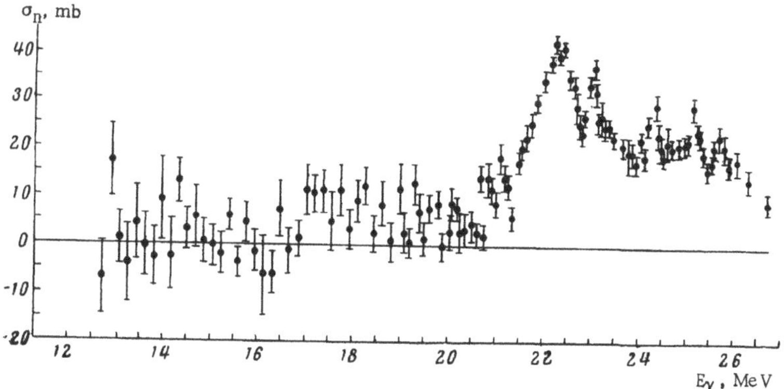

Fig. 20. Nuclear γ-ray absorption cross section in O^{16} obtained with a single-channel spectrometer.

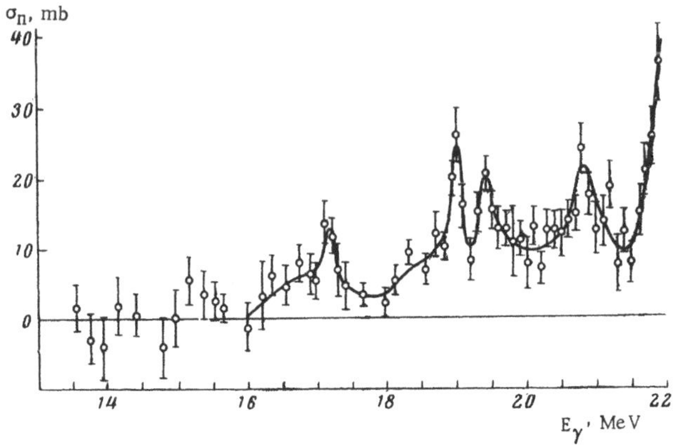

Fig. 21. Nuclear γ-ray absorption cross section in O^{16} below 22 MeV obtained with the 9-channel spectrometer.

To do this, a γ-ray energy region was selected where the nuclear cross section was negligibly small in comparison with the absorption cross section near the maximum of the giant resonance (or, in any case, considerably less than it). In practice, this means that the measurements must be made for γ-ray energies less than the thresholds for the main photodisintegration processes – the (γ, p) and (γ, n) reactions – or in their neighborhood. The nuclear cross section in this region is either assumed to be zero, or is estimated from other data if it exists. Since the magnitude of the nuclear cross section in this energy range is a few percent of the cross section near the maximum of the giant resonance, even a considerable error in such an estimate is unimportant. Accordingly, the difference between the experimental curve and the curve for "non-nuclear" processes in the normalization region is assumed equal to a quantity taken from other data or equal to zero.

Normalization in some comparatively small region assumes that the relative energy dependence of the "non-nuclear" cross section is determined with greater accuracy than its absolute magnitude. There is reason for considering that this is actually so. In particular, one of the main sources of uncertainty is the calculation

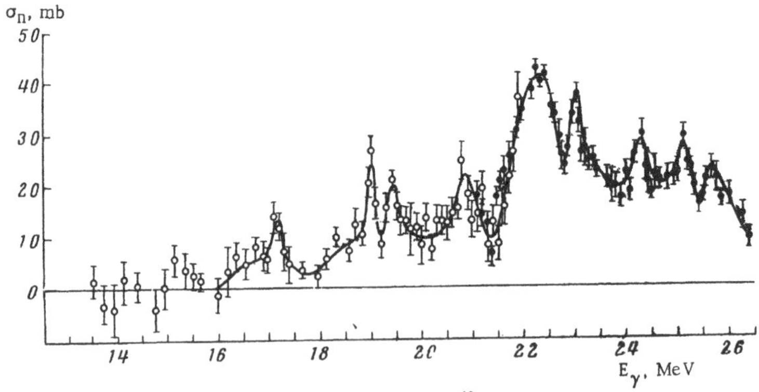

Fig. 22. Nuclear γ-ray absorption cross section in O^{16} in the 13.5-27 MeV energy range (combined data; below 21.5 MeV, the data are taken from Fig. 21).

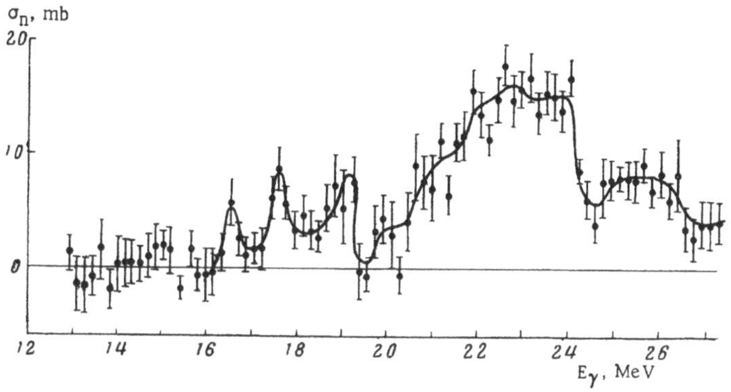

Fig. 23. Nuclear γ-ray absorption cross section for C^{12}.

of pair production in the field of an electron. Although the calculations of Borsellino [59] and Votruba [60] are considerably different in absolute magnitude, the cross sections obtained from them are similar in shape.

In the present instance, normalization was carried out in the following manner.

Oxygen O^{16}. Measurements were made almost to the threshold of the (γ, p) reaction (E_p = 12.1 MeV). Up to the threshold for the (γ, n) reaction (E_n = 15.6 MeV), the nuclear absorption cross section is made up of the cross sections for the (γ, p) and (γ, α) reactions and for elastic and inelastic γ-ray scattering. From existing data [66-68], the contribution of the last three processes is small in comparison with that of the first, and they can be neglected. Consequently, the nuclear absorption cross section in this energy region practically coincides with the cross section for the (γ, p) reaction. The 13.5-15.6 MeV energy range, in which the cross section for the (γ, p) reaction has no resonances, was selected for normalization. Therefore, possible inaccuracies in the energy scale found in various papers will have no important effect on the magnitude of the cross section. The average cross section for the (γ, p) reaction in this range is 1 mb [7, 69-71]. It was assumed that the average nuclear absorption cross section in this same range was also equal to 1 mb.

Carbon C^{12}. Below the threshold for the (γ, p) reaction (E_p= 15.95 MeV), the (γ, α) reaction and elastic and inelastic γ-ray scattering are possible. Measurements with the carbon sample were made down to 13 MeV excitation energy. Normalization was carried out in the 13-16 MeV energy range. The cross section for the (γ, α) reaction is about 0.05 mb in this range [72]. Neglecting this quantity as well as the contribution from γ-ray scattering [48, 68], which is considerably less than experimental error, it was assumed that the nuclear absorption cross section was zero in the 13-16 MeV range.

After normalization, the nuclear absorption cross sections were determined.

The nuclear γ-ray absorption cross section for oxygen in the 12.5-27 MeV excitation energy range, which was measured with the single-channel pair spectrometer, is shown in Fig. 20. Scale zero was determined with an accuracy of ± 1.5 mb. Below 21 MeV, the experimental errors are too large to permit any conclusions about the behavior of the cross section. One can only point out that its magnitude in this region is considerably less than in the 21-26 MeV range.

The nuclear γ-ray absorption cross section for oxygen in the 13.5-22 MeV range obtained with the 9-channel pair spectrometer is shown in Fig. 21. Scale zero was determined with an accuracy of ± 1 mb. The rapid increase in the cross section above 21.5 MeV is in good agreement, both in shape and absolute magnitude, with the behavior of the cross section in the preceding figure. Consequently, the results of these two measurements can be combined, using the data shown in Fig. 21 below 21.5 MeV. The combined results for the measurement of the γ-ray absorption cross section in oxygen in the 13.5-27 MeV excitation energy range are shown in Fig. 22. The errors are the standard deviations, and the curve has been drawn "by eye." The γ-ray absorption cross section in O^{16} has a distinct structure. The integrated absorption cross section over the measured range is 170 ± 20 MeV-mb.

The nuclear γ-ray absorption cross section for carbon in the 13-27 MeV excitation energy range is shown in Fig. 23. Errors are the standard deviations. Scale zero has an uncertainty of about ± 0.5 mb. There is structure in the cross section curve. The integrated absorption cross section is 84 ± 10 MeV-mb.

CHAPTER V

Analysis of O^{16} Data

The nuclear γ-ray absorption cross section for O^{16} in the 13.5-27 MeV energy range is shown in Fig. 22.

The cross section is close to zero for γ-ray energies below 16 MeV. As indicated above, measurements in the 13.5-16 MeV range were made for the purpose of normalization. Above 16 MeV, the cross section has a series of distinctly resolved peaks. A noticeable rise in the cross section begins at about 17 MeV where the first small maximum is located. Farther on – in the 18-21 MeV region – there are three peaks of approximately equal size; the average cross section in this region is about 12 mb. Above 21 MeV, there is a group of large peaks which also form the "giant resonance" in the O^{16} γ-ray absorption cross section. The peak parameters obtained from the experimental curve are given in Table 3. These parameters are estimates. The majority of the maxima overlap, leading to ambiguity in their separation. In this connection, the error in the quantities given, with the exception of E_m, is 30-50% depending on the magnitude of Γ.

The general behavior of the cross section in the 20.4-22.0 MeV energy range agrees with the results of recent measurements by Tessler and Stephens [42]. The nuclear absorption cross section for this energy range was obtained by them with a similar method, but monochromatic lines from the T(p, γ) reaction were used as the γ-ray source. This made it possible to use a high-efficiency detector (scintillation spectrometer) and to have good resolution at the same time (calculated value ~ 50 keV). In the opinion of the authors [42], peaks were found at energies of 20.6 and 21.0 MeV, while the present measurements give a single maximum at 20.9 MeV in this region. The sum of the integrated cross sections for these peaks is 14 MeV-mb. This quantity is in good agreement with the value of 15 MeV-mb obtained for the maximum at 20.9 MeV (Table 3).

In making measurements by this method, the energy location of the maxima is determined rather accurately making it possible to obtain reliable data on the energies of the corresponding nuclear levels. In the first instance, therefore, a comparison will be made of the location of peaks in the absorption cross section with structure in the cross sections for the main partial reactions (γ, n) and (γ, p) and in photoneutron and photoproton spectra.

TABLE 3. A) Parameters for Resonances in the O^{16} γ-Ray Absorption Cross Section

E_m, MeV	σ_m, mb	Γ, MeV	$\int\sigma dE$, MeV-mb	I, %
17.2	10	0.3	5	3
19.0	20	0.3	7	5
19.4	20	0.6	10	6
20.9	20	0.7	15	10
22.3	40	1.0	50	33
23.1	20	0.3	10	6
24.3	25	0.7	20	12
25.2	25	0.7	20	12
25.8	20	1.0	15	10

TABLE 4. Structure in Absorption Cross Section, (γ, n) Reaction Cross Section, and Neutron Spectra

Absorption cross section	Cross section for (γ, n) reaction			Neutron spectra			
	[76]	[77]	[74]	[75]	[45]	[39]	[73]
			16.0				
						17.1	
17.2	17.3	17.5	17.1				
						17.25	17.28
				18.5			
19.0			19.0			19.0	19.08
	19.3	19.4					
19.4			19.3			19.4	19.52
						20.1	
20.9		21.1	20.8			20.9	20.98
			21.8				
				21.8			
22.3	22.3	22.2	22.1		22.4	22.3	22.28
				22.8			
23.1		(23)	23.1			23.1	23.15
24.3	24.1	24.2	24.30	24.1	24.4	24.3	24.30
						25.0	
25.2		25.0					
						25.4	
25.8			26.2	26.0			

1. Structure in the Cross Section for the (γ, n) Reaction and in Neutron Spectra

Experimental data on structure in the absorption cross section and in the cross section for the (γ, n) reaction, obtained by direct measurement, are given in Table 4. Good agreement is noted with the results of Firk [39], Yergin et al. [73], and Bramblett et al. [74].

In the first two cases, the neutron spectrum was measured by the time-of-flight method at a linear accelerator. The energy resolution with which these results were obtained are the following:

E, MeV	18	20	22	24	26
ΔE, keV	13	42	68	95	125

The excellent resolution and high statistical accuracy of the measurements led to the uncovering of a considerably larger number of peaks in the neutron spectrum in these experiments than in those of Milone, et al. [75] and Fuchs, et al. [45]. The first of these was done with photographic plates, and in the second, the neutrons were detected by means of recoil protons in a stilbene crystal.

The maximum at 18.5 MeV, found in [75], is probably connected with an incorrect identification of the corresponding proton group in the spectrum, which in fact resulted from transitions to excited states in O^{15}. Such an explanation is also verified by [39] where it was established that additional maxima at 18.25, 18.70, and 19.9 MeV appeared in the neutron spectrum with target irradiation by a γ-ray bremsstrahlung spectrum with an upper limit of 32.5 MeV.

The refinement of the data for the location and number of maxima in the cross section for the (γ, n) reaction is apparent from a comparison of the results given in columns 2-4 of Table 4. In all these experiments, neutron detection was accomplished with BF_3 counters located in paraffin cubes.

Column 2 gives the results of measurements of cross section structure which were obtained with a γ-ray bremsstrahlung spectrum [76]. As is apparent from a comparison of these values with the peak locations in the absorption cross section, maxima corresponding to groups of peaks were observed in [76]. The width of the maxima is 1 MeV, or more, which indicates the order of magnitude of the resolution that can be obtained when

TABLE 5. Structure in γ-Ray Absorption Cross Section and O^{16} Photoproton Spectra

Absorption cross section	Proton spectra					
	[9]	[7]	[11]	[78]	[40]	[79]
		14.5				
						16,8
17.2		17.2			17,3	17,3
					18,1	18.1
				18,8		
19.0					19,0	19,1
19.4	19.6			19.3	19,6	19,7
20.9	20.6	20.7	20.7	20.6	20.6	20,8
		21,9				
			21,9			
22.3				22.2	22,3	22,0
23.1					23,1	
24.3			24.0	24.0	24,3	24.0
25.2				25.5		25,0
25.8						

calculating the cross section from a yield curve. In order to resolve such a coarse structure, very careful measurements of the yield curve are necessary (the standard error at an energy of approximately 25 MeV was roughly 0.5%). The cross section calculated by the Penfold-Leiss method [13] had an error of about 25%. The problem of attaining high accuracy for the cross section further limits the possibility of studying structure by this method at points removed from the reaction threshold.

Columns 3 and 4 give the results of measurements of the cross section for the $O^{16}(\gamma, n)$ reaction which were obtained with quasimonochromatic γ-ray beams [74, 77]. The resolution achieved in these experiments was ~ 3%. However, the results differ rather markedly. The structure in [77] resembles the "coarse" structure which was obtained with the use of a bremsstrahlung spectrum [76]. This is apparently associated with the fact that the measurement spacing was 0.5 MeV in [76]. Bramblett et al. [74] made measurements with a spacing of approximately 100 keV. The error in the measurement of individual points in the cross section did not exceed 5%. Thus they succeeded in establishing the complex nature of a number of maxima.

As is apparent from Table 4, "coarse" structure, which represented the result of averaging several neighboring maxima, was found in the experiments [45, 75-77]. The main peaks at γ-ray energies of 22.3 and 24.3 MeV were observed in all the experiments, with the exception of the work of Milone et al. [75], with a maximum deviation of 200 keV.

The cross section for the (γ, n) reaction was also determined from yield curves by the method of "second differences" [41]. The capabilities of this method are still not clear (details have not been published). Although the calculation of the cross section in [41] was carried out with a "resolution" of 68 keV up to 17.5 MeV and of 136 keV above that energy, the number of resonances found was considerably greater than in [39], which was done with better resolution and accuracy, and it is close to the number of breaks (see Table 7).

We shall discuss the slight discrepancies between the structure in the absorption cross section and the results of [39, 74]. In [39], two peaks were observed – at 17.10 and 17.25 MeV – instead of a single maximum at 17.2 MeV. Since the resolution in [39] was approximately 10 keV in this energy region, one can consider that the peak in the absorption cross section is the result of an averaging of two peaks with energies of 17.10 and 17.25 MeV.

There is an indication in the absorption cross section that there is also a broad maximum ($\Gamma \sim 1$ MeV) in this energy region.

TABLE 6. Structure in Absorption of Cross Section and (p, γ_0) Reaction Cross Section

Absorption cross section	Reaction $N^{15}(p, \gamma_0)O^{16}$		Absorption cross section	Reaction $N^{15}(p, \gamma_0)O^{16}$	
	[38]	[80]		[38]	[80]
	16.2				21.8
	17.1		22.3	22.3	
17.2			23.1	23.1	
	17.3		24.3	24.4	
19.0	19.1				24.7
19.4	19.5		25.2	25.1	
20.9	21.0		25.8		

The absence of a peak in the absorption cross section at 16.0 MeV, where one was found by Bramblett et al. [74], is associated with its smallness. The cross section at the maximum of the (γ, n) reaction is 0.2 mb (the integrated cross section is 0.2 MeV-mb) while the error in the absorption cross section is of the order of 3 mb in this region (this range was not covered in [39]).

The maximum at 20.1 MeV [39] is apparently connected with transitions to excited states in O^{15}, because it is absent in all the other experiments.

What has been said leads one to conclude that the structure in the absorption cross section and in the cross section for the (γ, n) reaction is practically the same. It should be emphasized that the agreement is maintained even for the considerably improved energy resolution which obtained in [39] and [73].

2. Structure in Proton Spectra and the (p, γ_0) Reaction Cross Section

Sources of information concerning structure in the cross section for the (γ, p) reaction are photoproton spectra and the inverse (p, γ_0) reaction. The location of peaks found in the absorption cross section and in proton spectra are given in Table 5.

Columns 2-5 and 7 give the positions of resonances obtained by the photographic plate method for target irradiation by a γ-ray bremsstrahlung spectrum. The results given in columns 2-4 were the first confirmations of the existence of structure in the cross section for the (γ, p) reaction. They have been described in detail in Chapter I. As can be seen in the table, only a small part of the data on structure in the cross section for the (γ, p) reaction was obtained in these experiments.

A considerable increase in accuracy and resolution was achieved in the work of Dodge and Barber [40] (column 6) where they measured the $(e, e'p)$ reaction. The protons were detected by a spectrometer with a resolution of $\sim 2.3\%$. In calculating the cross section, it was assumed that all the resonances in the spectrum were associated with proton groups whose emission left the N^{15} nucleus in the ground state. The existence of a resonance at 18.1 MeV is an obvious consequence of this assumption. For example, it could be produced by the emission of protons from an excited state in O^{16} located at 24.3 MeV to a level in N^{15} located at 6.3 MeV. The remaining resonances are in good agreement with the locations of maxima in the absorption cross section.

The last column gives the location of peaks found in proton spectra by Finck et al. [79]. In this work, the protons were detected with a scintillation spectrometer (CsI crystals); the resolution was 250 keV for a proton energy of 5 MeV. The statistical accuracy of the results is considerably less than that in the work of Dodge and Barber [40], and is close to the accuracy which is obtained with the use of photographic plates. A substantial reduction in the error of measurement is desirable for dependable separation of a number of maxima in the spectrum (for example, those at 19.1 and 19.7 MeV). Calibration of the energy scale was made on the basis of a peak to which the authors erroneously assigned an energy of 22.0 MeV. The location of the peak at 24.0 MeV instead of 24.3 MeV is also an indication of a shift in scale. For that reason, the agreement of the peak positions at 17.3, 19.1, and 20.8 MeV with data from the absorption cross section is not understandable.

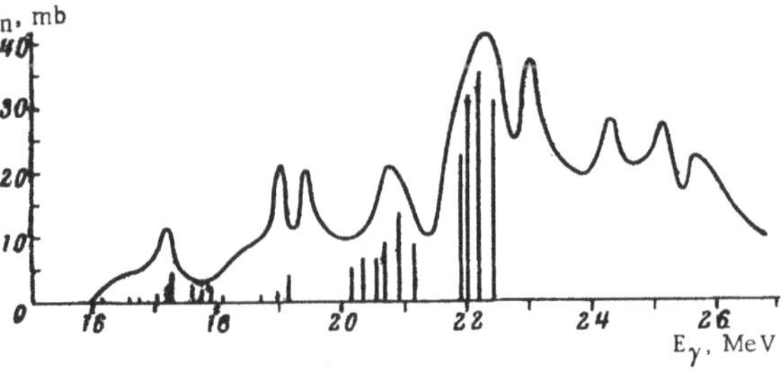

Fig. 24. O^{16} nuclear absorption cross section and interpretation of breaks in the yield curve for the O^{16} (γ, n) reaction [3].

As already mentioned, the maximum at 18.1 MeV is associated with proton emission leading to an excited state of N^{15}, and it is not a peak in the cross section for the (γ, p) reaction. It is clear from a comparison with absorption cross section that the resonance at 16.8 MeV must have the same sort of origin.

Data on structure in the (γ, p) reaction was also obtained by studying the inverse (p, γ_0) reaction which is connected with the (γ, p_0) reaction by the principle of detailed balance. Structure in the (p, γ_0) reaction is compared with structure in the absorption cross section in Table 6.

The table reveals the excellent agreement between structure in the absorption cross section and structure in the cross section for the (p, γ_0) reaction found in the work of Tanner et al. [38]. An additional peak at 16.2 MeV was observed in [38] which is not present in the absorption cross section, and a splitting of the maximum at 17.2 MeV into two resonances at 17.1 and 17.3 MeV was found in agreement with the data obtained for the (γ, n) reaction (see Section 1).

The method of studying structure with the help of the (p, γ_0) reaction has a great advantage insofar as resolution is concerned. The protons obtained from a tandem generator have an energy spread of about 5 keV. The resolution is practically determined by target thickness alone. On the average, the resolution was 30 keV in [38].

The measurements of Cohen et al. [80], which were made in the 21-26 MeV O^{16} excitation energy range, are also given in Table 6. They found only two peaks in the cross section for the (γ, p_0) reaction with the location of the main maximum at 21.8 MeV indicating considerable error in the determination of the energy scale. There were indications of a complex structure in the peaks at 21.8 and 24.7 MeV. However, the measurements were made in steps of approximately 0.3 MeV, making it difficult to arrive at more definite conclusions.

Thus, the structure in proton spectra and in the (p, γ_0) reaction agrees with the location of the maxima found in the absorption cross section. The agreement held up despite the fact that the cross section for the (p, γ_0) reaction was measured with considerably better resolution than was the absorption cross section. This result confirms the conclusions reached in the discussion of structure in neutron spectra.

3. Breaks

The structure found in the absorption cross section and the locations of the breaks observed in yield curves are compared in Table 7. We recall that breaks were considered as manifestations of narrow, isolated maxima in the absorption cross section. Papers [1, 3, 4] have been discussed in detail previously (see Chapter I). The work in [81, 82], performed more recently, is close in precision to that in [3, 4].

Above all, one is struck forcibly by the disagreement between the number of resonances in the absorption cross section and the number of breaks. The sole exception is the first results obtained by Katz et al. [1]; in this case, each break corresponds to a maximum in the absorption cross section, starting at 17 MeV. In later work [3, 81], it was indicated that the accuracy of the results in [1] was poor. For that reason, our main atten-

TABLE 7. Structure in Absorption Cross Section and Location of Breaks in (γ, n) Reaction Yield Curves

Absorption cross section	Breaks			
	[1]	[3, 4]	[81]	[82]
		15.60	15.60	15.65
				15.74
	15.9	15.99	15.85	15.92
		16.17	16.14	16.12
	16.4		16.45	16.41
		16.59		16.52
	16.7			16.70
	16.9	16.86	16.88	16.84
		17.05	17.05	
	17.1	17.12	17.15	17.15
17.2		17.23	17.21	
		17.27		17.33
		17.62		17.55
		17.74		17.68
		17.89		
		18.08		18.07
		18.69		18.72
19.0	18.9	18.97		18.89
		19.12		19.18
19.4	19.3			
		20.14		
		20.36		20.25
		20.55		20.58
20.9	20.7	20.67		
		20.91		
		21.17		
				21.59
	21.9	21.87		
		22.01		
22.3		22.19		
		22.41		

tion will be focused on a comparison with the results obtained in [3, 81, 82]. It is clear from the table that, with few exceptions, the location and number of the breaks shown in columns 3-5 agree with one another up to 19 MeV. A disagreement in the number of breaks is observed close to the region of maximum cross section values.

With respect to their location, the breaks can be divided into four groups, with each group corresponding to a maximum, or a group of maxima, in the absorption cross section. The cross section curve and breaks [3] are shown in Fig. 24. The ordinates are straight lines whose length is proportional to the value of the integrated cross section.

The first group extends from the threshold for the (γ, n) reaction (15.6 MeV) to approximately 18 MeV. It is clear from Fig. 24 that the breaks are grouped so that their envelope is the peak located near 17.2 MeV. It has been demonstrated that the maximum at 17.2 MeV consists of two resonances at 17.13 and 17.29 MeV [38] by comparison of the structure in neutron spectra and in the cross section for the (p, γ_0) reaction. In addition, there is another maximum at 16.2 MeV. Breaks corresponding to this resonance were found in all the experiments whose results are given in the table with the exception of the first paper by Katz et al. [1]. The width of the level at 16.2 MeV is 24 keV, as was established by a study of the (p, n) reaction [83]. The width determined by Penfold and Spicer [3] for the maximum in the cross section for the (γ, n) reaction corresponding to the break at 16.17 ± 0.05 MeV was 18 ± 5 keV. Therefore the peak at 16.2 MeV actually does correspond to a definite break in the yield curve.

It is more difficult to establish a connection with the breaks for the resonances at 17.13 MeV (Γ = 44 keV) and 17.29 MeV (Γ = 90 keV) because the breaks should appear below the resonance energy by an amount roughly

TABLE 8. O^{16} Energy Levels Excited by γ-Rays in the 16-26 MeV Range

Absorption cross section	(p, γ) cross section [38]	Neutron spectra [39]	Proton spectra [40]	(γ, n) cross section [74]
	16.2			16.0
	17.1	17.1		
17.2			17.3	17.1
	17.3	17.25		
19.0	19.1	19.0	19.0	19.0
19.4	19.5	19.4	19.6	19.3
20.9	21.0	20.9	20.6	20.8
		(21.6)		(21.8)
22.3	22.3	22.3	22.3	22.1
23.1	23.1	23.1	23.1	23.1
24.3	24.4	24.3	24.3	24.3
25.2	25.1	25.0		25.0
		25.4		
25.8				26.2

equal to its width (see Chapter I). Breaks corresponding to these resonances should be found at energies of 17.09 and 17.20 MeV. In the work of Penfold and Spicer [3], they may be among the breaks at 17.05, 17.12, and 17.23 MeV; in the work of King and Katz [81], they may be among those at 17.05, 17.15, and 17.21 MeV; there is only one break at 17.15 MeV (Table 7) in the paper by Sadch [82].

The next group of breaks is located close to 19 MeV. Here there are two maxima in the absorption cross section and two [1] or three breaks in the yield curves. On the basis of the value of the integrated cross section, one can assume that the resonances in the absorption cross section correspond to the breaks at 18.97 and 19.12 MeV. In this case, the first coincides with the peak in the cross section, and the second is shifted by 0.3 MeV toward the low-energy side. This indicates that the widths of the resonances are much different. There is no data on which one can base a more definite conclusion.

The third group of breaks is located in the 20-21 MeV range where the absorption cross section has only a single resonance (only two breaks were found in this region in [82]). There is no data with respect to the possibility of a comparison of any one of them with the peak in the absorption cross section.

The last group of breaks is associated with the principal maximum in the absorption cross section at 22.3 MeV. The integrated cross section corresponding to these breaks is considerably larger than that for breaks located at lower energies. (In [82], there is only one break in this region, and it corresponds to the maximum rise in the cross section.) The fact that the resonances corresponding to these breaks were not resolved in the absorption cross section, not even partially, indicates, at the very least, that their parameters differ from those reported in [3].

The cross section for the (p, γ_0) reaction was studied especially carefully [38] in the following O^{16} excitation energy ranges: 20.8-21.5 and 21.8-22.4 MeV. There were no indications of any splitting of the maxima at 20.8 and 22.3 MeV corresponding to "fine structure" (breaks).

In summary, it can be said that the peaks in the absorption cross section at 16.2, 17.1, and 17.3 MeV correspond to individual breaks in the yield curves. One can find a connection between peaks and breaks for the resonances at 19.0 and 19.4 MeV. For the majority of the breaks, one can only manage to establish an association between groups of breaks and maxima in the absorption cross section. The delineation of the groups is rather clear because they are separated by intervals of 0.8-1 MeV.

There arises a question as to why there was no detection of the other resonances whose manifestation in the yield curves are breaks according to assumptions. For a long time, this was explained by the insufficient resolution and low accuracy of the more direct measurements. Now the situation has changed in this respect.

TABLE 9. O^{16} Levels Determined by Absorption Cross Section and by Charged-Particle Reactions

Absorption cross section	Reaction cross section				
	(p, n), [83]	(p, p), [84]	(p, α_1), [85]	(p, α_0), [85]	(p, γ_0), [38]
			16.09		
	16.22	16.28	16.24		16.2
			16.68		
		16.79	16.86	16.73	
17,2	17.13	17.13			17.1
	17.29	17,34		17.31	17 3
		17.58	17.50		
	17.63		17.69		
	17.84	17.86	17.85		
	17.97		17.92		
	18.05	18.06	18.04	18.04	
		18.29	18.23		
		18.51	18.44		
		18.84	18.81		
19.0		19.19	19.28		19.1
19.4		19.49	19.40		19.5
		19.87	19.82		
		19.95	19.98		
		20.45	20.52	20.52	
		20.78	20.74		
20.9		21.02	21.06		21.0
		21.25			
		21.54	21.53		
		21.71	21.63	21.63	
		21.92			
22.3		22.32	22.28	22.28	22.3

The excellent agreement between structure in the absorption cross section and in the cross sections for the principal partial reactions, as well as their shapes, has demonstrated the adequacy of the various methods discussed above for the establishment of the O^{16} energy levels which are excited by the absorption of γ-rays. The absence of resonances in the cross sections for the (γ, p) and (γ, n) reactions presupposes their absence in the absorption cross section. In [38] (cross section for the (p, γ) resonance), the shape of the cross section in the neighborhood of 16.2 MeV was determined, and the width of the resonance was established. The integrated cross section for this peak is less than that for other peaks in this range. With that in mind, resonances corresponding to breaks at 16.86, 17.62, 17.74, 17.89, and 18.08 MeV (from the data of Penfold and Spicer [3, 4]) should have been observed. These resonances were not found. As mentioned above, neither were there found resonances corresponding to breaks in the 20.8-21.5 and 21.8-22.4 MeV regions.

Presently existing data are in agreement with the conclusion that each break is not necessarily associated with an isolated resonance in the absorption spectrum. Basically, peaks in the cross section are associated with groups of breaks. This can be explained, for example, by the fact that a group of breaks reflects either a group of overlapping resonances or a sharp change in the derivative at an individual maximum.

4. O^{16} Energy Levels Excited by Absorption of γ-Rays in the 16-26 MeV Energy Range

The considerations presented have shown that there exists good, and practically complete, agreement between the structure observed in the absorption cross section and in the cross sections for the principal partial reactions. This means that presently existing data represent reliable information about highly excited states in O^{16}. Summarized in Table 8 are experimental data on the energy levels in this nucleus which are excited by γ-rays in the 16-26 MeV energy range. The table does not include breaks because it was shown above that the collection of experimental data do not confirm a unique correspondence between breaks and resonances in cross sections.

The absence of a level at 25.0 MeV in the work of Dodge and Barber [40] can be explained by the reduction in accuracy in the "tail" of the experimental curve. Energy level values shown in parentheses [39, 74] correspond to changes in slope on the rising portion of the cross section curve in the region of the large maximum at 22.3 MeV.

The agreement between the various kinds of data can be considered as an indirect indication of the existence of charge symmetry in nuclear forces. It also makes it possible to assume that the resonances found in the absorption cross section correspond to individual O^{16} levels or to closely-packed, overlapping groups of levels.

5. O^{16} States Excited by Charged Particles

Up to now, O^{16} levels excited by photons have been discussed. The single exception was the (p, γ_0) reaction. However, it is the inverse of the (γ, p_0) reaction by virtue of the principle of detailed balance.

The locations of O^{16} levels observed in N^{15} + p reactions are shown in Table 9. A study of these reactions reveals fundamental information about O^{16} states in the 16-22.3 MeV excitation energy range. Reliable data are lacking above 22.3 MeV.

The levels printed in boldface in Table 9 correspond to strong resonances which are definitely associated with the N^{15} + p reaction. Gaseous nitrogen, enriched in N^{15}, was used as the target in such experiments. The amount of N^{14} was different in the different experiments; for example, it was 25% in [83]. Consequently, weak resonances might be associated with the N^{14} + p reaction. On this basis, a number of small peaks were omitted from the results in [83].

For purposes of comparison, O^{16} levels obtained from the cross section for the (p, γ_0) reaction [38] are given in the last column of the table; this paper was discussed in Section 2.

It is clear from Table 9 that a considerably larger number of states are excited in charged particle reactions (with the exception of the (p, γ_0) reaction) than in photon absorption. It can be assumed that this is not associated in any important way with insufficient resolution of the method by which the absorption cross sections were obtained. Such an assumption is based on a comparison with the results obtained from the (p, γ_0) reaction. As far as experimental possibilities are concerned, this method of studying excited nuclear states is similar to other methods where one studies reactions produced by protons: (p, p), (p, n), and (p, α) reactions. At the same time, considerably fewer levels are found in the (p, γ_0) reaction than in the (p, p) or other reactions. Thus the data in Table 9 indicate the existence of more rigid selection rules for level excitation by γ-ray absorption.

A comparison of the levels determined by the absorption cross section and by the (p, γ_0) reaction cross section with the levels obtained from the (p, α) reaction [85] shows that there is practically no correlation between them. This fact confirms the isotopic spin as a good quantum number. In this case, the emission of α particles is forbidden for a $T = 1 \rightarrow T = 0$ transition. In fact, none of the strong resonances in the cross section for the (p, α) reaction coincide with the peaks in the absorption cross section or in the cross section for the (p, γ_0) reaction. A well-known exception is the ($J^\pi = 1^-$, $T = 1$) level at 13 MeV, which decays by α-particle emission because it contains an admixture of a $T = 0$ state. It has been established that the source of this admixture is the ($J^\pi = 1^-$, $T = 0$) state at 12.4 MeV [86]. The results of a study of the O^{16} (γ, α) reaction [87] also are evidence for the low probability of a $T = 1 \rightarrow T = 0$ transition for α particles. The cross section curve for the (γ, α) reaction does not duplicate the giant resonance curves; the maximum of the curve is above 30 MeV where $T = 1 \rightarrow T = 1$ transitions become possible (the lowest $T = 1$ level in C^{12} is above 20 MeV). Nevertheless, a maximum corresponding to the giant resonance is also observed, which hints at some admixture of a $T = 0$ state.

There is no such limitation in the (p, p') and (p, n) reactions, and therefore levels which appear in reactions produced by γ-rays may also be excited. This fact was established in [38] for the levels located at excitation energies of 16.2, 17.1, and 17.3 MeV. Note also the agreement of levels in the absorption cross section with strong resonances in proton elastic scattering cross section [84] at 19.0, 20.9, and 22.3 MeV.

6. Theory

The existence of structure in the absorption cross section received a theoretical explanation on the basis of the shell model.

The process of γ-ray absorption by nuclei leading to the giant dipole resonance was first described in the language of the collective model [88]; according to this model, an oscillation of the protons with respect to the neutrons is produced by the absorption of photons. Different types of models [89, 90] lead to an energy dependence of the maximum of the giant resonance E_m on mass number A

$$E_m \sim A^{-1/3} - A^{-1/6}.$$

For heavy nuclei, the value of E_m agrees with experimental data, but it is much greater than experimental data for light nuclei. (For oxygen, $E_m \sim 30$ MeV.)

In the collective model, the nucleus is essentially considered as a harmonic oscillator. The shape of the γ-ray absorption cross section for such an oscillator is described by the Lorentz curve

$$\sigma = \frac{\sigma_0}{\left(\frac{E^2 - E_0^2}{E\Gamma}\right)^2 + 1}$$

with an experimentally introduced damping which is a consequence of the interaction of the collective dipole state with other degrees of freedom. The hydrodynamic model [90] was modified by Danos [91] and Okamoto [92] for aspherical nuclei. The Danos-Okamoto model explained the increase in the width of the giant resonance for aspherical nuclei and its possible splitting for nuclei with high eccentricity.

The assumptions of the hydrodynamic model are valid for heavy nuclei, and possibly medium nuclei, in which the level densities are very large at the excitation energies under consideration. From the point of view of this model, the giant resonance is a smooth resonance curve for spherical nuclei and a combination of two (or three) such curves for aspherical nuclei.

An opposite approach to the explanation of the giant resonance was proposed by Wilkinson [93], who used the shell model for this purpose. According to the single-particle model of Wilkinson, nucleon transitions into higher shells take place as the result of γ-ray absorption. If the nuclear potential is chosen in the form of an isotropic harmonic oscillator, all the dipole transitions will have an identical energy equal to the spacing between neighboring shells. Consequently, this approximation leads to the result which was obtained with the collective model. The transition energy obtained is [94]

$$E_m = 42\,A^{-1/3}\ \text{MeV}\ (r_0 = 1.2\cdot 10^{-13}\ \text{cm}).$$

This value for E_m is approximately twice as small as that found experimentally for medium and heavy nuclei. For light nuclei, the difference between experiment and theory is somewhat less for E_m; for example, the theoretical value of $E_m \simeq 17$ MeV for O^{16}.

For a nuclear potential approximated by a rectangular well, the high degree of degeneracy inherent in a harmonic oscillator is removed, and the E1 transitions have different energies. As Wilkinson has shown, the spread is 2-3 MeV. Transitions of the $nl \to n(l+1)$ types, where n is the principal quantum number and l is the orbital quantum number, have the highest probability.

In the Wilkinson theory, the width of the giant resonance is made up of three main parts:

$$\Gamma = \Delta E + \Gamma_{\text{part}} + 2W.$$

The first term includes the spread in transition energies mentioned above; the second is the single-particle width for direct decay of the excited state:

$$\Gamma_{\text{part}} \simeq \frac{\hbar^2}{MR} 2k_n P_l.$$

Here, k_n is the nucleon wave number; P_l is the transparency; and $\hbar^2/MR$ is the single-particle reduced width. Since the transitions are caused by nucleons with large orbital momentum values the single-particle width is small, of the order of 0.5 MeV. The third term, 2W, is associated with the smearing of each transition energy caused by "trapping" of the nucleon in the nucleus and by the distribution of its energy among all the other nucleons. The magnitude of 2W (imaginary part of the potential) was estimated by Wilkinson from nucleon scattering experiments and is 1-4 MeV. Therefore, the total width of the giant resonance for spherical nuclei should be 3-5 MeV, which is in agreement with experiment.

The energy at the maximum of the giant resonance in rectangular well calculations continued to remain close to the energy obtained with a harmonic oscillator potential and considerably less than the experimental value. Wilkinson increased it by assuming that a nucleon inside a nucleus has an effective mass equal to approximately half the mass of the free nucleon. It was shown later that the underestimate of the energy of the giant resonance could be explained by the neglect of residual pairing interactions between nucleons [95-99].

The role of the residual interaction was discussed by Brown and Bolsterli [96]. They showed that the particle-hole interaction mechanism mixed shell configurations, leading to removal of the degeneracies in particle-hole levels. Nuclear levels are described by a linear combination of the latter, with one of them, described by the most symmetric combination, possessing an energy considerably greater than the others. A large portion of the dipole sum also falls to its part. Contributions from all the particle-hole levels enter into the formation of this state giving it a collective nature resembling the coherent particle motion which is described by the collective model. This effect, which is connected with the off-diagonal portion of the residual interaction, is significant in heavy nuclei.

The discussion of Brown and Bolsterli indicates the existence of a connection between the shell-model and collective-model descriptions of the giant resonance. With photon absorption, states are excited as the result of shell configuration mixing that resemble the collective states produced by oscillation of protons with respect to neutrons. Brink [100] showed that the collective and shell models were identical for the particular case of a harmonic oscillator potential.

In the majority of absorption cross section calculations using the shell model, the following approach, first used by Elliott and Flowers [95], is employed:

1. The Hamiltonian is written as the sum of the single-particle Hamiltonian and the potential for particle-hole pairing forces:

$$H = H_0 + V, \quad V = \sum_{i<j} V_{ij}.$$

In the paper of Brown et al. [101], it was shown that the interaction force between particles and holes in cases where the isotopic spin T = 1 is a repulsive force, and that its consideration leads to an increase in the energy of the states under consideration.

2. Energy levels, without consideration of residual interactions (zeroth approximation), are determined from the experimental values of the corresponding single-particle levels in neighboring nuclei. For example, it is necessary to know the energy of the corresponding particle and hole states in nuclei with A = 15 and 17 in order to determine the energy of nucleon transitions from the 1p shell into the (1d, 2s) shell in O^{16}.

TABLE 10. Experimental and Theoretical Energy Levels and Relative Transition Probabilities for T = 1, 1^- States for O^{16}

This work			Theoretical papers											
			Elliott and Flowers [95]		Brown et al. [101]		Gillet and Vinh-Mau [102]				Eichler [103]		Yudin [104]	
E, MeV	J^π	I, %	E, MeV	I, %	E, MeV	I, %	E, MeV	I, %	E, MeV	I, %	E, MeV	I, %	E, MeV	I, %
			13.1	1	13.7	1	13.6	3	13.5	4	13.55	1		
17.2	1^-	3	17.3	1	17.6	1	18.1	1	18.1	1	17.38	2	18.0	1
19.0 }	$2^+, 1^+$	5												
19.4 }		6												
			20.4	1	20.0	1	19.6	2	19.6	2	19.96	1		
20.9	1^-	10											21.5	29
22.3	1^-	33	22.6	67	22.2	68	22.7	68	22.2	73	22.42	74		
23.1	1^-	6											23.0	19
													23.3	6
													23.7	10
24.3	1^-	12											24.8	16
25.2	1^-	12	25.2	32	25.0	29	25.4	26	25.2	20	24.31	22		
25.8	1^-	10												

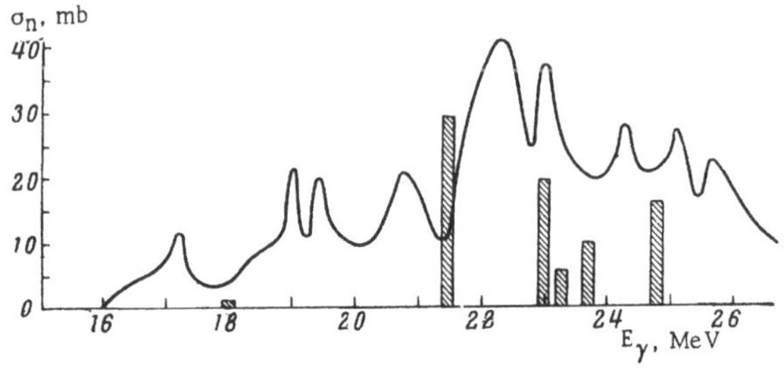

Fig. 25. Comparison of experimental and theoretical [104] γ-ray absorption cross sections for O^{16}.

3. The residual interaction of nucleons is taken into account. Its diagonal part, which corresponds to a correction for the change in the magnitude of the potential when a nucleon is removed from a closed shell, increases the transition energy by 2-3 MeV. The off-diagonal part of the residual interaction mixes particle-hole configurations and also increases the energy of the state (by 1.0-1.5 MeV in light nuclei).

Configuration mixing does not play an important role in the O^{16} nucleus. It has been shown [101] that the most intense transition corresponds to the $p^{-1}_{3/2}\, d_{5/2}$ configuration. Its energy is about 18 MeV. After consideration of the residual interaction, the most intense transition has an energy of about 22 MeV and contains 90% of the $p^{-1}_{3/2}\, d_{5/2}$ state. The remaining configurations, $p^{-1}_{3/2}\, s_{1/2}$, $p^{-1}_{1/2}\, d_{3/2}$, $p^{-1}_{1/2}\, s_{1/2}$, and $p^{-1}_{3/2}\, d_{3/2}$, contribute from 4 to 35% of their amplitude as admixtures to this state.

Values are given in Table 10 for transition energies and intensities in dipole absorption of γ-rays by O^{16} based on theoretical data [95, 101-104] and on experimental data obtained from the absorption cross section.

All the theoretical papers, with the exception of [104], predict that the O^{16} γ-ray absorption cross section should consist of five resonances with energies in the neighborhood of 13.5, 17.5, 20.0, 22.5, and 25.0 MeV with the predominant portion of the dipole sum occurring in the latter two.

The calculations made in [95, 101-103] were carried out in accordance with the general scheme given above. The slight differences in location of levels and in relative transition intensities indicate that a difference in the number of specific assumptions is of no great significance. In particular, the excellent agreement of the results in [95] and [101] shows that the calculations are insensitive to the simplifying assumptions made in the latter. In the first one, intermediate coupling and a Yukawa potential for the particle-hole interaction were used; in the second, they used j – j coupling and a zero-range force with two types of forces being considered: an ordinary force and a Soper mixture of forces

$$V = (0.3 + 0.43\,P_M + 0.27\,P_B)\,V(\mathbf{r}_1 - \mathbf{r}_2),$$

where P_M and P_B are the Majorana and Bartlett exchange operators. The results of calculations with the Soper forces are given in the table. In the calculation with ordinary forces, the main transition was located at 21.5 MeV, i.e., below the experimental value.

The results of Gillet and Vinh-Mau [102] which are given in Table 10 were obtained with j – j coupling and finite-range forces. In addition, they made calculations using the so-called RPA approximation, in which the excitation of an odd number of particle-hole pairs is considered (in this case, three pairs), and they took into account correlations in the ground state. In contrast to this calculation, the ground state of the O^{16} nucleus in the other papers was represented by the $(1s)^4(1p)^{12}$ configuration, i.e., by a system of 16 nucleons occupying the lowest single-particle states up to the Fermi surface and not interacting with one another. Both approaches lead to practically identical results for states with T = 1, but are greatly different for levels with T = 0.

TABLE 11. Radiation Widths of O^{16} Levels

Experimental data		Single-particle Weisskopf width [109] Γ_γ, eV	Theoretical data, Elliott and Flowers [95]	
E, MeV	Γ_γ, eV		E, MeV	Γ_γ, eV
			13.1	60
17.2	145	3 500	17.3	140
19.0	250	4 750		
19.4	375	5 000		
			20.4	20
20 8	650	6 200		
22.3	2500	7 700	22.6	12 000
23.1	530	8 500		
24.3	1200	10 000		
25.2	1260	11 000	25.2	5 800
25.8	1000	12 000		

All the calculations mentioned above were carried out with oscillator wave functions. Eichler did a calculation with a rectangular-well potential [103]. He used zero-range Soper forces for the particle-hole interaction. Four sets of values for transition energies and intensities were obtained depending on well depth and nuclear radius. The energy levels in the various sets differed little from one another. The third set of values is given in the table.

A comparison of the results of the theoretical calculations [95, 101-103] with the experimental data shows, first of all, that theory predicts too few transitions. The experimental level at 17.2 MeV is predicted nicely. This level is found in the 17.3-18.1 MeV energy range in the various theoretical papers. At higher energies, it is not clear whether a one-particle, one-hole state corresponds to one level, or to a group of levels, found experimentally.

Since only dipole transitions were considered in the theoretical papers, the disagreement may possibly be explained by the fact that a part of the experimentally observed resonances corresponds to transitions of other multipolarities. In particular, indications with regard to the quadrupole nature of the 24.3 MeV transition were obtained from experiments on inelastic electron scattering [105]. It was later established [40] that the contribution to the cross section for the (γ, p) reaction from quadrupole transitions did not exceed 1.8% in the neighborhood of the large maxima at 22.3 and 24.3 MeV. The asymmetry in angular distribution corresponding to such a contribution is possible without assumptions about the existence of a 2^+ state in the 22 and 24 MeV region [103]. The angular distribution also agrees with the assumption of a dipole character for the transition in the neighborhood of 20 and 17.2 MeV [38]. It is possible that the state near 19.0 MeV is excited by M1 and E2 transitions [7, 106, 107]. The assumed level characteristics are given in Table 10.

Consequently, all the transitions found, with the possible exception of the two at 19.0 and 19.4 MeV, should be considered to be dipole on the basis of the latest data. Hence one can conclude that consideration of only the lowest one-particle, one-hole configurations is not a sufficiently accurate approximation for carrying out theoretical calculations.

The paper of Yudin [104] showed that enhancement of the level spectrum in the giant resonance region could be obtained by consideration of an admixture of higher configurations. In the paper, a determination was made of the redistribution of dipole transition intensities resulting from the interaction of the usual shell-model levels with surface oscillations of the nucleus. The calculation was carried out in the one-phonon approximation. Instead of five levels, 18 states were obtained which were located in the region up to 40 MeV. Four sets of calculations were made for various values of phonon energy $\hbar\omega$ and of pair interaction amplitude V_0.

The results closest to the experimental data are shown in Table 10; these were obtained with the parameters $\hbar\omega = 10$ MeV and $V_0 = 45$ MeV. Only the transitions with energies up to 26 MeV are given; about 20% of

TABLE 12. Integrated Cross Section for (γ, n) Reaction, MeV-mb

	Source							
	[14]	[74]	[76]	[77]	[113]	[114]	[115]	[45]
Upper limit, MeV	31	30	30	25	30	33	32.5	30
$\int \sigma_{\gamma n} dE$	46±7	46±7	51	13	65±2	61±7	53±5	52±13
$\int^{27} \sigma_{\gamma n} dE$	38	37	40					40

the total integrated cross section is above this energy. In the energy range above 21 MeV, where the theoretical calculation did not take into account higher configurations, two transitions were predicted, and five transitions were obtained in [104] with more uniform distribution of intensities, which is considerably closer to the experimental results (Fig. 25). At the same time, Fig. 25 shows that the energy locations of the main transitions obtained in [104] do not agree completely with experiment. A transition with maximum intensity is predicted at 21.5 MeV, 0.8 MeV below the energy of the principal maximum found experimentally. As indicated in the paper, a small increase in the amplitude of the pairing interaction V_0 is necessary to improve agreement with experiment.

As far as the distribution of relative transition probabilities is concerned, the calculations indicate that about 95% of the dipole sum falls to the share of the two transitions at 22 and 25 MeV. Experiment indicates that 75% of the integrated cross section found is located in the 21-26 MeV range. The theoretical analysis underestimates the relative transition probabilities to levels below 21 MeV.

The relative intensities of the dipole transitions at 22 and 25 MeV vary from 2 : 1 to 3 : 1 for the various theoretical papers shown in Table 10. The experimentally determined integrated cross sections for the 22.3 and 25.2 MeV resonances in the absorption cross section are in the ratio 2.75 : 1. It is not clear whether this ought to be considered confirmation of the theory because a larger number of maxima were found experimentally. The suggestion about the transformation of each of the states as the result of considering interactions with higher configurations in the level structure of the compound nucleus [95] leading to the association of the 22.3 and 23.1 MeV resonances in the absorption cross section with the 22 MeV transition, and the remaining portion of the cross section in the 23.5-26.5 MeV range with the 25 MeV transition, is incorrect because the ratio of the integrated cross sections for the two resonance groups mentioned is approximately 1 : 1. Evidently the particle-hole state at 22 MeV is split in such a way by additional interactions that it is responsible for a considerable portion of the cross section at lower and higher energies.

The data obtained makes it possible to estimate the radiation widths of the states observed in the absorption cross section. The estimate was made by using the Breit-Wigner single-level formula [108]. In the present case, the radiation width Γ_γ is connected with the integrated cross section of a resonance in the following manner:

$$\Gamma_\gamma = \frac{1}{\pi^2 \lambda^2} \frac{1}{2s_0} \int \sigma \, dE,$$

where $\lambda = \hbar c/k$ is the photon wave length; $s_0 = [2J+1]/[2(2I+1)]$ is a statistical weighting factor; I is the spin of the initial state; and J is the spin of the excited state. For dipole transitions, $s_0 = {}^3/_2$. Values of Γ_γ for the O^{16} levels found are given in Table 11. For comparison, the single-particle radiation widths for E1 transitions [109] determined by

$$\Gamma_\gamma^W = 0.11\, A^{2/3} E^3, \text{ eV},$$

are also included along with the radiation widths obtained in the calculations of Elliott and Flowers [95].

TABLE 13. Integrated Cross Section for (γ, p) Reaction, MeV-mb

	Source					
	Absorption cross section	[7]	[113]	Proton spectra [40]	Proton spectra [79]	(γ, p_0) reaction [38]
Upper limit, MeV.	27	25.5	30	27	27	27
Integrated cross section.	105±20	76	100±4	56±11	57.9	33±8

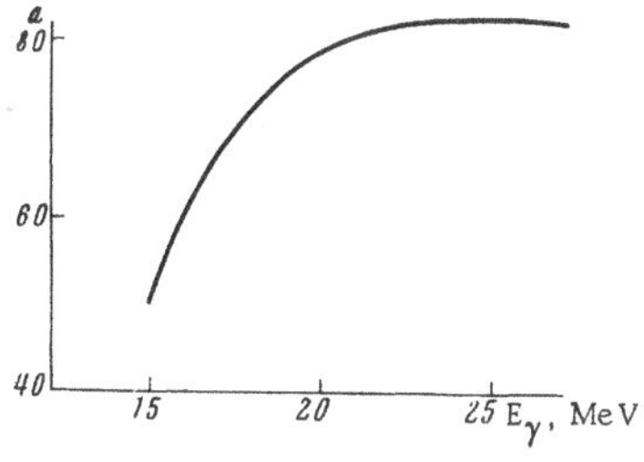

Fig. 26. Coefficient for calculating the cross section for the (γ, p_0) reaction from the cross section for the (p, γ_0) reaction.

The experimental values for Γ_γ indicate the dipole nature of the transitions under discussion because they are considerably larger than the single-particle widths for E2 and M1 transitions (for example, the single-particle Weisskopf width [110] of the E2 transition for the 23.1 MeV state is $\Gamma_\gamma = 41$ eV).

Wilkinson [109] did shell-model calculations of E1 transition intensities for a number of light nuclei. The transition intensity is expressed by the ratio of the experimental and single-particle radiation widths $|M|^2 = \Gamma_\gamma / \Gamma_\gamma^W$. The $|M|^2$ values calculated by Wilkinson roughly lie in the range from 1 to 10^{-4}, and the maximum of the distribution corresponds approximately to 0.05. The experimental values of $|M|^2$ vary from 0.04 to 0.3, and are therefore in agreement with the Wilkinson calculations. However, as noted in the discussion of relative transition probabilities, there is no quantitative agreement of experiment with calculations of specific transition intensities. This can be seen from a comparison of our data with the radiation widths calculated by Elliott and Flowers [95].

In the work of Cohen et al. [80], in which the differential cross section for the (p, γ_0) reaction was measured, an estimate was obtained for the radiation width of the state at 21.8 MeV, which corresponds to unresolved levels at 22.3 and 23.1 MeV, and of the state at 24.7 MeV, which obviously also corresponds to the two levels at 24.3 and 25.2 MeV. The value $\Gamma_\gamma = 2.2$ keV obtained for the first one can be considered in agreement with the value $\Gamma_\gamma \approx 3.0$ keV obtained here for the two levels at 22.3 and 23.1 MeV. An overestimate ($5 \leq \Gamma_\gamma \leq 14$ keV) was obtained [80] for the state at 24.7 MeV. This was apparently connected with an incorrect evaluation of the probability distribution for transitions between the ground state and excited levels in N^{15}.

Estimates of the widths of dipole levels were made in [103] and [111]. A value of 0.5-1.2 MeV was obtained in the first paper (on the assumption $\Gamma_n \simeq \Gamma_p$) for the various kinds of calculations we have mentioned above. The widths of the primary states at 22.2 and 25.0 MeV, determined in [111], were 0.75 and 1.0 MeV. The calculated values are in agreement with experimental values in the range 21-26 MeV equaling 0.3-1 MeV.

7. Integrated Cross Section

According to the sum rule [112], the integrated cross section for electric dipole absorption in oxygen is

$$\sigma_0 = \frac{2\pi^2 e^2 \hbar}{Mc} \cdot \frac{NZ}{A}(1 + 0.8\,x) = 336 \text{ MeV-mb}$$

(for the exchange force fraction x = 0.5). The experimental value for the O^{16} integrated γ-ray absorption cross section in the 13-27 MeV range is 170 ± 20 MeV-mb, or about 50% of the theoretical value. Consequently,

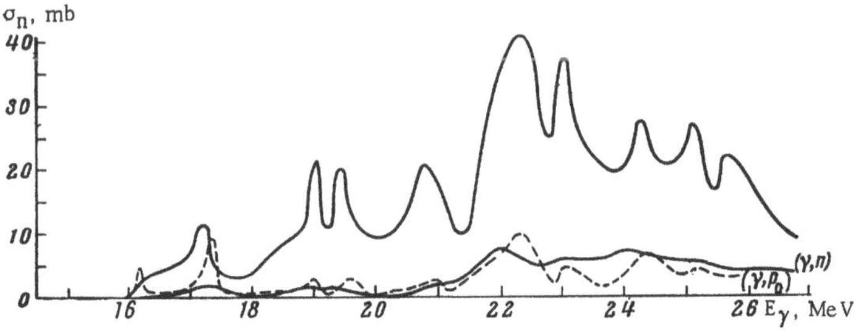

Fig. 27. Nuclear γ-ray absorption cross section and cross sections for the (γ, p_0) and (γ, n) reactions (O^{16}).

approximately 50% of the integrated cross section is above the giant resonance region if the sum rule is correct. This conclusion is confirmed by the results of Gorbunov and Asipova [113], who studied the photodisintegration of oxygen with a Wilson chamber. It was found [113] that the integrated cross section for photodisintegration of O^{16} up to 170 MeV was 410 MeV-mb. The value of 183 MeV-mb was obtained for the γ-ray energy region below 30 MeV, which is in good agreement with the integrated absorption cross section found. This same paper confirmed that the primary reactions in the giant resonance region, which make up about 90% of the integrated photodisintegration cross section, were the (γ, p) and (γ, n) reactions.

The integrated cross section for the (γ, n) reaction has been measured rather accurately while that for the (γ, p) reaction is poorly known. Using the data for the absorption cross section and for the (γ, n) reaction cross section, we shall estimate the integrated cross section of the (γ, p) reaction and compare it with existing experimental data.

The results of various experiments in which the integrated cross section for the (γ, n) reaction was measured are given in Table 12. (In [45], the cross section was calculated from the neutron spectrum on the assumption that all the transitions went only to the ground state.)

The last line gives the integrated cross section to 27 MeV obtained from curves given in these papers. In the majority of cases, with the exception of [77] where the results are clearly too low, the integrated cross section values agree with one another. On the basis of the data obtained in [14, 45, 74, 76], we shall assume that the integrated cross section for (γ, n) reaction from threshold to 27 MeV is 40 MeV-mb. The contribution of (γ, α), (γ, 4α), and (γ, pn) reactions is 18 MeV-mb for the energy region below 30 MeV [113]. Hence it follows that the integrated cross section for the (γ, p) reaction from threshold to 27 MeV is 115 MeV-mb. This value is an upper limit since it includes (γ, 2p), (γ, pα), and (γ, d) reactions, whose thresholds are 22.3, 23.1, and 20.7 MeV, respectively. Setting the contribution from these processes at 10 MeV-mb [113], we obtain a value of 105 MeV-mb for the integrated cross section for the (γ, p) reaction, which we shall use for comparison with other experimental results. The accuracy of this estimate is about 20%.

Data for the integrated cross section for the (γ, p) reaction is presented in Table 13. It is clear from the table that the integrated cross section for the (γ, p) reaction calculated from the integrated absorption cross section agrees with the data of Gorbunov and Asipova [113]. The results of [40, 79], in which it was assumed that the N^{15} nucleus remained in the ground state after proton emission, show an underestimate of the integrated cross section by a factor of two.

The last column gives the integrated cross section for the (γ, p_0) reaction calculated from the cross section curve for the (p, γ_0) reaction [38].

According to the principle of detailed balance

$$k_\gamma^2 \left(\frac{d\sigma}{d\Omega}\right)_{\gamma,\, p_0} (2I_\gamma + 1)(2I_0 + 1) = k_p^2 \left(\frac{d\sigma}{d\Omega}\right)_{p,\, \gamma_0} (2j_p + 1)(2j + 1).$$

TABLE 14. Integrated Cross Sections for Various Processes in O^{16}, MeV-mb

Energy range, MeV	Absorption cross section	$\sigma(\gamma, n)$	$\sigma(\gamma, p)$	$\sigma(\gamma, p_0)$	$\sigma(\gamma, p^*)$	$\frac{\sigma(\gamma, p_0)}{\sigma(\gamma, p)}$, %
20—21.5	16	2	14	3	11	21
21.5—23.5	57	16	41	12	29	30
23.5—27	65	18	48	15	38	30

Here, k_γ, k_p are the wave numbers of the γ-ray and proton; I_γ, I_0 are the spins of the photon and of the target nucleus; and j_p, j are the spins of the proton and of the final nucleus. Consequently,

$$\left(\frac{d\sigma}{d\Omega}\right)_{\gamma,\, p_0} = a\,(E_\gamma)\left(\frac{d\sigma}{d\Omega}\right)_{p,\, \gamma_0}.$$

A curve for $a(E_\gamma)$ is given in Fig. 26. In calculating the total cross section for the (γ, p_0) reaction from the differential cross sections, the results of angular distribution measurements [38, 40] were used.

From a comparison of the integrated cross sections for the (γ, p) and (γ, p_0) reactions (Table 13), it is clear that about 70% of all transitions with energies below 27 MeV go to excited states of N^{15}. The cross sections calculated from proton spectra differ greatly from the cross section for the (γ, p_0) reaction apparently because of the difficulty of differentiating transitions to ground and excited states in the residual nucleus. It is clear from the table that they lead to some intermediate result.

The ratio of the integrated cross sections for the (γ, p) and (γ, n) reactions is about 2.5. From the viewpoint of the hypothesis of charge-independent forces, the cross sections for these reactions should be identical for given individual states in this case. Consideration of the various thresholds and transparencies changes the relationship between them. The theoretical ratio of the cross sections for the (γ, p) and (γ, n) reactions [7], under conditions where d-particles are emitted is close to the experimental value. Hence it also follows that the corresponding characteristics of O^{16} states should be explained from the viewpoint of the shell model rather than from the statistical concepts of the compound nucleus.

The nuclear absorption cross section, the cross section for the (γ, p_0) reaction calculated from the cross section for the inverse (p, γ_0) reaction [38] as indicated above, and the cross section for the (γ, n) reaction [74] are shown in Fig. 27. Note that the magnitudes of the cross sections for the (γ, p_0) and (γ, n) reactions practically coincide. On the other hand, Fig. 27 makes it clear that the cross section for the (γ, p^*) process, in which the N^{15} nucleus remains in an excited state after emission of the proton, is considerably greater than the cross section for the (γ, p_0) reaction.

The probability of proton emission to a ground-state level from states or groups of states in O^{16} can be found by considering the distribution of integrated cross sections over appropriate energy ranges. The 20-27 MeV range was divided into three parts separated by the minima in the absorption cross section curve at 21.5 and 23.5 MeV. The difference between the integrated absorption cross section and the integrated cross section for the (γ, n) reaction was set equal to the integrated cross section for the (γ, p) reaction. The ratio of the integrated cross section for the (γ, p_0) to this quantity gave an estimate of the probability for transitions to the ground state of N^{15}. The results of this treatment are given in Table 14.

The results make it clear that the probability for transition to the ground state of N^{15} from the energy ranges considered is 20-30%. This value is a lower limit because the integrated cross section for the (γ, p) reaction also includes processes such as (2p), (pn), etc. Allowing for this, one can consider that our data are close to the results in [78, 116] where a value around 40% was obtained for the probability of transitions to the ground state. The result of Johansson and Forkman [7] (~ 14%) is an underestimate.

Fuller and Hayward [117] estimated the integrated absorption cross section from the sum of the integrated cross sections for the (γ, p) and (γ, n) reactions. They calculated the cross section for the (γ, p) reaction from

TABLE 15. C^{12} Energy Levels in the 16-28 MeV Range

Nuclear absorption cross section	Photoproton spectra [40]	Inverse reaction cross section [44]	Photoneutron spectra [45]	Charged particle reaction cross section [118]
16.5				16.58
				17.23
17.6				17.77
				18.40
				18.85
19.1		19.2		19.26
				19.42
				19.67
				19.88
				20.27
				20.49
				20.65
				21.34
				21.80
23.0	22.55	22.5	23.0	
25.6	25.5	25.5	26.0	
	27.5			

data in [7, 78] where proton spectra were measured for several values of maximum bremsstrahlung spectrum energy. The cross section for the (γ, n) reaction was taken from [14], but its absolute value was increased in accordance with the data in [114]. However, Table 12 makes it clear that the integrated cross section obtained in this work is approximately 20% greater than the average value so that such an increase is unjustified.

The distribution of the integrated cross section for the sum of the (γ, p) and (γ, n) reactions in each energy interval [117] is compared below with the analogous results obtained from the absorption cross section curve. (Note that the subdivision into intervals in [117] does not correspond to division into groups of resonances.)

Energy interval, MeV	16-18	18-21	21-23	23-26	16-26
Absorption cross section, MeV-mb . .	9	30	53	66	158
Sum of (γ, p) and (γ, n) reactions, MeV-mb	6	17	44	93	160

It is evident from this that the integrated cross sections agree over the entire 16-26 MeV energy region where the comparison was made. However, this agreement may be regarded as accidental because the ratio of the integrated cross sections varies from 1.7 to 0.7 in the separate intervals. The disagreement characterizes the possible error in obtaining the absorption cross section by summation of the cross sections for the partial reactions. Especially important is the disagreement in the 23-26 MeV range, which apparently is the result of an incorrect estimate of the probability of transitions to excited levels of N^{15} in the (γ, p) reaction. It led the authors [117] to the conclusion that the ratio of the integrated cross sections in the 21-23 and 23-26 MeV ranges was 1 : 2 and the inverse of the theoretical predictions by Elliott and Flowers [95]. This conclusion is not confirmed by the results of absorption cross section measurements.

CHAPTER VI

Discussion of C^{12} Results

1. Structure

The C^{12} nuclear γ-ray absorption cross section in the 13-27 MeV energy range is shown in Fig. 23. It makes clear that the measured cross section consists of several maxima with very different parameters. The main part of the cross section, which is associated with the C^{12} giant resonance, is located above 20 MeV. The absorption cross section in this excitation energy region (up to 27 MeV) consists of two broad maxima centered at 23 and 25.6 MeV. The integrated cross sections for these maxima are roughly in the ratio 3 : 1. It is evident that a comparatively slight worsening (by a factor 1.5-2) of instrumental energy resolution in comparison with the resolution used in this experiment (~ 200 keV) would lead to a rather symmetric "giant resonance" with a maximum close to 23 MeV, a width of approximately 3.5 MeV, and a high-energy "tail." This conclusion is confirmed by the results of Zeigler [28], who measured the γ-ray absorption cross section by a similar method but with poorer resolution (~ 300 keV at 22 MeV). The absolute values of the cross sections are in good agreement with each other.

The broad peak located at 20-24 MeV makes the main contribution to the absorption cross section. The value of the cross section at the maximum of the peak σ_m is ≃ 16 mb, the halfwidth $\Gamma \simeq 3$ MeV, and the integrated cross section is 54 MeV-mb. The rather flat, broad peak (~ 2 MeV) drops sharply at 24 MeV, and the irregular nature of the curve gives reason for supposing that this maximum consists of narrower, overlapping peaks.

The second maximum, which is located in the 24.5-26.5 MeV range, has the following parameters: maximum cross section $\sigma_m = 8$ mb, halfwidth $\Gamma \sim 2$ MeV, and integrated cross section about 17 MeV-mb.

There are indications [40, 77] that both the first maximum in the 20-24 MeV range and the second maximum in the 24-5-26.5 MeV range have a more complex structure.

Three relatively narrow peaks are observed below 20 MeV in the curve obtained for the C^{12} nuclear absorption cross section. The energies corresponding to the maxima of these peaks are 16.5, 17.6, and 19.1 MeV. The halfwidths of the peaks (without correction for spectrometer resolution) are 250, 400, and 650 keV, respectively.

Experimental data for levels in the C^{12} nucleus in the 16-27 MeV excitation energy range are shown in Table 15. The peaks observed in the absorption cross section are compared with recent, more accurate data on photoproton [40] and photoneutron [45] spectra and with results obtained from a study of the inverse reaction $B^{11}(p, \gamma_0)C^{12}$ [44]. The last column gives C^{12} levels observed in reactions produced by charged particles (other than the (p, γ_0) reaction) [118].

As is clear from the table, the location of the two resonances in the energy region above 20 MeV in the absorption cross section observed in the present work agrees with the location of maxima in the cross sections

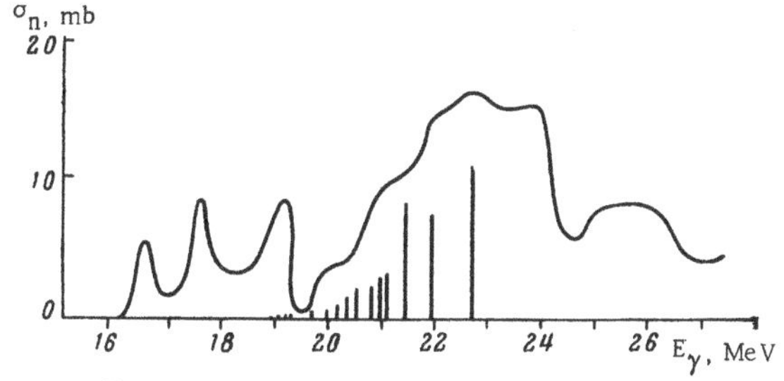

Fig. 28. C^{12} nuclear γ-ray absorption cross section and interpretation of breaks in the $C^{12}(\gamma, n)$ reaction yield curve.

TABLE 16. Giant Resonance Parameters in C^{12}

	Absorption cross section		$\sigma(\gamma, p)$							$\sigma(\gamma, n)$					
	This work	[28]	[125]	[9]	[127]	[126]	[128]	[46]	[40]	[16]	[124]	[123]	[77]	[45]	[14]
Location of maximum E_m, MeV	23.0	23.0	21.5	23.0	22.1	22.5	22.5	22.5	22.5	22.5	22.8	23.0	23.4	23.0	23
Cross section at the maximum σ_m, mb	16	17	34	14.7 *	8.1 *	24.0 *	12.7	12	11	8.3	10.4	7.9	6.3	7.5	8
Halfwidth Γ, MeV	3.2	4.0	1.7		3.6		3.3	3.1	3.1	4.3	3.5	3.2	3.2	4	4.2
Integrated cross section $\int \sigma dE$, MeV-mb	84 (27 †)	100 (27)	63 (24)	46 (24)			57 (27)		43 (27)	39 (27)	34 (25)	22 (24)	24 (26)	35 (27)	34 (27)

* Values given were corrected in [128] for nonisotropic angular distribution.

† Upper limit of integration is given in MeV in parentheses.

TABLE 17. Experimental Energy Levels and Relative Transition Probabilities for C^{12} and Theoretical Dipole Transition Characteristics

This experiment		Theoretical papers									
		Brown [132]		Gillet and Vinh-Mau		Mikeska [111]		Nilsson [129]		Mihailovic and Rosina [130]	
E, MeV	I, %	E, MeV	I, %	E, MeV	I, %	E, MeV	I, %	E, MeV	I, %	E, MeV	I, %
16.5	3										
17.6	6			17.7	7	18.3	5				
19.1	6	18.7	6.5							19.7	3
		22.2	75	21.9	73	21.7	60	22.2	31		
								23.0	9		
23.0	64							23.7	9	23.6	9
		23.9	0.5	24.2	1					24.6	40
						25.0	15	26.3	11	25.4	1
25.6	20									27.4	6
										28.0	3
								29.5	9	29.2	2
										30.7	7
		34.3	18	33.8	19	34.3	20	31.9	30	31.7	1
										40.7	28

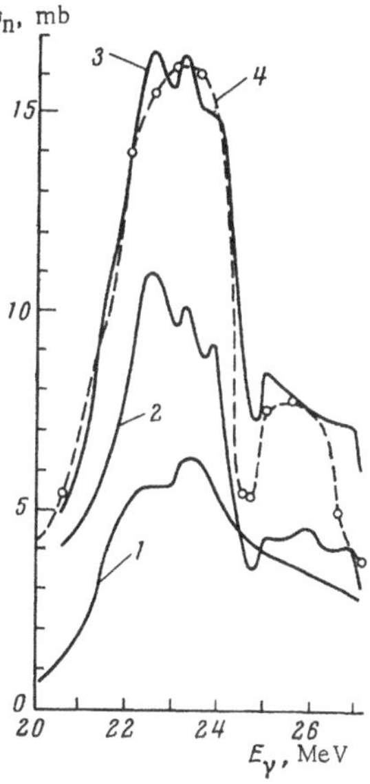

Fig. 29. Comparison of C^{12} nuclear γ-quanta absorption cross section with the summed cross sections for the (γ, n) and (γ, p) reactions. (1) $\sigma(\gamma, n)$; (2) $\sigma(\gamma, p)$; (3) $\sigma(\gamma, n) + \sigma(\gamma, p)$; (4) absorption cross section.

for the (γ, p) and (γ, n) reactions which were obtained from analysis of photoproton and photoneutron spectra and from the cross section for the inverse (p, γ_0) reaction.

Resonances found in the absorption cross section curve below 20 MeV are observed at the same excitation energies in the cross sections for reactions with charged particles.

The maximum at 19.1 MeV was observed in cross sections for the (p, γ_0), (p, p'), and (p, n) reactions [10, 44, 118]. The level widths of 450 and 500 keV given in [10] and [118] agree with our results.

The maximum located at 17.6 MeV can be associated either with the 17.23 MeV level (1^-, Γ = 140 keV) or with the 17.77 MeV level (0^+, Γ = 140 keV), which were obtained in a study of the cross sections for the (p, γ) and (p, p') reactions [118]. Since the $0 \rightarrow 0$ transition is forbidden, the association with the 17.77 MeV level can only be made if its characterization as (0^+) is erroneous.

The resonance observed for a γ-ray energy of 16.5 MeV possibly corresponds to the 16.58 MeV level (2^-, Γ = 295 keV) found in a study of the $B^{10}(He^3, p)$, $B^{11}(p, \gamma)$ and $B^{11}(p, \alpha)$ reactions [118], although the probability of exciting a 2^- level must be small.

The fact that a large number of the levels observed in charged particle reactions do not appear in the γ-ray absorption cross section can be explained by the low probability of their excitation in nuclear photodisintegration.

Three peaks in the absorption cross section at 20.15, 20.46, and 20.92 were found in [47], which also used the absorption method but with monochromatic γ-rays from the T(p, γ) reaction (the energy resolution in the 20.0-21.2 MeV region where the measurements were made was 40-70 keV). They were not observed in the partial reactions. A large number of breaks were found in the $C^{12}(\gamma, n)$ yield curves [46], just as happened in the study of O^{16}. They are shown in Fig. 28 together with the absorption cross section curve.

As is clear from the figure, the spacing between breaks is from 60 to 350 keV with the exception of the two breaks at 22.02 and 22.88 MeV. Since the energy resolution of the spectrometer was about 200 keV in making these measurements, and the experimental points were 175 keV apart, the corresponding resonances in the absorption cross section should not be clearly apparent regardless of their parameters. Note that no sort of correlation is observed between the breaks and the maximum at 19.1 MeV.

Data for photoproton [9, 119] and photoneutron [120] spectra from carbon, obtained in earlier work, was considered as confirmation that the absorption cross section is a grouping of narrow, isolated peaks. However, such a structure for carbon was not observed in more recent work carried out with more modern techniques [40, 45, 128].

A very careful study of the shape of the cross section for the $B^{11}(p, \gamma_0)C^{12}$ reaction in the C^{12} 21-22.6 MeV excitation energy range was made by Gove et al. [44] in order to clear up the question of the existence of fine structure. Although the resolution was about 30 keV, no sort of structure was observed outside of a small irregularity in the behavior of the cross section curve (not more than 10%). The measurement of the cross section for this reaction was repeated by Becker and Fox [121] for the 21-28 MeV C^{12} excitation energy range. The shape of the cross section was similar to that obtained by Gove et al. [44].

Therefore, the experimental data for the giant resonance in C^{12} do not confirm the idea that the absorption cross section consists of narrow, isolated peaks. Rather, it is in accord with the hypothesis that several broad levels, whose spacing is comparable with their width, are to be found in C^{12} in the 20-24 MeV range. Further experiments with higher accuracy and resolution are needed to fix the parameters of these states.

2. Parameters of the Giant Resonance in C^{12}

The integrated nuclear absorption cross section in the 13-27 MeV range is 84 ± 10 MeV-mb. This quantity is in agreement with the integrated cross section of 130 ± 20 MeV-mb which was obtained by a similar technique for the 10-30 MeV range [28]. Somewhat poorer is the agreement with a value for the integrated cross section in the 10-27 MeV range equal to 120 MeV-mb which was obtained in [122] (by the absorption method with a scintillation counter as detector); this is apparently associated with the insufficient accuracy of those results.

The integrated cross sections found in the present work and in [28] and [122] are less than half the magnitude of the 252 MeV-mb integrated absorption cross section predicted by theory in accordance with the "sum rule." Evidently, this is evidence of the fact that in carbon (as in oxygen), the giant resonance region below 30 MeV includes only approximately half the integrated cross section for dipole transitions.

As already mentioned above, the (γ, p) and (γ, n) reactions are the principal processes in the energy range under consideration. In this connection, it is interesting to make a comparison of the absorption cross section found with the sum of the cross sections for these reactions. Experimental data for the giant resonance parameters in C^{12} are given in Table 16.

The cross section for the $C^{12}(\gamma, n)C^{11}$ reaction was measured by using the method of residual activity [14, 16, 123] and by direct detection [77, 124]. In [45], the cross section was calculated from photoneutron spectra. The cross section for the $C^{12}(\gamma, p)B^{11}$ reaction was obtained by direct detection of protons [125], from the cross section for the inverse, (p, γ_0) reaction [44, 126], and from photoproton spectra [9, 40, 127, 128]. In calculating the cross sections for the (γ, n) and (γ, p) reactions from photoneutron and photoproton spectra, it was assumed that the emitted nucleons left the product-nucleus in the ground state. It was shown that such an assumption is sufficiently accurate for protons up to an excitation energy of 30 MeV [127]. This conclusion is also confirmed by a comparison of the absolute values of the cross sections for the (γ, p_0) and (γ, p) reactions.

As can be seen in Table 16, the maximum values of the cross sections for the (γ, p) and (γ, n) reactions average out to approximately 12 and 8 mb (the values for $\sigma_0(\gamma, p)$ obtained in [125] and [126] are clearly overestimated). Within limits of experimental error, the sum of these cross sections is not in contradiction with values of 16 and 17 mb for the absorption cross section at the maximum of the giant resonance. The sum of the average integrated cross sections for the (γ, p) and (γ, n) reactions (about 50 and 30 MeV-mb, respectively; see above) are in good agreement with the integrated absorption cross section found in the present work.

The curve for the absorption cross section and the summed cross sections for the (γ, p) and (γ, n) reactions are compared in Fig. 29 over the 20-27 MeV range. The partial cross sections $\sigma(\gamma, p)$ and $\sigma(\gamma, n)$ were taken from [40] and [77]. Both the absolute values and the general behavior of the curves are in completely satisfactory agreement. This indicates the insignificant contribution of transitions to excited states of the residual nucleus, confirming the results of [127].

Use of the cross sections $\sigma(\gamma, n)$ obtained in [16] and [124] leads to poorer agreement because the chief maximum in the 20-24 MeV range is found to be sharper upon summation with the cross section $\sigma(\gamma, p)$. Evidently, this indicates that a displacement actually exists between the maxima in the cross sections for the (γ, p) and (γ, n) reactions.

3. Comparison with Theory

Table 17 gives transition energies and intensities for dipole absorption of γ-rays by C^{12} nuclei as determined by theoretical calculations and analogous results obtained from the absorption cross section. All the calculations were based on the shell model. The assumptions on which such calculations are based have been discussed previously during a similar comparison of the theoretical and experimental data pertaining to oxygen. Here, only the main features will be noted.

The calculations in [132, 102, 111] were carried out within the framework of the approximation made by Elliott and Flowers [95]. In addition to the results presented in the table, Gillet and Vinh-Mau [102] also presented data for calculations taking into account correlations in the ground state. They practically coincide. Calculations were made with a deformed Nilsson potential [129]. In this case, the C^{12} dipole resonance is split into two groups of transitions corresponding to excitation of oscillations along the symmetry axis of the nucleus and perpendicular to it. These groups of transitions partially overlap because of strong spin-orbit interactions. Mihailovic and Rosina [130] considered the effect of two-particle, two-hole configurations. In their opinion, such configurations have an important effect on the shape of the giant resonance in nuclei with unfilled shells.

In contrast to oxygen, where the energies of all the calculated transitions were less than 30 MeV, a significant part of the dipole sum for C^{12} is located in the 30-40 MeV range. In this connection, it is necessary to measure the nuclear absorption cross section up to energies of 35-40 MeV in order to study the giant resonance. Such measurements may perhaps make it possible to obtain some information about the quadrupole moment of the nucleus although the overlap of transitions associated with oscillations along the symmetry axis and perpendicular to it, according to theoretical data [129], makes it difficult to obtain such data and requires preliminary identification of resonances in the absorption cross section.

For energies below 30 MeV, the levels making the main contribution to the absorption cross section are located in the neighborhood of 23 MeV for all calculations, i.e., they agree with the position of the principal maximum in the measured absorption cross section. The shape of the experimental curve above 20 MeV is in better agreement with the results of [129]. The transition energies obtained by Mihailovic and Rosina [130] are overestimated by approximately 2 MeV from a comparison with the experimental value of the principal maximum at 23 MeV.

Below 20 MeV, the calculations presented do not give the required number of levels. (The data of [131], in which the existence of three transitions below 21 MeV was predicted, are closer to the experimental results.) This can be explained either by insufficient accuracy of the calculations or by the fact that part of the transitions are of other multipolarities. Calculations made in [132] for other types of transitions indicate the possibility of identifying the 16.5 and 19.1 MeV transitions found in the present work with 2^+ and 2^- transitions. However, the probability of a 2^- transition for γ-ray absorption in this energy region must be relatively small.

Level widths were considered in [111] and [133]. As can be seen from Table 17, there is only qualitative agreement between theory and experiment, which makes it impossible to compare individual transitions. One can compare the 23 MeV resonance and the theoretical transition with maximum intensity. The width of this state was found experimentally to be about 3 MeV; calculation gives values of 1 MeV [111] and 2.9 MeV [133].

Conclusions

1. The method employed gives quantitative information about the structure and absolute value of the γ-ray absorption cross section in light nuclei. The results obtained are the first direct, reliable data on the structure of the nuclear absorption cross section.

2. A comparison of the curves for the nuclear γ-ray absorption cross sections in O^{16} and C^{12} indicates, on the one hand, that there are some common features in the giant resonances of these nuclei; for example, the integrated cross section is concentrated in the 21-25 MeV region to a considerable extent (apparently, this is explained by the fact that the $1p_{3/2} \rightarrow 1d_{5/2}$ transition makes the main contribution), and a considerable fraction of the dipole sum ($\sim 50\%$) is found above 27 MeV. On the other hand, the resonance structure of the O^{16} and C^{12} cross sections is very different; this conclusion can be made at this time despite the lack of complete clarity with respect to C^{12}. The explanation of the structure, and of the differences, evidently requires theoretical consideration of higher configurations.

3. Resonances in the absorption cross section, corresponding to levels or closely spaced groups of levels, were found at the following excitation energies: O^{16} – 17.2, 19.0, 19.4, 20.9, 22.3, 23.1, 24.3, 25.2, and 25.8 MeV; C^{12} – 16.5, 17.6, 19.1, 23.0, and 25.6 MeV.

4. The correspondence of breaks in the yield curves to narrow, isolated resonances in the absorption cross section was not confirmed except for particular cases. In O^{16}, groups of breaks basically correspond to an individual peak. In C^{12}, the breaks are more uniformly distributed, and, at the same time, structure such as is found in O^{16} is absent.

5. The representation of γ-ray-excited dipole states in the form of one-particle, one-hole configurations which mix the residual interaction between nucleons only leads to qualitative agreement with experimental data:

(a) The theory predicts fewer transitions (five for O^{16} and four for C^{12}). Existing data for transitions of other multipolarities only partially remove this discrepancy. Consideration of interaction with the surface or of excitation of two particle-hole pairs increases the spectrum of states, improving agreement with experiment;
(b) the contribution from the low-energy portion of the giant resonance is underestimated. In O^{16}, the integrated cross section for energies below 21 MeV is about 25%, which considerably exceeds the value of 3-6% obtained from the calculations.

6. In O^{16}, the practically complete agreement of the location and number of maxima on the absorption cross section and in the cross sections for the (γ, p) and (γ, n) reactions can be considered as an indirect indication of the charge symmetry of nuclear forces.

7. The radiation width of O^{16} states is 3-30% of the single-particle Weisskopf widths, which agrees with the predictions of the shell model for E1 transition probabilities.

8. The value of the estimated integrated cross section for the O^{16} (γ, p) reaction is 105 ± 20 MeV-mb, which is considerably larger than the results obtained from proton spectra on the assumption that the product nucleus remains in the ground state. The ratio of the integrated cross sections for the (γ, p) and (γ, n) reactions is 2.5. The probability of proton transitions to excited N^{15} states is approximately 70%.

9. The results obtained furnish a basis for supposing that the large maximum in the C^{12} absorption cross section located in the 20-24 MeV region consists of narrower, unresolved peaks. Clarification of this question requires further measurements with greater accuracy and better resolution.

10. A comparison of the absorption cross section and the cross sections for the (γ, p_0) and (γ, n) reactions in C^{12} shows that the contribution of the (γ, p*) reaction in carbon is considerably less than in oxygen.

In conclusion, the author wishes to thank L. E. Lazareva, scientific director for candidates in the physical and mathematical sciences, for suggesting the topic and for constant consideration and assistance. The author is grateful to F. A. Nikolaev, V. I. Korin, V. A. Zapevalov, and N. S. Kozhevnikov for a great deal of help in carrying out this investigation, to B. A. Tulupov for numerous valuable discussions, to the director of the Laboratory for Photomeson Processes, FIAN, for affording the opportunity to work at the FIAN 260-MeV synchrotron, and to the operating groups of the FIAN 30-MeV and 260-MeV synchrotrons. The author extends his appreciation to all his associates who were of help in carrying out various stages of this work.

Appendix

We shall calculate the value of the quantity A for the 15.11 MeV resonance in C^{12}:

$$A = \int_{k_0}^{\infty} (1 - e^{-n\sigma(k)})\, \Delta (k - k_0)\, dk.$$

Let

$$\sigma = \frac{\sigma_0 (\Gamma/2)^2}{(k-k_p)^2 + \Gamma^2/4} \text{ and } (k_0 - k_p) \sim \Delta k.$$

We divide A into three parts:

$$A = \int_{k_0}^{k_1} (1 - e^{-n\sigma(k)}) \Delta (k - k_0)\, dk +$$

$$+ \int_{k_1}^{k_2} (1 - e^{-n\sigma(k)}) \Delta (k - k_0)\, dk + \int_{k_2}^{\infty} (1 - e^{-n\sigma(k)}) \Delta (k - k_0)\, dk; \qquad (1)$$

k_1 and k_2 are chosen so that for energies $k < k_1$ and $k > k_2$, $e^{-n\sigma(k)}$ can be represented in the form

$$e^{-n\sigma(k)} \simeq 1 - n\sigma(k). \qquad (2)$$

This can be done with an accuracy $\leq 0.5\%$ if $k_{1,2} = k_p + a\Gamma$, where $a \geq 25$. Consider each of the terms in equation (1) separately:

$$\mathcal{I}_1 = \int_{k_0}^{k_1} (1 - e^{-n\sigma(k)}) \Delta (k - k_0)\, dk \simeq n \int_{k_0}^{k_1} \sigma(k) \Delta (k - k_0)\, dk \simeq$$

$$\simeq n\Delta (k^* - k_0) \int_{k_0}^{k_1} \sigma(k)\, dk.$$

In the range under consideration, $\Delta(k - k_0)$ can be represented sufficiently accurately by a linear function of the form

$$\Delta (k - k_0) \simeq \frac{1}{(\Delta k)^2} (k - k_0).$$

We can set

$$\Delta (k^* - k_0) \simeq \frac{1}{2(\Delta k)},$$

and then obtain

$$\mathcal{I}_1 \simeq n \frac{\sigma_0 \Gamma^2}{2(\Delta k)} \int_{k_p - \Delta k}^{k_p + a\Gamma} \frac{dk}{4(k-k_p)^2} = \frac{n}{2(\Delta k)} \cdot \frac{\sigma_0 \Gamma^2}{4} \left(\frac{1}{a\Gamma} - \frac{1}{\Delta k} \right) =$$

$$= \frac{n}{2(\Delta k)} \cdot \frac{\sigma_0 \Gamma}{4} \left(\frac{1}{a} - \frac{\Gamma}{\Delta k} \right). \qquad (3)$$

The evaluation of the third term in expression (1) is similar and leads to the result:

$$\mathcal{I}_3 \simeq \frac{n}{2(\Delta k)} \cdot \frac{\sigma_0 \Gamma}{4a}. \qquad (4)$$

Then

$$\mathcal{I}_1 + \mathcal{I}_3 \simeq \frac{n\sigma_0 \Gamma}{4\Delta k} \left(\frac{1}{a} - \frac{\Gamma}{2\Delta k} \right);$$

$$\mathcal{J}_2 = \int_{k_1}^{k_2} (1 - e^{-n\sigma(k)})\, \Delta(k - k_0)\, dk \simeq \frac{1}{\Delta k} \sum_{-a\Gamma}^{+a\Gamma} (1 - e^{-n\sigma(k)})\, \Delta l =$$

$$= \frac{\Delta l}{\Delta k} \sum_{-a\Gamma}^{a\Gamma} (1 - e^{-n\sigma(k)}) = \frac{\Gamma}{\Delta k} b(a). \qquad (5)$$

To evaluate b, we used the well-known value $\sigma_0 = 29.7 \times 10^{-24}$ cm^2; it was assumed that $\Delta l = 2.5\ \Gamma$. Inserting the expressions obtained, we have:

$$A = \mathcal{J}_1 + \mathcal{J}_2 + \mathcal{J}_3 = \frac{n\sigma_0}{4} \cdot \frac{\Gamma}{\Delta k}\left(\frac{1}{a} - \frac{\Gamma}{2(\Delta k)}\right) + \frac{\Gamma}{\Delta k} b =$$

$$= n\sigma_0 \frac{\Gamma}{\Delta k}\left[\frac{1}{4}\left(\frac{1}{a} - \frac{\Gamma}{2(\Delta k)}\right) + \frac{b(a)}{n\sigma_0}\right]. \qquad (6)$$

Formula (6) is similar to relation (2.13); however, the parameter a appears in it. The dependence of the quantity $\beta = \frac{1}{4}\left(\frac{1}{a} - \frac{\Gamma}{2(\Delta k)}\right) + \frac{b}{n\sigma_0}$ on this parameter is given below.

a	10	17.5	20	22.5	25	30	40	50	60
$\beta \cdot 10^2$ . .	9.76	9.86	9.99	10.06	10.16	10.20	10.38	10.47	10.49

It is clear that β slowly increases with increasing a, approaching some value β_0. Take $\beta_0 = 10.5 \times 10^{-2}$. One can write for A:

$$A = n\sigma_0 \frac{\Gamma}{\Delta k} \beta_0 = n\tilde{\sigma}. \qquad (7)$$

For $\Delta k = 175$ keV, $\tilde{\sigma} = 1.19 \times 10^{-27}$ cm^2; for $\Delta k = 160$ kev, $\tilde{\sigma} = 1.30 \times 10^{-27}$ cm^2.

Relation (4.3), which is obtained by numerical computation, will have the form:

$$A/n = 0.067\ \bar{\sigma}. \qquad (8)$$

LITERATURE CITED

1. L. Katz, R. N. H. Haslam, R. J. Horsley, and A. G. W. Cameron, Phys. Rev., 95:464 (1954).
2. J. Goldemberg and L. Katz, Phys. Rev., 95:471 (1954).
3. A. S. Penfold and B. M. Spicer, Phys. Rev., 100:1377 (1955).
4. B. M. Spicer, Private Communication (1960).
5. C. Tzara, J. Phys. Radium, 17:1001 (1956).
6. S. A. E. Johansson and B. Forkman, Phys. Rev., 99:1031 (1955).
7. S. A. E. Johansson and B. Forkman, Arkiv. Fys., 12:359 (1957).
8. B. M. Spicer, Phys. Rev., 99:33 (1955).
9. L. Cohen, A. K. Mann, B. J. Patton, K. Reibel, W. E. Stephens, and E. J. Winhold, Phys. Rev., 104:108 (1956).
10. J. K. Bair, J. D. Kington, and H. B. Willard, Phys. Rev., 100:21 (1955).
11. D. L. Livesey, Canad. J. Phys., 34:1022 (1956).
12. B. M. Spicer, Austral. J. Phys., 10:326 (1957).
13. A. S. Penfold and J. E. Leiss, Phys. Rev., 95:637 (1954).
14. J. J. Carver and K. H. Lokan, Austral. J. Phys., 10:312 (1957).
15. P. Erdos, P. Scherrer, and P. Stol, Helv. Phys. Acta, 30:639 (1957).
16. W. C. Barber, W. D. George, and D. D. Reagon, Phys. Rev., 98:73 (1955).

17. B. B. Kinsey and G. A. Bartholomew, Canad. J. Phys., 31:537 (1953).
18. G. A. Bartholomew, P. J. Campion, and K. Robinson, Canad. J. Phys., 38:194 (1960).
19. V. I. Gol'danskii, A. V. Kutsenko, and M. I. Podgoretskii, Counting Statistics in Nuclear Particle Detection. Fizmatgiz (1959).
20. N. A. Burgov, G. V. Danilyan, B. S. Dolbilkin, L. E. Lazareva, and F. A. Nikolaev, Zh. Eks. i Teor. Fiz., 37:1811 (1959).
21. N. A. Burgov, G. V. Danilyan, B. S. Dolbilkin, L. E. Lazareva, and F. A. Nikolaev, Zh. Eks. i Teor. Fiz., 43:70 (1962).
22. N. A. Burgov, G. V. Danilyan, B. S. Dolbilkin, L. E. Lazareva, and F. A. Nikolaev, Zh. Eks. i Teor. Fiz., 45:1693 (1963).
23. N. A. Burgov, G. V. Danilyan, B. S. Dolbilkin, L. E. Lazareva, and F. A. Nikolaev, Izv. Akad. Nauk SSSR, 27:866 (1963).
24. B. S. Dolbilkin, V. I. Korin, L. E. Lazareva, and F. A. Nikolaev, Letters to the Editor, Zh. Eks. i Teor. Fiz. Pisma, 1:47 (1965).
25. B. S. Dolbilkin, V. A. Zapevalov, V. I. Korin, L. E. Lazareva, and F. A. Nikolaev, Compt. Rend. Congr. Internat. Phys. Nucl., Paris, p. 1060 (1964).
26. B. S. Dolbilkin, V. A. Zapevalov, V. I. Korin, L. E. Lazareva, and F. A. Nikolaev, Report to the XV Annual Conference on Nuclear Spectroscopy and Nuclear Structure, Minsk (1965).
27. B. Zeigler, Z. Phys., 152:566 (1958).
28. B. Zeigler, Nucl. Phys., 17:238 (1960).
29. J. Dular, G. Kernel, M. Kregar, M. V. Michailovic, G. Pregl, M. Rosina, and C. Supancic, Nucl. Phys., 14:131 (1959).
30. M. V. Michailovic, G. Pregl, G. Kernel, and M. Kregar, Phys. Rev., 114:1621 (1959).
31. U. Miklavzic, N. Bezic, D. Jamnik, G. Kernel, Z. Milavc, and J. Snajder, Nucl. Phys., 31:570 (1962).
32. G. Tamas, J. Miller, G. G. Schuhl, and C. Tzara, J. Phys. Radium, 21:532 (1960).
33. W. C. Barber, Nucl. Instr. and Meth., 28:220 (1964).
34. E. Melkonian, W. W. Havens, and L. J. Rainwater, Phys. Rev., 92:702 (1953).
35. G. V. Danilyan, Dissertation, ITEF (1965).
36. A. A. Rudenko, Pribory i Tekhn. Eksp., 6:60 (1958).
37. V. A. Zapevalov, Dissertation, FIAN (1964).
38. N. W. Tanner, G. C. Thomas, and E. D. Earle, Nucl. Phys., 52:45 (1964).
39. F. W. Firk, Nucl. Phys., 52:437 (1964).
40. W. R. Dodge and W. C. Barber, Phys. Rev., 127:1746 (1962).
41. K. N. Geller and E. G. Muirhead, Phys. Rev. Letters, 11:371 (1963).
42. G. Tessler and W. E. Stephens, Phys. Rev., 135:B129 (1964).
43. G. V. Danilyan, ITEF Preprint 254 (1964).
44. H. E. Gove, A. E. Litherland, and R. Batchelor, Nucl. Phys., 26:480 (1961).
45. H. Fuchs and D. Haag, Z. Phys., 171:403 (1963).
46. I. M. Thorson and L. Katz, Proc. Phys. Soc., A77:166 (1961).
47. E. E. Carrol and W. E. Stephens, Phys. Rev., 118:1256 (1960).
48. E. L. Garwin, Phys. Rev., 114:143 (1959).
49. S. Z. Belen'kii and I. P. Ivanenko, Usp. Fiz. Nauk, 69:591 (1959).
50. S. Z. Belen'kii, Shower Processes in Cosmic Rays, p. 150. Gostekhizdat (1948).
51. A. G. Frank, Thesis, FIAN (1958).
52. H. A. Bethe and W. Heitler, Proc. Roy. Soc., A146:83 (1934).
53. O. Klein and Y. Nishina, Z. Phys., 52:853 (1929).
54. L. M. Brown and R. P. Feynman, Phys. Rev., 85:231 (1952).
55. F. Mandl and T. H. R. Skyrme, Proc. Roy. Soc., A215:497 (1952).
56. H. Davies, H. A. Bethe, and L. Maximon, Phys. Rev., 93:788 (1954).
57. J. A. Wheeler and W. E. Lamb, Phys. Rev., 55:858 (1939).

58. J. Joseph and F. Rohrlich, Rev. Mod. Phys., 30:354 (1958).
59. A. Borsellino, Nuovo Cimento, 4:112 (1947).
60. V. Votruba, Phys. Rev., 73:1468 (1948).
61. J. M. Wyckoff and H. W. Koch, Phys. Rev., 117:1261 (1960).
62. J. M. Moffatt and G. C. Weeks, Proc. Phys. Soc., A73:114 (1959).
63. H. A. Bethe and F. Rohrlich, Phys. Rev., 86:10 (1952).
64. G. W. Grodstein, NBS Report 584 (1957).
65. K. Siegbahn, Ed., Beta- and Gamma-Ray Spectroscopy. Interscience, New York (1955).
66. M. E. Toms, Nucl. Phys., 54:625 (1964).
67. E. G. Fuller and E. Hayward, Phys. Rev., 101:692 (1956).
68. A. S. Penfold and E. L. Garwin, Phys. Rev., 116:120 (1959).
69. D. H. Wilkinson and S. D. Bloom, Phys. Rev., 105:683 (1957).
70. N. W. Tanner, G. C. Thomas, and W. E. Meyerhof, Nuovo Cimento, 14:257 (1959).
71. K. Shoda, J. Phys. Soc. Japan, 16:1841 (1962).
72. M. E. Toms, Nucl. Phys., 50:561 (1964).
73. P. F. Yergin, R. H. Auguston, N. N. Kausal, H. A. Medicus, W. R. Moyer, and E. J. Winhold, Phys. Rev. Letters, 12:733 (1964).
74. R. L. Bramblett, J. T. Caldwell, R. R. Harvey, and S. C. Fultz, Phys. Rev., 133:B869 (1964).
75. C. Milone and A. Rubino, Nuovo Cimento, 13:1035 (1959).
76. L. N. Bolen and W. D. Whitehead, Phys. Rev. Letters, 9:458 (1962).
77. J. Miller, G. Schuhl, G. Tamas, and C. Tzara, Phys. Letters, 2:76 (1962).
78. C. Milone, S. Milone-Tamburino, R. Rinzivillo, A. Rubino, and C. Tribuno, Nuovo Cimento, 7:729 (1958).
79. E. Finck, T. Kosiek, K. H. Lindenberger, K. Maier, U. Meyer-Berkout, M. Schechter, and J. Zimmerer, Z. Phys., 174:337 (1963).
80. S. G. Cohen, P. S. Fishee, and E. K. Warburton, Phys. Rev., 121:858 (1961).
81. H. King and L. Katz, Canad. J. Phys., 37:1357 (1959).
82. D. Sadch, Compt. Rend., 249:2313 (1959).
83. K. W. Jones, L. J. Lidofsky, and J. L. Weil, Phys. Rev., 112:1252 (1958).
84. G. Dearnaley, D. S. Gemmell, B. M. Hooten, and G. A. Jones, Phys. Letters, 1:269 (1962).
85. H. S. Adams, G. M. Temmer, and G. Roy, Annual Progress Report of Dept. of Physics, Florida State University (1961).
86. D. F. Hebbard, Nucl. Phys., 15:289 (1960).
87. W. K. Dawson and D. L. Livesey, Canad. J. Phys., 34:241 (1956).
88. A. B. Migdal, Zh. Eks. i Teor. Fiz., 15:81 (1945).
89. M. Goldhaber and E. Teller, Phys. Rev., 74:1046 (1948).
90. H. Steinwedel and J. H. D. Jensen, Z. Naturforsch., 5a:413 (1950).
91. M. Danos, Nucl. Phys., 5:23 (1958).
92. K. Okamoto, Phys. Rev., 110:143 (1958).
93. D. H. Wilkinson, Physica, 22:1039 (1956).
94. J. S. Levinger, Phys. Rev., 97:122 (1955).
95. J. P. Elliott and B. H. Flowers, Proc. Roy. Soc., A242:57 (1957).
96. G. E. Brown and M. Bolsterli, Phys. Rev. Letters, 3:472 (1959).
97. V. V. Balashov, Izv. Akad. Nauk SSSR, 26:1459 (1962).
98. A. B. Migdal, D. F. Zaretskii, and A. A. Lushnikov, IAE Preprint 613 (1964).
99. D. F. Zaretskii and A. A. Lushnikov, IAE Preprint 612 (1964).
100. D. M. Brink, Nucl. Phys., 4:215 (1957).
101. G. E. Brown, L. Castillejo, and J. A. Evans, Nucl. Phys., 22:1 (1961).
102. V. Gillet and N. Vinh-Mau, Nucl. Phys., 54:321 (1964).
103. J. Eichler, Nucl. Phys., 56:577 (1964).
104. N. P. Yudin, Izv. Akad. Nauk SSSR, 26:1222 (1962).

105. D. B. Isabelle and G. R. Bishop, J. Phys. Radium, 22:548 (1961).
106. G. R. Bishop and D. B. Isabelle, Phys. Letters, 1:323 (1962).
107. W. C. Barber, J. Goldemberg, G. A. Peterson, and Y. Torizuka, Nucl. Phys., 41:461 (1963).
108. D. C. Peaslee, Phys. Rev., 88:812 (1952).
109. D. H. Wilkinson, Phil. Mag., 1:127 (1956).
110. V. F. Weisskopf, Phys. Rev., 83:1073 (1951).
111. H. J. Mikeska, Z. Phys., 177:441 (1964).
112. J. S. Levinger and H. A. Bethe, Phys. Rev., 78:115 (1950).
113. A. N. Gorbunov and V. A. Asipova, Zh. Eks. i Teor. Fiz., 43:40 (1962).
114. R. Brix, H. Fuchs, K. H. Lindenberger, and C. Salander, Z. Phys., 165:485 (1961).
115. H. Breuer and W. Pohlit, Nucl. Phys., 30:417 (1962).
116. V. P. Denisov and L. A. Kul'chitskii, Report to the XV Annual Conference on Nuclear Spectroscopy and Nuclear Structure, Minsk (1965).
117. E. G. Fuller and E. Hayward, Nuclear Reactions, Vol. 2, Amsterdam (1962).
118. F. Ajzenberg-Selove and T. Lauritsen, Nucl. Phys., 11:1 (1959).
119. D. L. Livesey, Canad. J. Phys., 35:987 (1957).
120. V. Emma, C. Milone, and A. Rubino, Phys. Rev., 118:1297 (1960).
121. J. A. Becker and J. D. Fox, Nucl. Phys., 42:669 (1963).
122. J. Kockum and N. Starfelt, Nucl. Instr. and Methods, 5:37 (1959).
123. J. P. Roalsvig, I. C. Gupta, and R. Haslam, Canad. J. Phys., 39:643 (1961).
124. B. S. Cook, Phys. Rev., 106:300 (1957).
125. J. Halpern and A. K. Mann, Phys. Rev., 83:370 (1951).
126. D. S. Gemmerl, A. H. Morton, and E. W. Titterton, Nucl. Phys., 10:33 (1959).
127. S. Penner and J. E. Leiss, Phys. Rev., 114:1101 (1959).
128. V. J. Vanhuyse and W. C. Barber, Nucl. Phys., 26:233 (1961).
129. S. G. Nilsson, J. Sawicki, and N. K. Glendenning, Nucl. Phys., 33:239 (1962).
130. V. M. Mihailovic and M. Rosina, Nucl. Phys.,40:252 (1963).
131. V. G. Neudachin and V. N. Orlin, Nucl. Phys., 31:338 (1962).
132. N. Vinh-Mau and G. E. Brown, Nucl. Phys., 29:89 (1962).
133. E. Boeker and C. C. Jonker, Phys. Letters, 6:80 (1963).

γ-RAY ABSORPTION CROSS SECTIONS OF F^{19}, Mg^{24}, Al^{27}, AND Ca^{40} IN THE GIANT DIPOLE RESONANCE REGION*

F. A. Nikolaev

Introduction

A detailed investigation of the interactions of high-energy γ-rays with nuclei can give extensive and reliable information about the properties of highly-excited nuclear states. In principle, the theoretical interpretation of these studies is simpler than the interpretation of results obtained from studies of nucleon and nucleon system interactions with nuclei because the properties of the electromagnetic interaction are well known and reliably explained by theory. From the point of view of interpretation of results, studies of the nuclear γ-ray absorption section are of great interest because the requirement for inclusion of supplemental information about the properties of the channel along which the decay of the excited nucleus goes is eliminated in that case. However, the smallness of the cross section and the need for working with a continuous bremsstrahlung spectrum create serious difficulties in performing such experiments. These difficulties are compounded by the requirement for good energy resolution and high statistical accuracy because there is considerable interest in data on absorption cross section structure in addition to data on the integral characteristics of the cross section. In practice, such experiments can only be performed with light nuclei where the level densities are low. All this is why, at the start of the experiments to be discussed (middle of 1962), sufficiently accurate measurements of the absorption cross section which could be reliably compared with theoretical data had been completed only for C^{12} and O^{16}.

This paper contains a study of structure in the γ-ray absorption cross sections of F^{19}, Mg^{24}, Al^{27}, and Ca^{40} in the giant dipole resonance region. Interest in the study of the absorption cross section for the selected nuclei was determined, in addition to general considerations, by the fact that the Ca^{40} nucleus is doubly magic, and the F^{19}, Mg^{24}, and Al^{27} nuclei are strongly deformed, which might or might not be reflected in the structure of the absorption cross section.

One can distinguish two directions in which studies of this kind have proceeded: studies of partial photonuclear reactions and studies of the total nuclear photoabsorption cross section. Extremely interesting information about highly-excited nuclear states may also be obtained from experiments using electron beams.

The reliability and amount of information about the giant resonance are determined mainly by the method used. This requires a very careful procedure for the selection of a method and for the analysis of the experimental possibilities of setting up experiments to investigate the structure of the giant resonance.

*Dissertation offered in satisfaction of the requirements for the scientific degree of candidate in the physical and mathematical sciences. Defended on May 26, 1965, at the MGU Research Institute for Nuclear Physics (abridged version).

CHAPTER I

Methods of Investigating Structure in the Giant Resonance

1. Partial Reactions

Measurement of "Breaks" in Yield Curves and Structure in (γ, n) Reaction Cross Sections. The first "breaks" in yield curves for (γ, n) reactions were observed in C^{12} and O^{16} by Katz and his associates [1, 2] in 1954. They were observed while calibrating with the help of a newly-developed system for stabilizing the energy of a betatron which made it possible to maintain the energy of the upper limit of the bremsstrahlung spectrum E_γ^{max} with an accuracy of 20 keV over several hours of machine operation. During the performance of these experiments, irregularities were observed in the behavior of the yield curves for insignificant changes in E_γ^{max}.

Additional control experiments showed that the observed breaks in the yield curves did not result from instrumental effects. The authors interpreted the breaks as a manifestation of individual peaks in the cross section of the process under consideration. Experiments repeated in many laboratories in subsequent years showed excellent reproducibility of results. At the present time, there is no doubt as to their reality.*

Although the existence of breaks in yield curves has been established for a relatively large number of nuclei, the achievement of any sort of systematic analysis of this data is difficult because the results give very limited information – only that concerning the energy location of levels and extremely qualitative estimates of integrated cross sections. Attempts made to estimate the widths of these levels [1-4] showed large uncertainties; the width of a given level might vary between 30 and 400 keV. In addition to this deficiency, the method being discussed is further limited because it can only be used in the near-threshold region (the energy interval over which one can depart from the threshold is determined by the number and strength of transitions). The possibility of separating each successive break depends essentially on the ratio of the contributions to the reaction yield from a given level to those from all preceding levels and also on the accuracy of the measurement. Ordinarily, the energy range is limited to 2-6 MeV above threshold.

Unfortunately, measurements of the (γ, p) reaction cross section with accuracy sufficient to obtain spectroscopic information are made difficult because of the deficiencies of existing methods for proton detection; at high detection efficiencies, they do not permit reliable discrimination of low-energy protons (< 3 MeV) from the electron background. Use of the activation method, which would allow one to get around these difficulties is quite limited for (γ, p) reactions because the final nuclei which are produced as the result of (γ, p) reactions in light nuclei are stable in the overwhelming majority of cases. However, even in the case of an unstable product nucleus, half-lives unsuitable for measurement, low abundance of the parent isotope in a natural mixture, and small cross sections for the process hinder the making of measurements with the required accuracy.

Cross sections calculated from yield curves measured with great accuracy exhibit structure in a number of cases. However, a large error is introduced by the conversion from yield to cross section, making it difficult to perform an analysis. The difficulties in making such measurements and the complexities of the interpretation of the resulting data are easily seen in a study of the O^{16} (γ, n) O^{15} reaction [5]. The measurements were made in steps $\Delta E = 0.5$ MeV. At each value of E_γ^{max}, measurements were repeated as many as 30 times so that the root-mean-square error in the yield was 0.2-0.5%, and rose to 1% only in the near-threshold region (about 10^5 counts were obtained at each energy value). Despite the high accuracy of the yield curve measurements, the cross sections calculated from them had relatively large errors, amounting to 50% at energies above 25 MeV, which made it impossible to perform analyses in that energy region. Such uncertainties in the final results are associated with the necessity of working with bremsstrahlung beams. However, one can hope that continually improving techniques for monochromatization of high-energy γ-ray beams will make it possible to obtain extremely interesting results in the coming years from measurements of the cross sections for partial reactions.

*The energies of some breaks are very often used in laboratory practice as reference points for calibrating the energy stabilization system of an electron accelerator (for example, the break at $E_\gamma^{max} = 17.5$ MeV in the yield curve for the O^{16} (γ, n) O^{15} reaction).

Photoproton and Photoneutron Energy Spectra. The errors arising in the analysis of yield curves for the purpose of obtaining data on structure in the cross sections for partial reactions can be avoided if one measures the energy spectrum of the products of photonuclear reactions. Indeed, if there are maxima, or groups of closely-located maxima, in the reaction cross section, maxima should also be observed in spectra of the reaction products which can be determined from the relation:

$$E_\sigma = \frac{M_1}{M_2} E_x + Q,$$

where E_σ and E_x are the energies of the maxima in the reaction cross section and in the energy spectrum, respectively; M_1 and M_2 are the masses of the initial and final nuclei; and Q is the energy threshold for the process, equal to the sum of binding energy of the emitted particle and the excitation of the residual nucleus.

The most interesting data obtained by this method was in studies of photoproton spectra. Emulsions are often used to record these spectra; however, scintillation and semiconductor spectrometers have been coming into ever-increasing use in recent years. Any of the detection methods mentioned is capable of giving an energy resolution which would allow one to distinguish structure in a spectrum. Emulsions are capable of yielding measurements which are accurate to 100 keV; the resolution of scintillation spectrometers can be made 2-3%, and that of semiconductor detectors can be made even better than 1%. However, the actual resolution of measured proton spectra is usually considerably worse, and is determined by the effective target thickness, which cannot be made too small because of low photoproton yields. In the case of an aluminum target 15.3 mg/cm^2 thick ($\sim$50 μ), for example, the proton energy spread because of energy loss during penetration of a target half-thickness is 25% for 3 MeV protons, 7.5% for 6 MeV protons, and 4% for 9 MeV protons.

Photoneutron spectra can be measured by similar methods by detecting the recoil protons produced in the detector material itself (if it contains hydrogen) by the flux of incident neutrons or produced in a special radiator located ahead of the detector. However, the efficiency of the method is low, and it is not widely used.

In recent years, very successful attempts have made use of the time-of-flight method, so highly developed for neutron spectroscopy, for measuring photoneutron spectra. This became possible only after high-current linear electron accelerators (1 A for a 10^{-9} sec beam pulse) capable of producing the necessary neutron yield became operative.

The time-of-flight method for photoneutron detection has undergone extensive development in the work of the physics group at Harwell (England) which has obtained energy resolutions of 4 and 100 keV for 1 and 10 MeV neutrons, respectively, [6] over a flight path of about 40 m. At the present time, the resolution of this method is one of the best of all the methods used for studying highly-excited states (with excitation energies > 10 MeV). Unfortunately, the use of this method to measure photoproton spectra is exceedingly difficult (although possible in principle) because of the need to work with thin targets which do not produce the required photoproton fluxes even at the intensities generated by linear accelerators.

Despite important advantages (good resolution, and the capability of simultaneous measurements over a broad energy range), the compelled use of a bremsstrahlung spectrum introduces a large uncertainty in the information obtained by this method. This results from the fact that a large number of nuclear states are excited for a given maximum bremsstrahlung energy, and these states may decay either to the ground state or to excited states of the residual nucleus with the emission of a neutron or proton. Therefore, a number of maxima in the nuclear spectrum, associated with various states in the residual nucleus, may correspond to each excited state of the target nucleus. Calculation of a cross section from such a spectrum is only possible in cases where the transition probabilities to the various states of the final nucleus are known. Theoretical estimates of these probabilities are unreliable in most cases so that the inclusion of supplemental experimental data is required in order to perform an analysis of photonucleon spectra. The most reliable results can be obtained from an analysis of spectra measured at different values of E_γ^{max}. The performance of additional measurements considerably complicates an experiment, particularly when emulsion is used, and supplemental information is lacking in most cases. This compels one to make assumptions about the predominant probability of ground state transitions that are not always justified when making an analysis of photonucleon spectra.

The investigation of photoproton and photoneutron spectra can be carried out in a number of cases by using the Wilson cloud chamber in a magnetic field, a method recently developed by A. N. Gorbunov and associates [7]. This method can be used with great success in investigations of the lightest nuclei (He^3, He^4, etc.) where kinematic considerations are significantly simplified and the number of possible reactions is small because of the small number of nucleons participating in the process. However, Gorbunov et al. [7, 8] also made measurements on C^{12} [9], N^{14}, O^{16}, and Ne^{20} [10] in addition to measurements on the lightest nuclei.

Despite the obvious advantages (simultaneous measurement of energy and angular distribution of all reaction products, and also of their cross sections), this method has a number of additional deficiencies beside the usual ones resulting from the necessity of working with a bremsstrahlung spectrum – poor resolution, limitations on the choice of nuclei (gases or simple gaseous compounds), and laboriousness because it is necessary to examine and analyze a tremendous number of pictures in order to obtain the necessary statistical accuracy.

Inverse Reactions. Interesting data on the properties of highly-excited states in light nuclei was obtained by studying reactions that are the inverse of the photonuclear reaction – the (p, γ_0) reaction, to be more precise.

A study of the (n, γ_0) reaction for high-energy neutrons (1-20 MeV) could give very valuable information about nuclear structure, but the unfortunate lack of monochromatic neutron sources with energies varying smoothly in this range makes it impossible to set up such experiments.

The principle of detailed balance makes it possible to obtain complete information about the inverse (γ, p_0) process from data for the (p, γ_0) process. The reliability of this information is determined by the accuracy of the measurements made.

In studying proton radiative capture processes, sufficiently good energy resolution can be obtained; it is determined by the energy spread of the proton beam and by target thickness in most cases. Electrostatic generators with charge exchange (tandems) developed in the past 5-7 years make it possible to produce 10-12 MeV proton beams with currents of several microamperes and an energy spread no greater than a few kilovolts so that the error in energy is primarily determined by target thickness when a tandem is used as a source of energetic protons. The choice of the necessary target thickness depends on proton beam intensity and reaction cross section since only target thickness determines γ-ray yield for a given intensity and geometry. For light nuclei ($A \leq 40$) and reasonable γ-ray yields, the effective target thickness can be made 30-50 keV. It will also determine the experimental resolution (usually the resolution is 0.5-0.25% for excitation energies $E_\gamma > 15$ MeV).

Studies made with proton beams from circular and linear accelerators, which have considerably greater energy spread (several hundred kilovolts) than found in tandem beams, also make it possible to obtain results of interest [11]. However, this method also has a very serious deficiency which strongly limits its area of usefulness. It results from a flaw in the method of detecting high-energy γ-rays. Because of the small cross section for the (p, γ_0) reaction, one ordinarily uses as γ-ray detectors total-absorption scintillation spectrometers having high detection efficiency (~ 10%) and an energy resolution of about 15-20% for 15-20 MeV γ-rays. Such poor resolution does not allow one to distinguish proton radiative capture quanta produced in transitions to the ground state (these are of most interest) from those produced in transitions to the first excited state of the final nucleus where the latter is sufficiently close to the ground state. In order to separate these transitions reliably, it is necessary that the energy of the first excited state be more than 2-3 MeV, or that the transition probability to this state be negligibly small.

In the case of the (p, γ_0) reaction, the resulting information contains data about the final nucleus. Therefore it is necessary to measure the proton radiative capture cross section in the nucleus $A_Z^{(Z-1)+N}$ in order to make a comparison of the results obtained by this method and those obtained by studies of the (γ, p) reaction in a particular nucleus A_Z^{Z+N}. Such nuclei may not always exist in nature, or their abundance in a natural mixture of isotopes may be extremely low. For example, it is necessary to have the nucleus Na^{25}, which does not exist in nature, in order to study Mg^{26}, and to study F^{19}, one needs O^{18}, the abundance of which in the natural mixture of isotopes does not exceed 0.2%. This also limits the area of usefulness of the inverse reaction method for studying highly-excited nuclear states.

The development of techniques for detecting high-energy γ-rays with semiconductor counters having an energy resolution of about 1% and the developments in accelerator technology lead one to hope that the number of nuclei which can be studied by this method will increase, and that the region of excitation energies will continually expand.

2. Nuclear Absorption Cross Section

As already mentioned, it is necessary to know the characteristics of the channel involved in the decay of the excited nucleus in order to carry out theoretical calculations of cross sections for individual reactions and of the spectra of emitted particles. To do this, one must make use of nuclear models. The models which are used for consideration of excitation and decay may be basically different, creating serious difficulties in the interpretation of results. Consequently, measurement of total nuclear γ-ray absorption cross sections is of considerable interest.

The nuclear absorption cross section can be obtained by a simple addition of the cross sections for all nuclear reactions possible at a given excitation energy. However, even if these cross sections are known (actually, the cross sections for the principal reactions have been measured for only a small number of light nuclei), such a method of determining the total cross section is extremely tedious and leads to significant errors because the errors for all the individual reactions, which may be quite large, enter into the final result. One of the methods for measuring the absorption cross section is that of Fuller and Hayward [12], who measured the elastic γ-ray absorption cross section and then calculated the total absorption cross section by using dispersion relations. However, the extremely low counts in experiments of this kind (the counting rate in the experiments of Fuller and Hayward was 20 cts/hr) excludes their use for obtaining data related to structure in the absorption cross section.

The nuclear absorption cross section can be determined from direct measurements of the bremsstrahlung spectrum outside a thick absorber made of the material under investigation. In this situation, the number of quanta N(E) with energy in the range $E - E + \Delta E$ recorded by a detector is

$$N(E) = N_0(E) \exp\left[-\left(\sigma_T(E) + \sigma_C(E) + \sigma_p(E) + \sigma_n(E)\right) A\right],$$

where $N_0(E)$ is the number of quanta with energy $E - E + \Delta E$ incident upon the absorber; A is the number of nuclei per cm^2 of absorber; $\sigma_T(E)$, $\sigma_C(E)$, $\sigma_p(E)$, and $\sigma_n(E)$ are the cross sections for photoelectric effect, Compton scattering, pair production, and nuclear absorption, respectively. Knowing the cross sections for the non-nuclear absorption processes (σ_T, σ_C, and σ_p), one can separate out the nuclear part with sufficient accuracy. Because the energy dependence of the non-nuclear absorption processes is a smooth function of γ-ray energy, structure in nuclear absorption should be distinguished rather readily provided the energy resolution is appropriate.

Because the cross section for non-nuclear absorption processes is considerably greater than σ_n even in the lightest nuclei, the demands for measurement accuracy are increased considerably. For example, in order to determine the nuclear absorption cross section in O^{16} with an accuracy of 5%, it is necessary to make measurements of the total absorption cross section with a statistical accuracy no less than 0.5% ($\sigma_n/(\sigma_C + \sigma_p) \approx 0.1$-$0.15$ at the maximum of the giant resonance). Roughly speaking, the requirements for measurement accuracy increase proportionally to Z for heavier nuclei because the nuclear absorption cross section is proportional to Z and the cross section for non-nuclear processes is proportional to Z^2 (for elements with $Z > 10$ and for γ-ray energies greater than 10-15 MeV, the total cross section for non-nuclear processes is mainly determined by the pair production cross section).

The energy resolution of this method is entirely determined by the resolution of the detection system, which must satisfy two requirements: good resolution and high efficiency for detection of γ-rays in the energy region up to $E_\gamma \geq 30$ MeV. The first condition is satisfied by magnetic spectrometers, which have a resolution of about 1% at 10 MeV but an extremely low detection efficiency of 10^{-5}-10^{-6}. In contrast to the magnetic spectrometers, scintillation spectrometers, which have considerably higher detection efficiency (~10%), have low resolution – 10-15% for γ-rays with energies in the neighborhood of 20 MeV. Unfortunately, semiconductor counters, which combine the advantages of both types of spectrometers, cannot be used as yet for photon detection in this energy range.

Despite the measurement difficulties associated with low detection efficiencies, experiments have been set up in a number of laboratories to investigate the shape of the photointeraction cross section [13-15] by using magnetic spectrometers of varying resolution – from 2.0% [13] to 0.5% [15] – as detection systems. As will be shown later, this method makes it possible to obtain interesting data about the structure of the nuclear absorption cross section.

The development of accelerator technology, the construction of high-current accelerators, and the development of γ-beam monochromatization techniques create opportunities for the broader use of the absorption method. Operating with a monochromatic beam makes it possible to use a scintillation spectrometer as detector. Because the resolution of the method will be determined only by the degree of monochromatization in this case, the use of a scintillation spectrometer may prove to be more advantageous in a number of cases despite the considerable loss of intensity in a monochromator.

3. Interaction of High-Energy Electrons with Nuclei

High-current linear accelerators make it possible to use such methods as inelastic electron scattering and nuclear reactions produced by high-energy electrons for studying nuclear structure at high excitation energies.

The study of inelastic scattering of high-energy electrons makes it possible to obtain, in addition to data similar to that from experiments on nuclear absorption of γ-rays, supplementary information about form factors and the probabilities for excitation of monopole transitions which are not excited by photoabsorption because of the transverse nature of the electromagnetic field; it also makes possible more reliable identification of the type of transition.

The information which is obtained by the study of electron-nucleus interactions is considerably richer than the information obtained from γ-ray experiments. The photointeraction matrix element $\langle f | H | i \rangle$ is only a function of the energy, but the interaction of electrons with nuclei is a function not only of energy but also of momentum transfer, with $\Delta q \approx 2p$ (if the excitation energy $\varepsilon \ll E_e$). In all cases, therefore, the dependence of the cross section for a given process on momentum transfer can be determined for a given value of electron energy; consequently, the class of possible relationships describing the given process is sharply reduced.

The energy resolution in studies of inelastic scattering of electrons is determined, other than by the resolution of the detection system, by the energy spread of the beam itself, which amounts to 1-2.5 MeV at electron energies of 50-100 MeV in presently existing electron accelerators. The use of special analyzing systems leads to considerable loss of intensity. The overall resolution is normally about 0.5% for primary electron energies of the order of 100 MeV. Considerable difficulty also arises in the analysis of the data – the proper consideration of the contribution from ordinary Mott scattering and of radiation effects. The energy dependence of Mott scattering is a smooth function, and the problem of its consideration is similar to the problem of considering the non-nuclear part of the absorption in total absorption experiments. Consideration of radiation effects is more complex; besides, it may appear twice – the electron may emit low-energy γ-rays both before and after a nuclear interaction. The importance of this correction falls with improvement in the monochromaticity of electron beams and in the resolution of detection systems.

The high-intensity electron beams which are produced by linear accelerators make it possible to carry out on a higher level studies of nuclear reactions accompanied by the emission of nucleons. Data was recently published for the proton spectra produced in the (e, e'p) reaction [16]. The spectra were measured with a magnetic proton spectrometer. The use of this instrument permitted a considerable increase in the accuracy of measurement both because of the resolution and because of the statistics, which are usually clearly insufficient with the very frequently used photographic method (in rare cases, the number of analyzed tracks over the entire spectrum exceeds 4×10^3). The difficulties in performing and analyzing such experiments are similar to the difficulties which arise in the study of photoproton spectra.

Studies of (e, e'n) reactions for the purpose of revealing structure in the giant resonance are obviously not reasonable because a comparatively thick target is ordinarily chosen in this case in order to assure the required yield; a large flux of bremsstrahlung radiation will be produced with high probability in a thick target, and the neutron yield will be determined to a considerable extent by the (γ, n) reaction because the cross section for the reaction produced by the electrons is $^1/_{137}$ times smaller than the cross section for the photoreaction of the same type.

Information about highly excited nuclear states can also be obtained from a large number of reactions involving heavy particles such as (p, p'), (p, n), striping, pickup, etc. However, the interpretation of the results is considerably more complicated than the interpretation of data obtained from experiments with electromagnetic radiation and fast electrons.

Presently existing information about nuclear states with excitation energies in the giant resonance region is clearly insufficient for arriving at any kind of systematics or generalization. Only two nuclei have been studied very thoroughly – C^{12} and O^{16}, for which there is data obtained by almost all the methods enumerated, and for which theoretical studies have been made also. It is, therefore, of considerable interest to make systematic studies of the structure of highly excited states in nuclei with $A \leq 40$ because structure evidently appears rather clearly in those nuclei.

CHAPTER II

Formulation of Problem and Method

As indicated above, the study of structure in the nuclear absorption cross section in the giant resonance region places very stringent requirements on the resolution of the method used. As is clear from the review of the methods used for carrying out such studies, the best resolution can be achieved if the time-of-flight or inverse reaction methods are used, i.e., by the study of partial processes. Although these methods have "outstanding" resolution, they are characterized by serious deficiencies. A primary flaw is the fact that these methods only give information about a partial process, and also ambiguous information associated with the necessity for distinguishing ground state transitions. The latter circumstance seriously complicates the interpretation of results, leaving large possibilities for different sets of speculative conclusions.

The absorption method, which allows direct study of the structure of the nuclear absorption cross section, is free of these deficiencies. Although this method is also laborious, and modern experimental technique does not allow one to make measurements with it having a resolution better than 100 keV at an excitation energy $E_\gamma > 20$ MeV, it possesses a number of important advantages. Among them are: a) the total nuclear absorption cross section is measured directly; b) the result is unique; c) the result is free of the usually large errors associated with the use of a bremsstrahlung spectrum, i.e., the measurements are carried out under quasimonochromatic photon beam conditions; d) sufficiently good resolution (no worse than 1%) can be achieved over the entire energy range of interest.

Despite the fact that the procedure of subtraction, the non-nuclear portion of the cross section from the measured total absorption cross section, which is employed in this method, significantly increases the error in the measurement of the absolute magnitude of the nuclear cross section because of the smallness of the nuclear cross section (10% of the total), the accuracy of the determination of the energy location of resonances in the nuclear absorption cross section does not depend on this procedure and is influenced only by the resolution of the detection system used.

The idea of using the absorption method for studying giant resonance structure was put forward by L. E. Lazareva. (Recent publications indicated that this method was developed in Yugoslavia [13] and the FRG [14] at the same time.) The first experiments for measuring the γ-ray absorption cross section in C^{12} and O^{16} which were carried out with this method* made it possible to discover a number of interesting details in the structure of the nuclear absorption cross section and to make comparison with existing theoretical calculations.

*The method, the apparatus used and its characteristics, and the results are discussed in detail in the papers of G. V. Danilyan (ITEF, 1965) [20] and B. S. Dolbilkin (FIAN, 1965) [17].

The results which were obtained for C^{12} and O^{16} demonstrated the need for carrying out systematic studies of the cross section for absorption of γ-rays by nuclei. Beside revealing the structure of the nuclear absorption cross section, such measurements are of interest from the point of view of the study of general nuclear properties which are computed with the help of sum rules whose validity is not always verified by experimental material in the case of a large number of light nuclei ($A \leq 40$).

Experience acquired during measurements on O^{16} showed that systematic studies of the nuclear absorption cross section were only possible after increasing the transmission of the detector – a high-resolution, magnetic pair spectrometer. Such an increase in transmission can be achieved at the cost of a loss in spectrometer resolution and by using a multichannel detection system (see below). As indicated by measurements of the nuclear absorption cross section in carbon [19], the necessary loss of resolution does not lead to complete "smearing" of the structure in the nuclear absorption cross section.

Measurements were made on F^{19}, Mg^{24}, Al^{27}, and Ca^{40}. This choice was basically the result of two factors: a) the Ca^{40} nucleus is doubly magic, and the F^{19}, Mg^{24}, and Al^{27} nuclei are strongly deformed, which may be reflected in the shape of the nuclear absorption cross section curve; b) theoretical calculations have been made for Mg^{24}, Al^{27}, and Ca^{40}, and their comparison with experimental data is of considerable interest.

The absorption method has been rather extensively developed and used successfully both in nuclear spectroscopy and in neutron physics (transmission method), but the possibilities for using this method in studying giant resonance structure are not obvious.

The experimental arrangement for studying the structure of nuclear photon absorption is shown in Fig. 1 of the paper by Dolbilkin (p. 28, this volume). A collimated beam of bremsstrahlung radiation from an accelerator passes through a slab of the material under investigation where it is attenuated because of nuclear and non-nuclear absorption processes; it is collimated a second time and then is incident upon a detector which records the number of photons of a given energy. As in all experiments of this kind, the "good geometry" condition must be fulfilled, i.e., for a given aperture, the detector must be located at a sufficiently large distance from the absorber.*

It is most reasonable to make measurements not of the beam spectrum beyond the absorber but of the attenuation coefficient, i.e., the ratio of the number of photons of a given energy in the beam before and after the absorber. Indeed, if the detector counting rate, reduced to unit dose, in the direct beam (absorber "outside the beam") is $N_0(E) = \frac{1}{D_0(E)} \varepsilon(E) \Gamma(E, E^{\max}) \Delta E$, where $\varepsilon(E)$ is the detector efficiency, $\Gamma(E, E^{\max})\Delta E$ is the number of photons with energy between E and $E + \Delta E$ in the beam, and $D_0(E)$ is the dose, the counting rate behind the absorber (absorber "in the beam") is

$$N(E) = \frac{1}{D(E)} \varepsilon(E) \Gamma(E, E^{\max}) \Delta E \, [e^{-T\tau(E)}].$$

In this case, the attenuation is

$$N_0(E)/N(E) = \frac{D(E)}{D_0(E)} e^{T\tau(E)}. \tag{2.1}$$

Here, T and τ are the thickness and absorption coefficient, respectively. Thus a measurement of attenuation eliminates the need to calculate, or determine experimentally, such quantities as the efficiency of the detector and its energy dependence, the spectral distribution of photons in the beam, or to carry out absolute dosimetry

*As will be seen later, fulfillment of the "good geometry" condition is also very important for the reduction of effects from photon "transfer" out of higher-energy portions of the spectrum into the measured range because of shower-producing processes in the absorber.

of the beam, i.e., the measurements become relative. Although such a procedure slightly increases the time of measurement, the advantages gained in this way compensate for the relatively small additional expenditure of time.

The total absorption cross section can be calculated from known attenuation coefficients by means of simple relations. Indeed, we have from (2.1)

$$\sigma_n + \sigma_a = \frac{1}{n} \ln N_0/N, \tag{2.2}$$

where σ_n and σ_a are the nuclear and non-nuclear parts of the absorption, respectively; n is the number of nuclei per cm^2 of absorber, and N_0 and N are normalized to the same dose. One can show that σ_n obtained in this way will correspond to the nuclear absorption cross section when a number of conditions are fulfilled.*

If the nuclear absorption cross section consists of a number of resonances with width Γ, it is obvious that the most important factor determining the correct transmission of the shape of the measured cross section is the ratio between the quantity Γ and the detector resolution ΔE. Assigning a definite shape for the resonance in the absorption cross section and for the detector resolution curve, and assuming further that $n\sigma_n \ll 1$, one can make some rather valid estimates.

If the resonance being investigated has a Breit-Wigner shape, and if the resolution function is triangular, then

$$\bar{\sigma}_n \approx \sigma_0 \quad \text{when} \quad \Gamma/\Delta E \gg 1;$$

$$\bar{\sigma}_n \approx \sigma_0 \Gamma \frac{\pi}{2\Delta E} \quad \text{when} \quad \Gamma/\Delta E \ll 1;$$

$$\bar{\sigma}_n \approx 0.7\,\sigma_0 \quad \text{when} \quad \Gamma/\Delta E \approx 1;$$

$$\bar{\sigma}_n \approx 0.9\,\sigma_0 \quad \text{when} \quad \Gamma/\Delta E \approx 2.$$

The last result is of great interest because the condition $\Gamma \geq 2\Delta E$ can be used in the present measurements as a criterion for the applicability of expression (2.2). Indeed, if the measurements of σ_n are made with an accuracy of 10% (which is valid, as will be seen later on), the error introduced in the calculations will not exceed the error of measurement.

Similar conclusions are reached if it is assumed that the resolution curve has a rectangular form

$$\bar{\sigma}_n \approx \sigma_0 \quad \text{when} \quad \Gamma/\Delta E \gg 1;$$

$$\bar{\sigma}_n \approx \sigma_0 \Gamma \frac{\pi}{2\Delta E} \quad \text{when} \quad \Gamma/\Delta E \ll 1;$$

$$\bar{\sigma}_n \approx 0.8\,\sigma_0 \quad \text{when} \quad \Gamma/\Delta E \approx 1;$$

$$\bar{\sigma}_n \approx 0.9\,\sigma_0 \quad \text{when} \quad \Gamma/\Delta E = 1.8 \approx 2.$$

Consequently, expression (2.2) may be used for cross section calculations in all cases where the width of the resonance $\Gamma \gg 2\Delta E$, and the error introduced in this way will not exceed 10% in the limiting case $\Gamma = 2\Delta E$. For $\Gamma \lessdot 2\Delta E$, the cross section obtained from expression (2.2) will be underestimated by an amount depending essentially on the ratio of Γ and ΔE.

* The separation of σ_n from the total absorption cross section is discussed in detail in Chapter IV. It is assumed here that it can be performed properly.

CHAPTER III

Apparatus

As noted above, a single-channel magnetic pair spectrometer was used as a detector in the first work on the measurement of γ-ray absorption cross sections in C^{12} and O^{16} [15, 18, 19]. Nearly two years' experience in the operation of this instrument showed that its efficiency was insufficient for carrying out systematic studies of nuclear absorption cross sections even with a reduction of resolution (within reasonable limits).

The transmission of the method under discussion could not be increased by using a scintillation spectrometer as a detector because the fundamental requirement placed on a detector is the necessity for good energy resolution. The resolution of scintillation spectrometers does not exceed 10-15% in the 10-30 MeV γ-ray energy range. The use of a Compton magnetic spectrometer also leads to no solution of the problem because its efficiency, which is somewhat greater than that of a pair γ-spectrometer at $E \approx 10$ MeV [21], falls rapidly (like E^2_γ) with increasing γ-ray energy, while the efficiency of the pair spectrometer rises with energy, although slowly (like $\ln E_\gamma$).

The problem of increasing the efficiency of the method can only be solved by using a multichannel recording system. Therefore, in order to carry out systematic studies of structure in the absorption cross section of light nuclei, a five-channel, magnetic, pair γ-spectrometer was designed and built which made it possible to increase detection efficiency by almost an order of magnitude.

The efficiency of the magnetic pair spectrometer was increased because of an increase in the number of slits. This method was first used in 1948 by Walker and McDaniel [22]. Indeed, if the number of slits on one side of the radiator is l, and on the other side is k, the number of combinations of slit pairs is lk or n^2 if $k = l = n$. Neglecting the dependence of detection efficiency on γ-ray energy (this energy will be somewhat different for different pairs of slits if the magnetic field is constant over the entire working area of the spectrometer), we find that the efficiency of the instrument increases by n^2 in this case. If the distances between slits (located on one side) are the same, it is easily seen that such an instrument enables one to detect γ-rays simultaneously in $2n - 1$ energy intervals with a total efficiency n^2 times greater than the efficiency of a single-channel spectrometer with the same parameters.

For structural reasons, we selected $n = 3$ (it was impossible to install a larger number of photomultipliers in our magnet), which made it possible to increase the total efficiency of the spectrometer by a factor of nine for five energy detection ranges.

The calculations made for the single-channel spectrometer [17, 20, 23-25] remained valid for the five-channel spectrometer although the condition that the pairs are produced in the center of the radiator, i.e., at a point equidistant from the internal edges of a given pair of slits, will not be fulfilled, for the greater number of slit combinations (six of nine) (this condition was used in calculating spectrometer resolution). That conclusion is based on the results of experimental determinations of the dependence of resolution on beam position at the radiator, made on the single-channel spectrometer, which showed that the resolution changed by approximately 5% when the beam was displaced 6 cm from the center of the radiator (see Fig. 13 in Dolbilkin's paper, p. 38, this volume). Since the maximum asymmetry of the slits in the five-channel spectrometer is 2 cm, the difference in resolution for symmetric and asymmetric slits is hardly more than 2-2.5%. The fact that resolution was independent of slit symmetry also allowed summation of various channels with the same energy values.

The construction of the five-channel spectrometer was based on an electromagnetic whose principal parameters were the following: pole-tip diameter, 600 mm; inter-polar spacing, 120 mm; maximum magnetic field in the gap, 4.0 kilogauss; uniform magnetic field region, 400 mm (flat within 2%).

In order to improve the uniformity of the magnetic field in the spectrometer working region, discs of St-3 steel (600 mm in diameter, 10 mm thick) were installed in the top and bottom of the chamber which acted as pole-tip caps. The reduction of the magnet gap to 100 mm (a 20% reduction) brought about a considerable reduction in the nonuniformity of the magnetic field, which was no worse than 0.05% inside the chamber (in an area 40 cm in diameter).

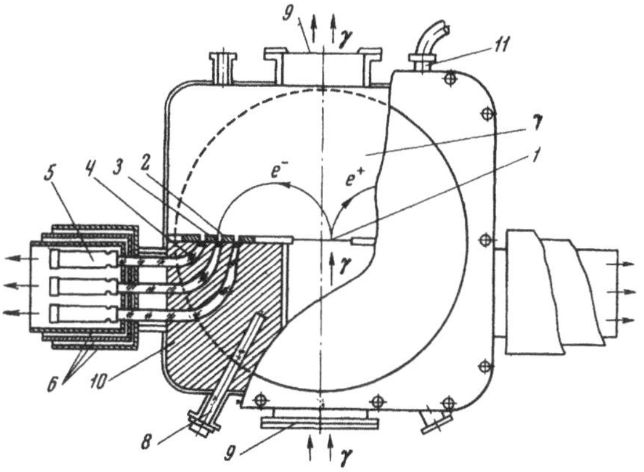

Fig. 1. Spectrometer chamber. (1) Radiator; (2) slit system; (3) scintillators; (4) light pipes; (5) FÉU-36 photomultipliers; (6) magnetic shielding; (7) uniform magnetic field region; (8) magnetic field stabilization sensor; (9) γ-ray beam entrance and exit windows; (10) lead shielding; (11) pumpout.

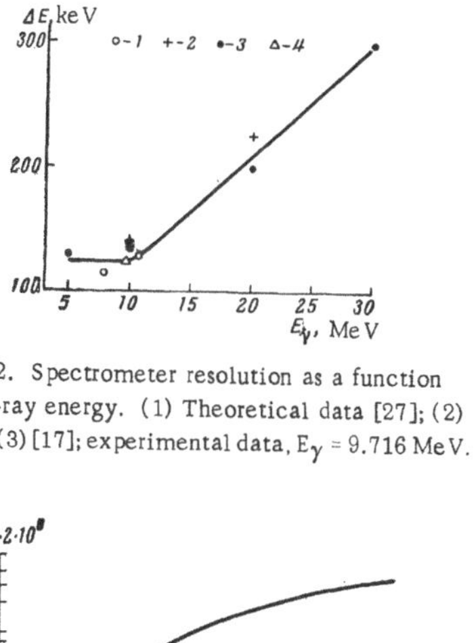

Fig. 2. Spectrometer resolution as a function of γ-ray energy. (1) Theoretical data [27]; (2) [23]; (3) [17]; experimental data, E_γ = 9.716 MeV.

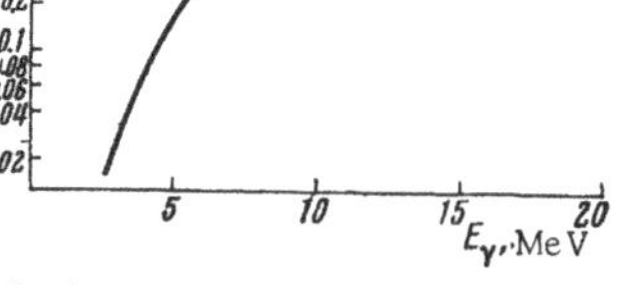

Fig. 3. Spectrometer detection efficiency as a function of γ-ray energy.

A sketch of the chamber is shown in Fig. 1. The chamber had outside dimensions of 750 × 750 × 120 mm and was made of stainless steel. For convenience, the top and bottom assemblies of the chamber were removable, and were fastened to the body against rubber gaskets by tie bolts.

The spectrometer radiator was made of gold foil 10 mg/cm^2 thick. The correct choice of radiator thickness is very important because the efficiency of the instrument rises with increasing radiator thickness. However, ionization loss and multiple scattering effects play an ever larger role as the target thickness increases. The second effect has a more important influence on the resolution because it distorts the angular distribution of the pair components. It is evident that equality of the multiple scattering and pair component emission angles can serve as a criterion in this case. Those angles are equal if the electrons and positrons pass through a layer of matter that has an average thickness of about 10^{-3} radiation lengths. Since the attenuation of a γ-ray beam in the radiator is negligibly small, the mean range of the pair components can be assumed to be half the target thickness. In that case, the value of the critical thickness is 2×10^{-3} radiation units (for gold, this corresponds to 11.2 mg/cm^2 or approximately 6 μ).

The construction of the radiator holder was such that it was possible to remove the radiator from the γ-ray beam without losing vacuum in the spectrometer chamber.

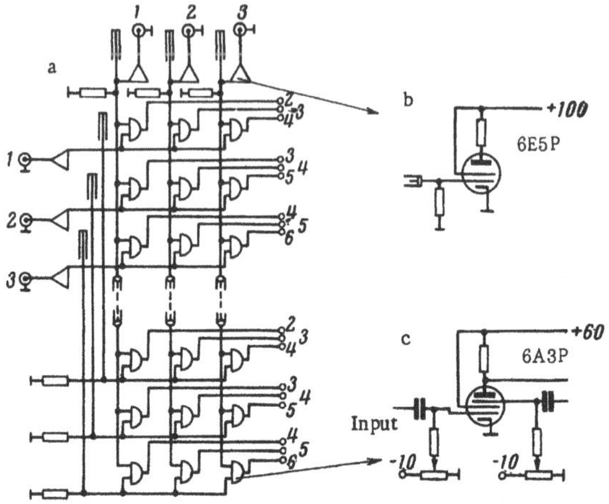

Fig. 4. Switching system (a) and basic pulse-shaping (b) and coincidence (c) circuits.

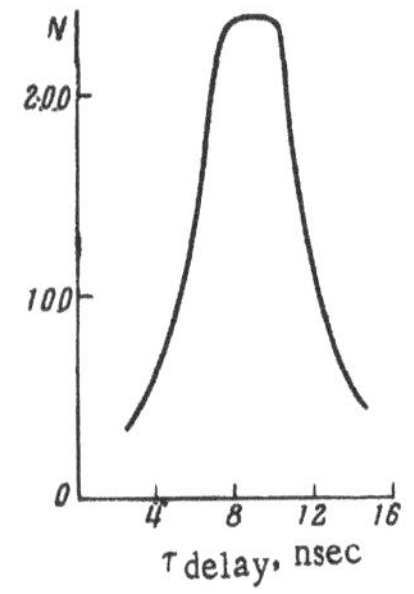

Fig. 5. Typical coincidence curve.

Because of this, it was possible to check operationally on the absence of spurious coincidences and of background contamination of the counters during spectrometer adjustments.

The spectrometer slits were 4 mm wide and located in the plane of the radiator. The distances between the inner edges of symmetric slits were 37.6, 39.6, and 41.6 cm. The distances between the inner edges of the slits located on one side of the radiator were 10 mm. The slits were made of lead 10 mm thick in the form of two blocks with three slits in each block. Preparation and installation of the slits was carried out with an accuracy of 0.1 mm, which insured an accuracy of 0.05% (because of errors in preparation and installation) in the energy determination for γ-rays detected by the spectrometer.

The energy dependence of the resolution and efficiency of a magnetic pair spectrometer with this construction is shown in Figs. 2 and 3. As the figures indicate, the spectrometer employed had a resolution of about 1% in the γ-ray energy range 10-30 MeV and an efficiency of 8×10^{-5} at $E_\gamma = 20$ MeV which fell to $\sim 2 \times 10^{-5}$ at $E_\gamma = 10$ MeV.

Plastic scintillators located behind the slits were fastened to the ends of curved, plastic light pipes. The scintillators were in the form of parallelepipeds $70 \times 6 \times 6$ mm in size. The light pipes were reliably isolated from one another, both optically and with respect to penetrating particles, by means of curved lead plates installed between them.

In order to reduce the background produced by electron (and positron) scattering, the chamber was lined with polythene and was pumped down to a pressure of 10^{-2} mm Hg.

The exit faces of the light pipes, which had a cylindrical shape at their far end, were held in place by rubber rings which fulfilled the role of vacuum gaskets at the same time. The cylindrical face of the light pipes was "glued" to the photomultiplier cathode with vacuum oil.

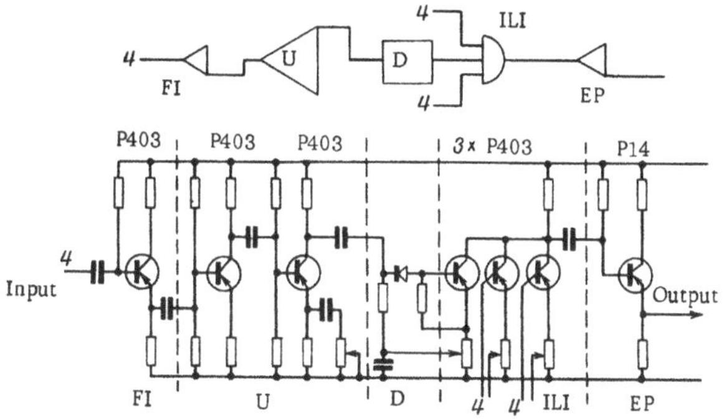

Fig. 6. Main circuit of output section.

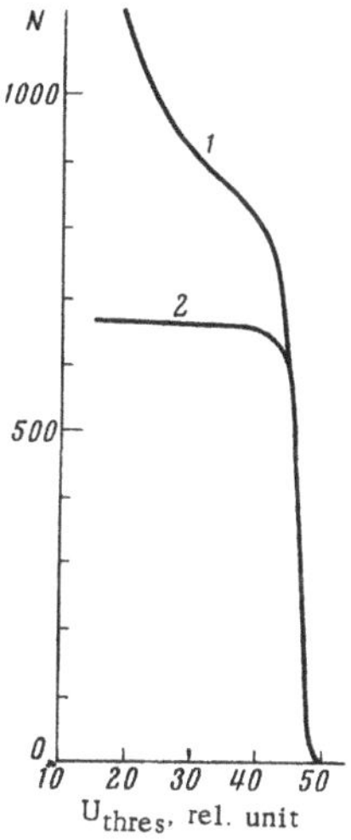

Fig. 7. Typical curve for the dependence of counting rate on discrimination threshold. (1) True and random coincidences; (2) true coincidences.

The type FÉU-36 photomultipliers used had a sufficiently high amplification factor of 10^7-10^8 and produced an output pulse with a rise time of $\sim 2 \times 10^{-9}$ sec. From the 25 photomultipliers at our disposal, two sets of three each were selected which had approximately the same operating voltage (difference no greater than 100 V) and amplification. The power for the photomultipliers was obtained from two VS-22 rectifiers (each of them fed a group of three photomultipliers).

Since the photomultipliers were located in the immediate vicinity (~ 10 cm) of the edge of the magnet pole tips, and were, therefore, in a region of considerable magnetic field, they were surrounded with a four-layered magnetic shield of soft steel, and the far portion of the light pipes passed through a thick magnetic shield (total thickness, 20 mm) of Armco iron arranged along the magnetic lines of force. It was only though the introduction of this magnetic shielding that it was possible to reduce the magnetic field in the photocathode region to the amount essential for normal operation of the photomultipliers.

With a 100 Ω load on all six counters, the scintillator-light pipe-photomultiplier cathode optical system and the magnetically shielded photomultiplier itself provided a ~ 10 V pulse with pulse length of $5\text{-}6 \times 10^{-9}$ sec and signal-to-noise ratio of 10. The pulse heights from all counters remained constant within about 5% for all magnetic field values from 0 to ~ 4.5 kilogauss.

Pulses from the photomultipliers were transmitted over 25 m of RK-2 cable to an instrumentation room and into pulse-shaping stages made up of 6E5P tubes, then into a matrix of nine coincidence circuits (Fig. 4). This same figure shows the basic coincidence stage circuit and a block diagram of its connections.*

Pulses shaped in a cable (0.4 m of RK-50 cable) in the plate circuit of the pulse-shaping stage were sent to the first and third grids of a coincidence circuit using 6A3P tubes having high transconductance on both grids. The use of a tube coincidence circuit made it possible to obtain a selection coefficient of about 7 at a

*The electronic equipment for the five-channel spectrometer was designed by V. A. Zapevalov and is described in detail in [26].

Fig. 8. General view of spectrometer.

Fig. 9. General view of recording and control systems.

relatively high coincidence signal pulse height (0.1 V) and provided the extremely important needed stability. Attempts to use diode coincidence circuits (type D2E crystal diodes) showed that temperature variations in the instrumentation room changed the parameters of these circuits significantly. Such a source of instability can be eliminated by thermostatic control or the introduction of special compensating circuits which would complicate the equipment considerably and lead to a number of drawbacks.

The coincidence circuits employed provided a resolution of nearly 3×10^{-9} sec under operating conditions with a detection efficiency close to 100%. A typical coincidence curve is shown in Fig. 5.

Signals from the coincidence circuits were fed through the corresponding emitter follower FI and amplifier U (Fig. 6) into a diode discriminator D operating on the integrator ILI and then through the output emitter follower EP into a univibrator which triggered a decatron scaling circuit with a capacity of 10^4 pulses.

Input numbering on the coincidence circuit matrix in Fig. 4 is the following: the internal symmetric slits correspond to number 1, the next pair to number 2, and the outermost pair to number 3. It is easy to see that electron-positron pairs produced by γ-rays of the same energy will be recorded (in order of increasing energy) by five groups of coincidence circuits: 1 – 1; 1 – 2 and 2 – 1; 1 – 3, 2 – 2, and 3 – 1; 3 – 2 and 2 – 3; 3 – 3. Thus, pulses at the output of the coincidence circuits with identical sum of input numbers correspond to γ-rays of the same energy. The energy channels of the spectrometer were of different efficiencies. Their efficiencies were in the ratio 1 : 2 : 3 : 2 : 1, i.e., like the number of coincidence circuits operating for a given energy channel. In this connection, only one integrator input was used in channels I and V, two were used in channels II and IV and all three were used in channel III.

A feature of the equipment employed was the possibility of simultaneous recording of chance coincidences. This was accomplished by the introduction of a parallel channel for counting delayed coincidences similar to the main channel. A delay of 2×10^{-8} sec was produced by a RK-50 cable in the arm of the second channel immediately after the pulse-shaping stage. The circuit connections of the chance coincidence and main matrices were identical and are shown in Fig. 4.

The use of semiconductors in the counting circuits made it possible to produce rather economical electronic equipment which operated under light thermal load and which had high stability with high detection efficiency (significant loss of amplification over the entire counting channel was not observed during nearly two years of operation).

In Fig. 7, there is a typical curve for the dependence of counting rate on discrimination threshold showing a rather broad "plateau" which cuts off sharply at high thresholds. Such a dependence is evidence of the proper operation of the entire counting channel (all the amplification stages before the discriminator operated in the linear portion of the pulse-height characteristic).

An important element of the spectrometer was the system for stabilization and measurement of the magnetic field by proton nuclear magnetic resonance (NMR) similar to the previously described system for the stabilization of the single-channel spectrometer [20]. The use of NMR provided magnetic field stability no worse than 0.01%, and, what is particularly important, made it possible to restore the value of the magnetic field with a high degree of accuracy (0.05%) determined by the precision of a wavemeter.

A general view of the spectrometer, and also of the detection and control systems is shown in Figs. 8 and 9. The basic parameters of the five-channel magnetic pair spectrometer are the following:

Radiator thickness (gold)	10 mg/cm^2
Radiator working area	60 x 50 mm
Distances between internal edges of symmetric slits	376, 396, 416 mm
Slit width	4 mm
Slit length	60 mm
Magnetic field uniformity in operating region . .	0.05%
Magnetic field stability (by NMR)	0.01%
Coincidence circuit resolution	3×10^{-9} sec
Chamber pressure	10^{-2} mm Hg
Resolution.........................	120 keV (at 10 MeV) 220 keV (at 20 MeV)
Detection efficiency	8×10^{-5} (at 20 MeV)

CHAPTER IV

Measurements and Analysis of Experimental Data

1. Absorbers

Measurements were made on fluorine, magnesium, aluminum, and calcium nuclei. Except for magnesium, these elements have monoisotopic or close to monoisotopic composition (Table 1).

In the case of magnesium, the results can be assigned to the principal isotope Mg^{24} because, if the shapes of the absorption cross sections for Mg^{25} and Mg^{26} are different from the shape of the absorption cross section for Mg^{24}, this difference would have to be quite considerable in order to appear in the background of the Mg^{24} cross section at the expected measurement accuracy of ~ 10-15%.

In the preparation of the absorbers, we used pure metallic magnesium (> 99.9%), aluminum (> 99.9%), and calcium (> 99%), particular attention being paid to the admixture of heavy elements, which were certified not to exceed 0.01%.

TABLE 1

Nucleus	Abundance in natural mixture, %	Absorber thickness		Nuclei per cm², 10^{23}	Expected attenuation (for E_γ = 20 MeV)
		g/cm²	cm		
$F^{19}(C_2F_4)$	100	105.3 (as F^{19})	62	33.2	6
Mg^{24} Mg^{25} Mg^{26}	78.6 10.1 11.3	112.4	64.5	27.4	10.3
Al^{27}	100	108.0	40	24.0	10.3
Ca^{40} $Ca^{42, 43, 44, 45, 46}$	96.9 3.1	70.8	45.5	10.6	7.8

Since metallic calcium oxidizes in air, it was placed in an oil bath. In that case, the beam passed through an oil layer 3 mm thick and the 2 mm-thick end walls of a plexiglas container in addition to the calcium itself. The total thickness of oil and container wall was no more than 0.5 mg/cm² or 0.5% of the absorber thickness.

In the case of fluorine, teflon was used as the absorber. Since the teflon molecule consists only of fluorine and carbon atoms (C_2F_4), the attenuation N_0/N will be

$$N_0/N = \exp\left[A_T(2\sigma_C + 4\sigma_F)\right],$$

where A_T is the number of teflon molecules per cm², and σ_C and σ_F are the total absorption cross sections for carbon and fluorine, respectively. In order to eliminate the absorption resulting from the presence of carbon in the absorber, a block of graphite was introduced into the beam during measurements of N_0 ("absorber out of the beam"), the thickness of which was based on the condition that the number of carbon nuclei per cm² was the same in the graphite and main absorbers. In this case, the attenuation coefficient is

$$N_C/N_T = \exp\left[-A_C\sigma_C\right]/\exp\left[-A_T(2\sigma_C + 4\sigma_F)\right],$$

and, since, $A_C\sigma_C = 2A_T\sigma_C$, then

$$N_C/N_T = \exp\left[4A_T\sigma_F\right] = \exp\left[A_F\sigma_F\right],$$

where A_F is the number of fluorine atoms per cm² and N_C and N_T are the coincidence counting rates with graphite and teflon absorbers, respectively. For a teflon absorber thickness of 138.6 g/cm², the graphite block was 33.3 g/cm² thick.

The selection of absorber thickness was determined by the minimal error in the measured absorption coefficient, by the condition $n\sigma_n < 1$, and also by practical considerations: the availability of a sufficient quantity of a given material and spatial limitations on absorber placement (because of lack of room in the experimental area, the absorber length could not be greater than 1 m).

It is easy to show, as was done in [27], for example, that the square of the relative fluctuation in a measurement of the absorption coefficient is

$$\delta^2 = A\left(\frac{1+\sqrt{x}}{\ln x}\right)^2, \tag{4.1}$$

where $x = N_0/N$ and $A = \text{const}$, and it is a minimum when $x \approx 13$. As can be seen from this relation, however, δ^2 does not exceed its minimum value by more than 20% over a wide range of N_0/N values (in any case, when $6 \leq N_0/N \leq 28$). Therefore, it is necessary to make measurements with an absorber that provides attenuation by at least a factor of six in order to obtain the minimum error in the absorption coefficint.

As can be seen from Table 1, all the absorber requirements enumerated were fulfilled.

2. Procedure for Measurement

The measurements were made at the FIAN synchrotron having a maximum energy of 260 MeV. The selection of this accelerator as the γ-ray source was based on the fact that an increase in the maximum energy of the bremsstrahlung spectrum to 260 MeV considerably raised the intensity in the excitation region under study (~20 times) in comparison with the intensity of the 30-MeV FIAN synchrotron present in the laboratory, and the relatively long beam duration (2-3 msec) provided a sufficiently small amount of chance coincidences (~20% for measurements without absorber).

The use of a synchrotron (with a maximum energy of 260 MeV) had its shortcomings. The chief one was that showers developed during the passage of high-energy bremsstrahlung through an absorber several radiation units thick, leading to photon "transfer" from the high-energy portion of the spectrum into the region under study (this effect is discussed in detail below). In order to avoid the effects of shower production in the absorber, it is necessary to operate at a maximum bremsstrahlung energy without exceeding some critical value. However, we had no opportunity to do this because the accelerator used was heavily scheduled. Therefore, all measurements were made in parallel with other work. Further, the maximum energy of the bremsstrahlung spectrum varied from 200 to 264 MeV.

In the selection of experimental geometry, particular attention was devoted to the fulfillment of "good geometry" conditions (see Fig. 1 in Dolbilkin's paper, p. 28, this volume): the solid angle from the center of the absorber to the spectrometer radiator was not greater than 10^{-4} sr. The absorber, which was fixed in a special holder allowing remote removal and introduction into the beam, was located 10 m from the synchrotron target; the beam incident on the absorber was defined by three lead collimators 50 mm in diameter. There was always a liquid–hydrogen target between the synchrotron target and the absorber, but the amount of material in the beam (structure of the hydrogen target and the liquid hydrogen itself) remained constant during all measurements.

The relative γ-ray dose incident on the absorber was measured with two standard ionization chambers developed at the FIAN Laboratory for Photomeson Processes, which were located directly at the first collimator and immediately before the absorber. The main chamber was the one placed in front of the absorber; the first chamber was used for monitoring and for control of the automatic counting system.

Measurements were made in the following manner. At a given value of the magnetic field, which was changed by changing the frequency of the high-frequency generator of the stabilization system, counts were measured for the absorber out of the beam (or in the beam). After the accumulation of a preset dose, automatic equipment stopped all 10 counters, the electronic clocks, and ionization chamber counter. After taking readings from the counters, the absorbers was remotely introduced into the beam (or removed from the beam), and "absorber in beam" (or "absorber out of beam") measurements were made. The ratio of the doses for measurements in and out of the beam was determined from consideration of the minimum error in the attenuation N_0/N, which is $\sqrt{N_0/N}$ as has been shown in [27], for example. The time for one such measurement was ordinarily not more than 15 min. Then the next value of the magnetic field was set, and the cycle was repeated, etc. The entire possible energy range was covered in this manner. The statistical accuracy at each point of measurement was no worse than 3%, as a rule. After going through the entire energy range, the magnetic field was returned to its original value. Measurements at the same magnetic field values were repeated not less than 8-10 times throughout the entire energy range.

Such cycling made it possible to eliminate systematic errors resulting from instrumental drift which are unavoidable during continuous operation covering a period of 100-300 h. All counter readings, times, and frequencies of the high-frequency generator in the spectrometer magnetic field stabilization system were recorded on special forms for each measurement, and later on punched cards for analysis with an electronic computer.

3. Processing of Experimental Data

Experimental data was processed on the FIAN electronic computer. The need to use an electronic computer for performing relatively simple computations grew out of the tremendous amount of material produced by the measurements. Estimates of the time needed to process the data indicated that one man, working six hours a day with a desk calculator, would require more than two months to process the results for each one of the elements.

1. Determination of γ-Ray Energy. The energy of γ-rays recorded is determined by the distance between slits and the magnitude of the magnetic field in the spectrometer gap, and the magnitude of the magnetic field is given by the frequency f_0 of the high-frequency generator in the system for measurement and stabilization of the magnetic field. Because the measurements were made with a continuous spectrum of bremsstrahlung radiation and yielded a quantity σ_n averaged over the resolution curve, the energy of most interest was that corresponding to the maximum of the coincidence peak of the spectrometer. It is most reasonable to assign the measured quantity σ_n to this energy because the cross section for E_{res} contributes to σ_n with the greatest weight [17]. Calculations that were performed showed that the γ-ray energy recorded by the five-channel spectrometer could be determined from the following expressions:

Channel I . . . . $1.3322\, f_0 + 0.0857$
Channel II . . . . $1.3676\, f_0 + 0.0857$
Channel III . . . . $1.4031\, f_0 + 0.0857$
Channel IV . . . . $1.4385\, f_0 + 0.0857$
Channel V . . . . $1.4739\, f_0 + 0.0857$

2. Counting Loss Corrections. The output scaling circuits of the pair spectrometer were made up of decatrons, which have a relatively long "dead" time (150 msec in the true coincidence channel and 400 msec in the chance coincidence channel). This time was comparable to the duration of the accelerator radiation pulse (~ 2000 msec) which could lead to an appreciable counting loss at counting rates of 50 cts/sec.

To allow for the effects of counting circuit "dead" time, a relation given in [27],

$$\frac{1}{m} = \frac{1}{n} + \tau, \tag{4.2}$$

can be used where m and n are the observed and actual count rate and τ is the resolving time. Hence, the number of counts in a time t, corrected for "dead" time, is

$$N_0 = \frac{N}{1 - \frac{N}{t}\tau}. \tag{4.3}$$

Considering the duty cycle of the accelerator, which equals 0.1, and the magnitude of the "dead" time, we have for the principal channel and for the chance coincidence channel, respectively,

$$N_0' = \frac{N'}{1 - \frac{N'}{t} 1.5 \cdot 10^{-3}}; \quad N_0'' = \frac{N''}{1 - \frac{N''}{t} 4 \cdot 10^{-3}}. \tag{4.4}$$

The magnitude of the correction for counting loss usually did not exceed 10% both for the true coincidence channel and for the chance coincidence channel.

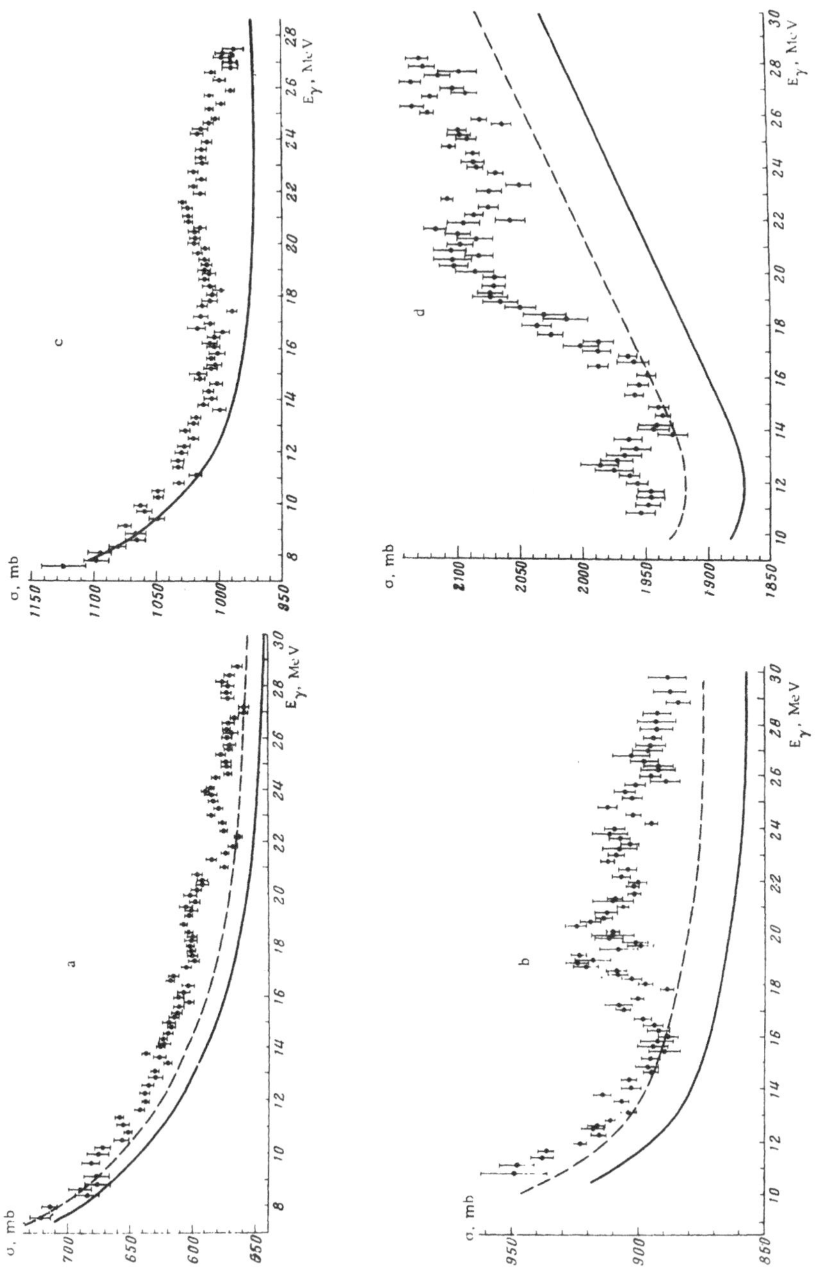

Fig. 10. Total absorption cross section for (a) F^{19}, (b) Mg^{24}, (c) Al^{27}, and (d) Ca^{40}. Solid line is the cross section for non-nuclear absorption processes from the data of [30]; dashed line is the cross section for non-nuclear absorption processes normalized to the experimental data.

3. Consideration of Chance Coincidences. The problem of correctly accounting for chance coincidences is extremely complex particularly when working with a pulsed radiation source whose intensity changes significantly with time. The principal method is the reduction of the number of chance coincidences to a negligibly small amount – by increasing the multiplicity of the coincidences or by reducing the resolving time, for example. An increase in the multiplicity of coincidences would have required the installation of additional counters in the working region of the magnet, which was impossible for structural reasons, and a reduction of resolving time by a factor of 2-3 (the scintillators and photomultipliers in our setup did not allow construction of a coincidence circuit with resolution better than 1×10^{-9} sec) would not have produced a significant reduction in the number of chance coincidences. Therefore, we chose the method of parallel counting of chance coincidences as described in Chapter III.

In adjusting the spectrometer, special attention was devoted to the equivalence of the coincidence circuits in the main channel and in the channel for counting chance coincidences. In addition, a check was made on this equivalence every 4-5 hours during the process of measurement. (A delay was introduced into one of the branches of the main coincidence circuit system so that the counts in both channel should have agreed within statistical error.) Such a system made it possible to take account of chance coincidences with sufficient accuracy and within the limits of statistical accuracy. The number of chance coincidences during measurements in the direct beam did not normally exceed 20-25%, and was negligibly small (~ 1%) during measurements with an absorber. The number of true coincidences was defined as the difference between the counts in the main and delayed coincidence circuits (after correcting for counting losses).

4. The Role of Cascade Processes. As already pointed out, electron-photon showers are produced during the penetration of a beam of bremsstrahlung radiation with $E_\gamma^{max} \approx 260$ MeV through a thick absorber (several t-units) because of cascade processes in the absorber. Therefore, the number of photons in the beam beyond the absorber which are in the measured energy interval 10 MeV < E_γ < 30 MeV, where $E_\gamma < E_\gamma^{max}$, will be determined to a considerable degree by these processes, leading to deviation from the exponential law for absorption.

One can show, as was done in [17], for example, that the number of cascade photons incident on the spectrometer radiator which were in the measured energy range was negligibly small (no more than 0.25% of the number of photons of the same energy in the direct beam) in the case of the actual geometry (solid angle from absorber to spectrometer radiator ~ 10^{-4} sr).

5. Consistency of the Computations. After the introduction of corrections for counting loss, we considered the ratio

$$N_0/N = \frac{N_0' - N_0''}{N_+' - N_+''} \frac{M_+}{M_0}, \tag{4.5}$$

where N'_+ and N''_+ are the counts in the main channel and in the chance coincidence channel with an absorber in the beam, and M_0 and M_+ are the relative doses for "absorber out of the beam" and "absorber in the beam" measurements. Counting loss corrections were not introduced for measurements with absorber because the loading was considerably less in this case (by a factor of approximately 10), and the magnitude of the corrections, in all cases, was very much less than the statistical accuracy of the measurements of N'_+ and N''_+.

The ratio N_0/N, calculated for the separate measurements in each of the five channels, was averaged by channel. Its standard deviation was computed. The corresponding energies were assigned to the $(N_0/N)_{av}$ ratios obtained in this way.

As an additional check on the correctness of card punching, all the values of N_0/N and $(N_0/N)_{av}$ were printed out by channel and inspected. Cases of large deviation of N_0/N from the average (more than 4-5 standard deviations) were checked. Such deviations did not exceed 3-5% for each of the elements measured, and were produced by punching errors in the majority of cases (most typical error – incorrectly punching the order of the digits). No sampling of points was made in doing this.

TABLE 2

Reaction	Isotope, content		
	Mg^{24} (78.6%)	Mg^{25} (10.1%)	Mg^{26} (11.3%)
(γ, n)	16.56	7.33	11.11
(γ, p)	11,69	12.06	14.3
(γ, α)	9.33	9.0	9,8

After correction of any errors found, the channel values of the ratio $(N_0/N)_{av}$ and the corresponding energies were put in the computer and were averaged, using weight,* over a 0.2 MeV interval in the energy range from 7 to 30 MeV. The 0.2 MeV interval corresponded to the resolution of the spectrometer at $E_\gamma \approx 20$ MeV. Then the total absorption cross section was calculated from the averaged N_0/N values:

$$\sigma = \varkappa \ln N_0/N, \tag{4.6}$$

where $\varkappa = n^{-1}$ (n is the number of nuclei per cm^2 of absorber), and the standard deviation of the total absorption cross section:

$$\Delta\sigma = \varkappa \frac{1}{N_0/N} \Delta N_0/N. \tag{4.7}$$

Errors connected mainly with the consideration of chance coincidences were taken care of in the following fashion: for each of the channels, average (weighted) values of N_0/N were determined in the 15-25 MeV range which were then compared with the average value of N_0/N in the third channel (statistical accuracy was always maximum in the third channel). The average values of N_0/N obtained in this way for the individual channels differed little from one another (~ 1%), testifying to the proper operation of the spectrometer. After normalization of all channels to the third, the total absorption cross section and its error was calculated once again. The resulting values of the cross section differed insignificantly from the original values (within the limits of error).

Total absorption cross sections for F^{19}, Mg^{24}, Al^{27}, and Ca^{40} are shown in Fig. 10.

6. Determination of Nuclear Absorption Cross Section. As already noted, the total absorption cross section $\sigma = \sigma_c + \sigma_{pair} + \sigma_{photo} + \sigma_n$ (σ_c, σ_{pair}, σ_{photo}, and σ_n are the cross sections for the Compton effect, pair production in the field of the nucleus and of electrons, photoelectric effect, and nuclear absorption processes, respectively). Separation of the nuclear portion from σ should be accomplished sufficiently correctly because the absolute magnitude $\sigma_n \ll \sigma$, and, therefore, a systematic error introduced in this way can be considered small in any case.

The absorption for γ-rays with energies of 10-30 MeV is mainly determined by the Compton scattering and pair-production processes. The absorption resulting from the photoelectric effect is small in this energy range and can be neglected.

One can show that the non-nuclear portion of the total absorption cross section can be calculated with an accuracy of 2-3%. Such an uncertainty is associated to a considerable extent with the complexity involved in considering exchange effects and screening effects in calculations of the cross section for pair production in the field of atomic electrons. (This problem has been discussed in more detail in [18] and also on p. 41, this volume.)

Numerical computations of the total absorption cross section (and of cross sections for individual processes) have been made by a number of authors. Evidently the best of these are the results obtained at the National Bureau of Standards (Washington) in 1952 [28] and 1957 [29]. The results obtained in 1952 are given in Siegbahn's monograph [30], and the more recent ones can be found in the book by Leipunskii et al. [31]. The differences in these calculations are quite insignificant and are associated with differing consideration of the "triplet" formation process.

* The inverse of the variance was used as the weight.

However, even a two-percent uncertainty in the total absorption cross sections leads to a very large uncertainty (~ 20%) in the "zero" of the nuclear cross section because the nuclear absorption cross section is about 10% of the total cross section at the maximum of the giant resonance. The error in the determination of the "zero" of the nuclear cross section can be reduced considerably if normalization of the calculated absorption cross section to the experimental data is carried out in an energy interval in which the nuclear absorption cross section from any source whatever is well known. Such a procedure was employed in treating the experimental data obtained for all four elements.

(a) Fluorine. It is obvious that such normalization can be performed most reliably in an energy region where the nuclear absorption cross section is small or equal to zero. The energy region below the energy thresholds for the main (γ, p) and (γ, n) reactions best satisfies this condition.

Presented below are the thresholds of energetically possible reactions in F^{19} for $E_\gamma < 15$ MeV, to which one should also add elastic and inelastic nuclear scattering.

Reaction	Thres., MeV	Reaction	Thres., MeV
(γ, α)	4.012	(γ, n)	10.442
(γ, p)	7.992	(γ, t)	11.669
		(γ, d)	13.814

Since the measurements were made down to an energy of 7.3 MeV, and the effective threshold for the (γ, p) reaction is 9.5 MeV, normalization of the calculated cross section was accomplished in the 7.5-9.5 MeV energy range. In this energy region, the nuclear absorption cross section is determined only by the (γ, α) reaction and by scattering processes (both elastic and inelastic). Unfortunately, there is no data on the cross sections for these processes; however, they can be estimated from existing data for O^{16} – the neighboring nucleus. It appears that such estimates will not lead to large error because a comparison of the α-particle yields from N^{14}(γ, α) and O^{16}(γ, α) reactions reveals that they are practically the same [32]. The integrated cross section for the O^{16}(γ, α) reaction at excitation energies $E_\gamma \leq 30$ MeV is no more than 4 MeV-mb [33]. Assuming that the integrated cross section for the (γ, α) reaction is also 4 MeV-mb in the case of fluorine and that 50% of this cross section occurs in the 7.5-9.5 MeV interval, we find that the average cross section for the reaction in this energy region is no more than 1 mb. The assumption that 50% of the integrated cross section for the F^{19} (γ, α) reaction is concentrated in the 7.5-9.5 MeV range is equivalent to the assumption that the maximum for this reaction is at $E_\gamma \simeq 8.5$ MeV and that it has a halfwidth $\Gamma \simeq 2$ MeV. As a matter of fact, the cross section for the (γ, α) reaction can have a complex shape and can be different from zero over a broad energy range [34]. Consequently, the assumption made leads to somewhat of an overestimate of the value of the average cross section.

There also have been no measurements made of the elastic and inelastic scattering cross sections in fluorine. Therefore, as in the case of the (γ, α) reaction, data for oxygen was used [35], an analysis of which shows that the total cross section for these processes at the maximum of the giant resonance where $E_\gamma = 22.0$ MeV is no more than 0.1-0.2 mb and drops sharply in the direction of lower energies, i.e., its contribution is negligibly small in the energy region of interest to us. Thus the average cross section for nuclear absorption does not exceed 1 mb in the 7.5-9.5 MeV range.

The measured total γ-ray absorption cross section in fluorine is shown in Fig. 10a. No calculations of the non-nuclear portion of the absorption cross section were made directly for fluorine; therefore the data for oxygen was used; it was converted with allowance made for the difference in Z. The error introduced by this conversion is negligibly small because the dependence of the cross section on Z is well known for Compton scattering and pair production.

The energy dependence of the non-nuclear portion of the absorption (dashed curve) which was obtained after normalization in the 7.5-9.5 MeV range differs from the calculated curve by 2%, which is in agreement with estimates of the accuracy of the calculations for cross sections of non-nuclear absorption processes.

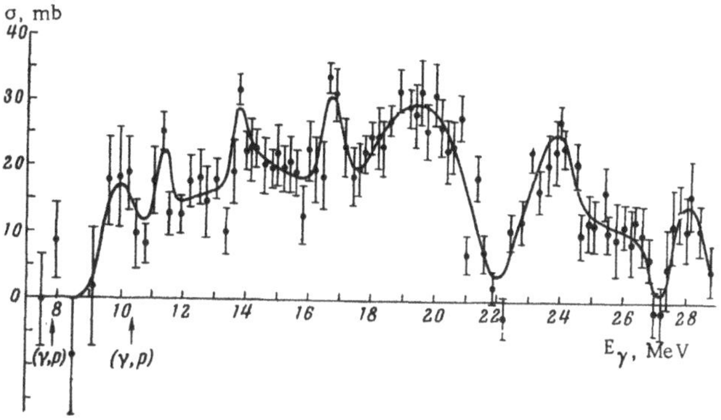

Fig. 11. Nuclear γ-ray absorption cross section of F^{19}.

The accuracy of the determination of the "zero" of the nuclear cross section can be found by an analysis of the experimental data. In fact, the error in the average nuclear absorption cross section in the 7.5-9.5 MeV region, which is ±4.0 mb, can be taken as the error in "zero." However, the points at 22 MeV introduce additional limitations; their consideration somewhat reduces the spread in the direction of lower cross section values and leads to an accuracy in the determination of the "zero" of the nuclear absorption cross section of (+4.0)-(−2.0) mb.

(b) Magnesium. In contrast to fluorine, normalization of the non-nuclear portion of the absorption cross section in the case of magnesium was made in the 14.5-16.0 MeV region. This is because a definite peak was observed in the nuclear absorption cross section for magnesium at the excitation energy E_γ = 11 MeV down to which measurements were made.

Thresholds of reactions (MeV) which can make a contribution to the absorption cross section for E_γ = 14.5-16.0 MeV are given in Table 2.

As is clear from this table, the (γ, n) reaction can only occur in the isotopes Mg^{25} and Mg^{26} at energies of 14.5-16.0 MeV. Spicer et al. [36] made measurements of the (γ, n) reaction cross section in a mixture of Mg^{25} and Mg^{26} which showed that they observed a peak at E_γ = 13.5 MeV with the cross section having a value of 2.6 mb at the maximum; the cross section then dropped sharply and was 0.5-0.6 mb on the average in the 14-16 MeV range. Earlier measurements by Yergin [37] gave a higher value for the cross section (~10 mb), which, as Spicer et al. showed [36], was connected with a contribution from oxygen activity (measurements in [37] were made by the induced activity method, and magnesium oxide was used for the samples). Since the content of Mg^{25} and Mg^{26} in the natural mixture of magnesium isotopes is about 20%, the contribution of the (γ, n) reaction in these isotopes to the total absorption cross section in the 14.5-16.0 MeV energy region is small and is about 0.15-0.2 mb.

The (γ, p) reaction makes a considerably larger contribution in this energy region. The cross section for the Mg(γ, p) reaction (in a natural mixture of isotopes) was measured recently by Ishkhanov et al. [38] who used a charged particle scintillation spectrometer which detected protons with energies $E_p \geq 3$ MeV. The results of this work make it possible to determine the average cross section of the (γ, p) reaction in the 14.5-16.0 MeV interval, which is approximately 2.5 mb. The fact that protons with energies $E_0 \geq 3$ MeV were measured in [38] should not significantly distort the cross section in this energy region because the threshold for the (γ, p) reaction in Mg^{24} is approximately 11.7 MeV and protons with $E_p \geq 3$ MeV should be completely detected (the height of the Coulomb barrier is about 4.5 MeV for the magnesium nucleus). Photoprotons from the Mg^{25} (γ, p) reaction – threshold 12.06 MeV – evidently should also be completely recorded, but of those from the Mg^{26}(γ, p) reaction – threshold, 14.3 MeV – practically none are detected. Thus the average cross section,

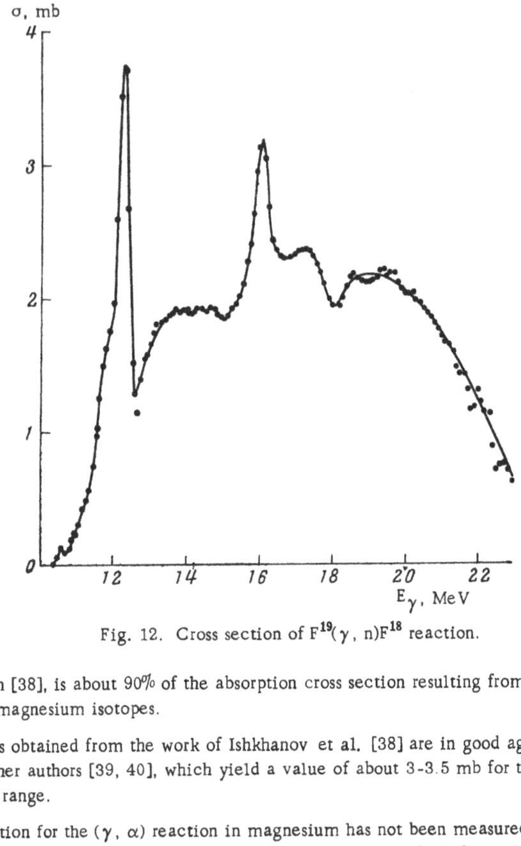

Fig. 12. Cross section of $F^{19}(\gamma, n)F^{18}$ reaction.

based on the data in [38], is about 90% of the absorption cross section resulting from the (γ, p) reaction in the natural mixture of magnesium isotopes.

The estimates obtained from the work of Ishkhanov et al. [38] are in good agreement with the estimates from the data of other authors [39, 40], which yield a value of about 3-3.5 mb for the average cross section in the 14.5-16.0 MeV range.

The cross section for the (γ, α) reaction in magnesium has not been measured, but data for Al^{27} (neighboring nucleus) [41] show that this cross section is no more than 1.1 mb at the maximum of the giant resonance (for α particles with energies $E_\alpha > 5$ MeV). In the energy region where normalization was performed, there is found only the near-threshold part of this cross section, i.e., its contribution to the absorption cross section is rather small, and is hardly more than 0.1-0.2 mb.

The processes of elastic and inelastic γ-ray scattering also make a small contribution to the average cross section at E_γ = 14.5-16.0 MeV. According to the data of Fuller and Hayward [12], the elastic scattering cross section in this energy region is close to zero (10^{-2} mb, or less than 0.1 mb in any case). The inelastic scattering cross section is usually of the same order of magnitude as the elastic, i.e., the total contribution of these processes is negligibly small. Consequently, the average cross section for absorption is about 3 mb in the energy region where normalization was done.

The accuracy of the determination of the "zero" of the nuclear absorption cross section can be estimated from the accuracy of the experimental data used, from the spread in the average cross section values at E_γ = 14.5-16.0 MeV which are obtained from different experimental data, and from the accuracy of the measurements discussed. Such estimates indicate that the error in setting the "zero" of the scale for the nuclear cross section in magnesium is (+2)-(−1) mb.

TABLE 3

Breaks [2]	σ (γ, n) [45]	Proton spectra from (γ, p) reaction [47]	Proton spectra from (γ, p) reaction [46]	Proton spectra from (e, e'p) reaction [16]	Absorption cross section
	10.6	10.4			(10.1)
11.0					
11.5		11.4		11.4	11.4
11.9		11.9	11.8	11.9	
	12.4		12.55		
		12.8		12.7	
		13.6			
	14.0		14.19		14.0
15.3		15.4	15.41	15.7	
	16.1		16.61	16.5	16.8
	17.2				
		18.1		18.7	
	19.3				19.5
					24.0
					28.0

The measured total absorption cross section for magnesium is shown in Fig. 10b. As in the case of fluorine, the absorption cross section has not been computed for magnesium; therefore, the solid curve was obtained by conversion from data for Al [30] with allowance for the difference in Z.

The cross section for non-nuclear processes normalized to the experimental data (dashed curve) differs from the original values by 2%, i.e., it is within the limits of accuracy of the calculations for the cross section of non-nuclear processes just as in the case of fluorine.

(c) Aluminum. As was done with fluorine, normalization was performed in the low-energy region – below the thresholds of the (γ, p) and (γ, n) reactions.

The energy thresholds for reactions occurring at energies $E_\gamma < 15$ MeV are the following:

Reaction	Thres., MeV
(γ, n)	13.07
(γ, p)	8.27
(γ, α)	10.3

It is clear from the table that the absorption cross section will be determined only by scattering (elastic and inelastic) in the energy region up to 10.5 MeV – the effective threshold for the (γ, p) reaction.

According to the data of Fuller and Hayward [12], the elastic scattering cross section in Al^{27} is about 0.25 mb in the 7.0-10 MeV region. It falls sharply to ~0.05 for $E_\gamma \simeq 12$ MeV. If it is assumed that $\sigma_{\gamma\gamma} \simeq \sigma_{\gamma\gamma'}$, the average absorption cross section in the energy range $E_\gamma < 10$ MeV is approximately 0.5 mb.

Normalization was carried out in the 7.5-9 MeV range. Error in the determination of the "zero" results only from the error of the average cross section in this region, which is ± 3.0 mb.

The measured total absorption cross section for Al^{27} is shown in Fig. 10c. The calculated values for the non-nuclear absorption cross section are in good agreement with the experimental values in the 7.5-9 MeV region; the normalization factor is one.

(d) Calcium. Normalization at low excitation energies was made difficult in the case of calcium because of the presence of a relatively strong peak in the nuclear absorption cross section at an energy of the order of 12.5 MeV, below the threshold for the (γ, n) reaction. Therefore, normalization was performed in the 14-16 MeV range close to the threshold for photoneutron reaction where the magnitude of $\sigma_{\gamma,p}$ is sufficiently well known and where the absorption cross section changes slowly with energy.

The thresholds of reactions making contributions to the absorption cross section for $E_\gamma < 16$ MeV are given below.

Reaction	Thres., MeV
(γ, p)	8.3
(γ, n)	15.8
(γ, α)	7.1

As is evident, the average value for the absorption in the 14-16 MeV range will be determined by the (γ, p) and (γ, α) reactions and by scattering (elastic and inelastic), the contribution to the average cross section from the (γ, α) reaction being 1 mb from an analysis of the α-particle yields made in [32]. The contribution from elastic and inelastic scattering processes (on the assumption that $\sigma_{\gamma\gamma} \simeq \sigma_{\gamma\gamma'}$) is no more than 0.02 mb. The total cross section for these processes is negligibly small in comparison with the cross section for the (γ, p) reaction at all excitation energies because the cross section for the Ca^{40} (γ, p) reaction is larger than that for any other reaction produced in Ca^{40} by γ-rays with $E_\gamma \le 30$-50 MeV. The average cross section in the 14-16 MeV range for protons with $E_p \ge 3$ MeV, which was obtained in [42], was 10-11 mb, and 3.5-4 mb for protons with $E_p \ge 5$ MeV. The latter result is in good agreement with the data of Ratner [43] (~4 mb), which considerably increases the reliability of the results for $E_p \ge 3$ MeV. Therefore, the average cross section in the energy range where normalization was done is 10-11 mb. The error in setting the "zero" of the Ca^{40} nuclear absorption cross section is determined by the accuracy of the $\sigma_{\gamma,p}$ measurements [42]. This accuracy can be assumed to be about 20-25% because considerable additional error is introduced in the calculation of the cross section from the yield curve. One can consider that the accuracy in fixing the "zero" of the nuclear cross section is no worse than ± 2.5 mb. The correctness of the normalization is confirmed by a comparison of the resulting nuclear absorption cross section (25 ± 3.0 mb) and the sum of the $\sigma_{\gamma,p}$ [42] and $\sigma_{\gamma,n}$ [44] cross sections, which are 22.0 ± 3.0 and 2.0 ± 0.5, respectively, in the 16-17 MeV range.

The total γ-ray absorption cross section for calcium is shown in Fig. 10d. The normalization factor for calcium is 1.025.

The nuclear absorption cross sections for all four elements, which were obtained by subtraction of the normalized values of the non-nuclear portion of the absorption cross section from the experimental data, are shown in Figs. 11, 13, 16, and 18. An attempt was made with an electronic computer to fit the resulting experimental data by the superposition of resonance curves, the number of which could be varied from 1 to 10. The resonance curves were assumed to be in the following forms:

$$\sigma = \frac{\sigma_0 (\Gamma/2)^2}{(E - E_p)^2 + (\Gamma/2)^2} \quad \text{(Breit–Wigner)},$$

$$\sigma = \frac{\sigma_0 (\Gamma/2)^4}{(E^2 - E_p^2)^2 + (\Gamma/2)^4} \quad \text{(Lorentz)},$$

$$\sigma = \sigma_0 \exp\left[-\left(\frac{E - E_p}{\Delta}\right)^2\right]; \qquad \Gamma = 2(\ln \Delta)^{1/2} \quad \text{(Gaussian)}.$$

The minimum functional method was used for the process of description. We chose σ_0, E_p, and Γ as the free parameters. Such attempts met with success only in the case of magnesium (see Fig. 13) where the experimental results were fitted accurately with Breit-Wigner curves. In all other cases, the reliability of the resulting curves was insignificant (the sum of functionals in the best case was of the order of 250 for 70 degrees of freedom).

The curves in Figs. 11, 13, and 18 for fluorine, aluminum, and calcium were drawn "by eye."

TABLE 4

E_p, MeV	σ_0, mL	Γ, MeV	$\int \sigma dE$, MeV-mb	E_p, MeV	σ_0, mL	Γ, MeV	$\int \sigma dE$, MeV-mb
11.2	21	1.5	48	20.4	29	1.3	57
13.8	13	0.3	7	22.9	26	2.6	106
17.1	20	0.4	14	24.9	20	0.6	18
18.9	37	0.8	44	27.0	19	4.9	146

CHAPTER V

Discussion of Results

1. Fluorine F^{19}

The nuclear absorption cross section for F^{19} is shown in Fig. 11, where the errors shown are standard deviations. The principal features of the measured nuclear absorption cross section are the very broad, asymmetric maximum (halfwidth $\Gamma \simeq 10$ MeV) at $E_\gamma = 19.5$ MeV (the cross section rises by a factor of two for a change in E_γ from 10 to 19.5 MeV, falling sharply to a minimum at 22 MeV), and the existence of a second maximum at $E_\gamma = 24$ MeV ($\Gamma \simeq 2$ MeV).

The cross section of the $F^{19}(\gamma, n)F^{18}$ reaction, measured by King et al. [45] (Fig. 12), has a similar shape in the energy range $E_\gamma < 22$ MeV. The cross sections for the partial photonuclear reactions in fluorine at excitation energies $E_\gamma > 22$ MeV have not been measured, and it is, therefore, impossible to carry out any sort of comparison in that energy range. Qualitatively, the shape of the absorption cross section can be explained by differences in transition energies of p- and d-nucleons and by considerable deformation of the F^{19} nucleus (the question of the shape of the resonance will be discussed more thoroughly later).

Four maxima at excitation energies of 16.8, 19.5, 24.0, and 28.0 MeV can be distinguished with reasonable confidence in the F^{19} nuclear absorption cross section within the energy range investigated. The three maxima at E_γ = (10.1); 11.4, and 14.0 MeV can only be considered as indications of the possible existence of transitions (or groups of transitions). The degree of confidence in the existence of maxima at $E_\gamma = 11.4$ and 14.0 MeV increases considerably if one considers that peaks in photoproton spectra (from both (γ, p) [46, 47] and $(e, e'p)$ [16] reactions) and in the (γ, n) reaction cross section [45] were observed at the same excitation energies. The results of these measurements, and also the location of "breaks" in photoneutron yield curves observed by Goldemberg and Katz [2], are given in Table 3.

The total photoneutron yield was measured in [2] and [45]. Photoproton spectra were determined by means of emulsion [47], semiconductor counters [46], and a magnetic proton spectrometer [16]. As is clear from a comparison of the results shown in Table 3, the use of various methods for investigating the structure of the giant resonance led to consistent results, verifying the objectivity of the resulting information. The energy location of the maxima observed in the nuclear absorption cross section is also in satisfactory agreement with the results obtained in studies of partial reactions in fluorine (within the limits of the errors associated with the resolutions of the various methods).

The peak which is clearly visible in photoproton spectra for $E_\gamma = 18.1$ MeV [47] and in proton spectra from the $F^{19}(e, e'p)O^{18}$ reaction [16] at $E_\gamma = 18.7$ MeV is not observed in the nuclear absorption cross section at excitation energies 18.1-18.7 MeV. This maximum is also absent in the cross section for the $F^{19}(\gamma, n)F^{18}$ reaction [45]. There are two reasons why this might occur: (a) the maximum at $E_\gamma = 18.5$ MeV corresponds to a "pure" proton transition; its integrated cross section is small, which makes it difficult to distinguish this peak in the background of the broad maximum at $E_\gamma = 19.5$ MeV; (b) the maximum at $E_\gamma = 18.5$ MeV is produced by transitions of the O^{18} nucleus in an excited state.

However, in case (b), it should not have been observed in [47] because spectral measurements were made with $E_\gamma^{max} = 19$ MeV, and the energy of the first excited state of O^{18} $(2)^+$ is 1.98 MeV. (The threshold of the (γ, p) reaction is 7.96 MeV, and the energy of the corresponding peak in the spectrum is $E_p = 10$ MeV.)

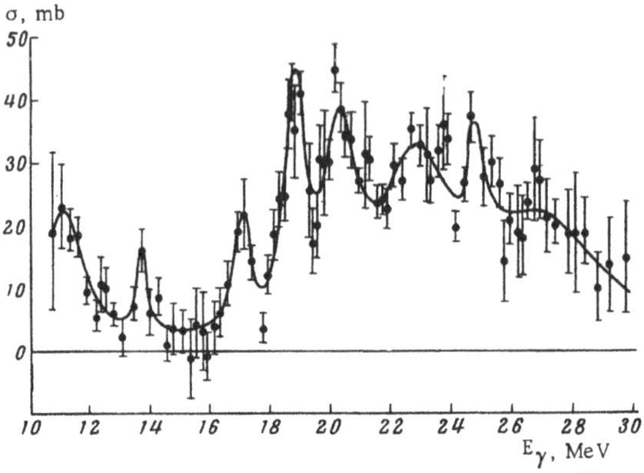

Fig. 13. γ-Ray nuclear absorption cross section of Mg^{24}.

Also absent in the nuclear absorption cross section is the maximum at E_γ = 12.4 MeV which is clearly observed in the cross section for the $F^{19}(\gamma, n)F^{18}$ reaction (Fig. 12) and in photoproton spectra (at E_γ = 12.55 MeV [46], 12.8 MeV [47], and 12.7 MeV [16]). This is apparently because the integrated cross section of the peak at E_γ = 12.4 MeV is relatively small (~ 1.5 MeV-mb), and it is, therefore, impossible to pick it out at the level of error achieved in the present measurements. (To separate this maximum, it is necessary to increase the accuracy of measurement by a factor of three at the very least.)

Analyzing the shape of the cross section for the $F^{19}(\gamma, n)F^{18}$ reaction in which Taylor et al. [48] observed a maximum at $E_\gamma \simeq$ 12.5 MeV (similar measurements were made in [45] but were less accurate), Fujii interpreted this peak as the result of E1 photon absorption by a neutron in the $d_{5/2}$ state and not coupled to the nuclear core. His calculations showed that a peak at E_γ = 14 MeV with an integrated cross section of about 10 MeV-mb ($\sigma_0\Gamma$ = 15 MeV-mb) and a halfwidth $\Gamma \sim 0.5$ MeV should be observed in the (γ, n) reaction cross section for direct excitation of this neutron into the $f_{7/2}$ state. (The weak peak at $E\gamma \simeq$ 18 MeV corresponds to a transition to the $f_{5/2}$ state.) Similar parameters in the experimental nuclear absorption cross section give a peak at E_γ = 14.0 MeV (integrated cross section 10-15 MeV-mb and $\Gamma \simeq 1$ MeV).

Fujii neglected the coupling between the external neutron and the nuclear core in his calculations. In fact, the neutron can be in the excited state sufficiently long to manage to interact with core nucleons. Consideration of this coupling leads to a reduction in neutron emission probability, i.e., to a reduction in the cross section for the (γ, n) reaction and to the appearance of a group with the corresponding energy in the proton spectrum. However, the integrated absorption cross section should remain unchanged in this case. Indeed, a weak proton group at E_p = 5.83 MeV is observed in the photoproton spectrum, which corresponds to an excitation of 14.19 MeV in the case of transitions to the ground state. On the basis of all the facts, the maximum in the nuclear absorption cross section at E_γ = 14 MeV can be interpreted as the result of a single-particle, electric dipole absorption of a γ-ray by a neutron in the $d_{5/2}$ state. The reliability of this interpretation is determined by the validity of the extremely qualitative ideas developed by Fujii.

No measurements were made of the cross sections for partial reactions in fluorine in the energy range E_γ > 22 MeV; further, any sort of information about the level structure in F^{19} in this energy region is lacking. The present measurements eliminate this gap and allow one to assert that there are at least two levels in the fluorine nucleus (or two groups of closely-spaced levels) located 24 and 28 MeV above the ground state.

As shown by measurements of the reduced probability for Coulomb excitation of the $5/2^+$ state (198 keV), which were made by Sherr et al. [50], the F^{19} nucleus is strongly deformed, and the electric quadrupole moment of this nucleus is $Q_0 = 0.24 \times 10^{-24}$ cm^2 (the accuracy of these measurements are estimated to be ± 25%).

TABLE 5

Absorption cross section	Photoneutron cross section $\sigma_n = \sigma_{\gamma,n} + 2\sigma_{\gamma,2n}$ [56]	(p, γ_0) cross section [57]	(γ, p) cross section [39]
11.2 (13.8)			14.3
17.1	17	16.0 16.6 17.6	16 17.2
18.9	18.9	18.5 19.2	18.5 19.1
		19.7	19.7
20.4	20.5	20.1 20.7	20.3
22.9 24.9 27.0	23 25	21.2 22.0 22.4	21.0

According to Okamoto [51] and Danos [52], who considered the shape of the giant resonance in deformed nuclei, the electric quadrupole moment of such nuclei is

$$Q_0 = \frac{2}{5} Z (a^2 - b^2) = Z R_0^2 A^{2/3} (x - 1)/x^{1/3},$$

where x is the ratio of the large and small axes of the nucleus, as determined by the ratio $E_b/E_a = 0.911x + 0.089$ (E_a and E_b are the resonance energies of the maxima in the absorption cross section).

Based on the results of Sherr et al., and on the theory developed by Okamoto and Danos, Spicer [53] pointed out the possibility of a splitting of the fluorine giant resonance with a ratio of resonance energies of the maxima $E_b/E_a = 1.31$ ($E_a \simeq 15$ MeV and $E_b \approx 20$ MeV). King et al. [45], who measured the cross section for the $F^{19}(\gamma, n)F^{18}$ reaction from threshold to approximately 22 MeV, assumed maxima at $E_\gamma \simeq 14.0$ and 19.3 MeV as the maxima corresponding to such a splitting. The value $Q_0 = 0.3 \times 10^{-24}$ cm^2 obtained as the result of these assumptions is in good agreement with the data from Coulomb excitation.

It would be logical to expect that a similar splitting would be no less clearly observed in the nuclear absorption cross section as well. However, as Fig. 11 makes clear, if a split is also present in the giant resonance, then the maxima at $E_\gamma = 19.5$ and 24 MeV may be associated with it. In that case, as shown by analysis of the integrated cross sections for those peaks, the electric quadrupole moment should be negative and of absolute magnitude $|Q_0| = 0.5 \times 10^{-24}$ cm^2, contradicting the data obtained from Coulomb excitation.

A comparison of the value of the electric quadrupole moment obtained by King et al. [45] with the value determined by the shape of the nuclear absorption cross section shows that the selection of the maxima corresponding to a splitting of the giant resonance in light nuclei is extremely subjective, at least in the case of fluorine, because there are no sufficiently rigid criteria for determining this selection. The criteria usually used in the analysis of photonuclear cross sections in deformed medium and heavy nuclei (the integrated cross section of one of the maxima is twice as large as that of the other) is not applicable in the case of light nu-

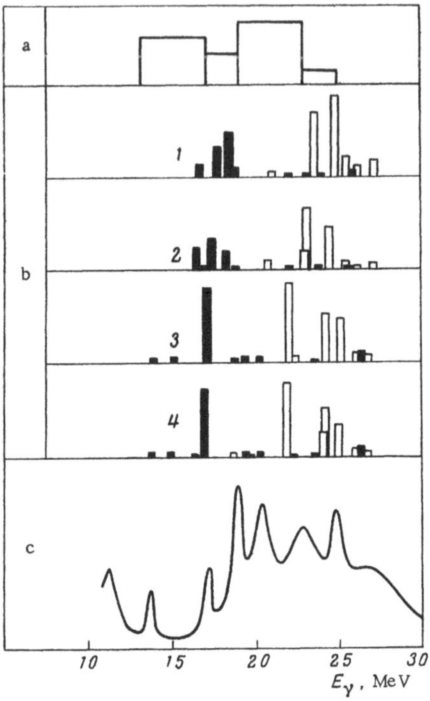

Fig 14. Comparison of measured Mg^{24} absorption cross section with various calculations (a) absorption cross section computed by Neudachin and Orlin [60]; (b) energy and strength of the transitions obtained in [61] for various types of forces [Ferrell and Visscher (1, 2) and Wigner (3, 4)] with (2, 4) and without (1, 3) consideration of ground-state correlations; solid rectangles are K = 0 transitions, open rectangles are K = 1 transitions; (c) nuclear absorption cross section.

clei because the transitions corresponding to K = 0 and K = 1 may overlap considerably. In addition, they are localized in energy than in the case of heavy nuclei. The arbitrariness in the choice of resonance energies leads to inconsistency in the value of Q_0, and thus the satisfactory agreement obtained by King et al., is obviously accidental. In addition, the use of the Okamoto-Danos concept for such a relatively few-nucleon system as F^{19} is hardly justified. The accuracy of the approximations used in the Danos-Okamoto theory falls roughly like $A^{-1/3}$ [54] and is no better than 40% in the case of fluorine; the considerable overlap of transitions corresponding to different values of K, which are more localized in energy than in the case of heavy nuclei, can also lead to error in the determination of the sign of Q_0.

The integrated absorption cross section of fluorine is 335^{+80}_{-40} MeV-mb, and is 85^{+80}_{-40} % of the value predicted by sum rules for dipole transitions (398 MeV-mb) with an exchange force fraction x = 0.5.

It should be pointed out that a sharp increase is observed in the contribution to the integrated cross section from the energy region $E_\gamma < 30$ MeV during the transition from O^{16} to F^{19}. In the case of oxygen, this con-

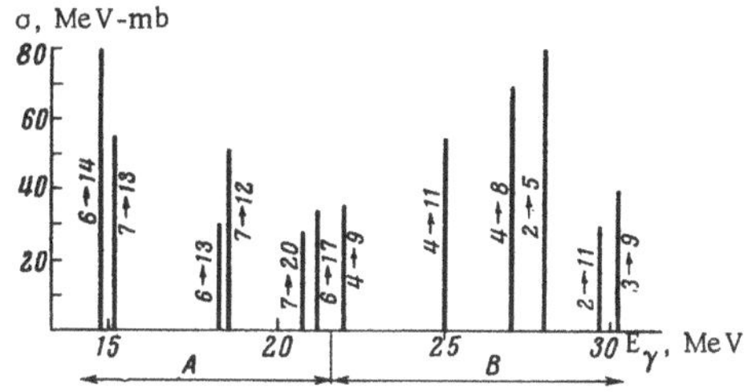

Fig. 15. Calculated energy and strengths of transitions in Mg^{24} [62].

tribution is about 50% [18, 33], and it is more than 80% in the case of fluorine, i.e., the center of gravity of transition strengths is shifted significantly toward lower energies when the d-shell starts filling.

The sum of the integrated cross sections for the fluorine (γ, n) and (γ, p) reactions in the energy range from threshold to about 25 MeV is no more than 115 MeV-mb (37 ± 11 MeV-mb for the (γ, p) reaction [16] and 77 MeV-mb for the (γ, n) reaction [55]), which is considerably less than the integrated absorption cross section for that same energy region amounting to 295 MeV-mb. Such a large difference may be associated with either a large scattering contribution or with inaccuracy in the measurement of the partial reaction cross sections, which is more likely.

2. Magnesium Mg^{24}

The nuclear absorption cross section of Mg^{24} (Fig. 13, errors shown are standard deviations) has a clearly expressed resonance structure.

A fit of the measured cross section by the superposition of eight maxima made it possible to obtain an estimate of the basic parameters of these peaks – the resonance E_r, the halfwidth Γ, the cross section at the maximum σ_0, and the integrated cross section; numerical values for these quantities are given in Table 4. The values of σ_0 and Γ were not corrected for the finite resolution of the spectrometer because this correction is no greater than the error in the determination of σ_0 and Γ even for the narrowest resonances.

The parameters given in Table 4 were calculated on an electronic computer by the minimum functional method, all parameters being considered free. The curve drawn through the experimental points in Fig. 13 was obtained on the basis of these parameters (the shape of each peak was described by the Breit-Wigner formula). Although the Breit-Wigner curves best described the results (when compared with Lorentz and Gaussian curves), one must consider this only as a successful attempt at a qualitative fit to the shape of the absorption cross section obtained for magnesium.

A similar structure in the giant resonance (with allowance for differences in resolution) was also observed in partial reaction studies, the results of which are shown in Table 5. Column 2 shows the maxima in the cross section for the emission of photoneutrons $\sigma_n = \sigma_{\gamma,n} + \sigma_{\gamma,np} + 2\sigma_{\gamma,2n}$, which was measured by Miller et al. [56] in the energy range 16.4-26.2 MeV on a linear accelerator producing monochromatic γ-ray beams (monochromatization by the method of positron annihilation in flight). The resolution was about 2% for these measurements (spectral distribution of photons). In the entire energy range, the location of the observed maxima are in good agreement with the location of the peaks in the total absorption cross section (column 1).

Column 3 gives the location of the maxima observed by Gove in the cross section for the $Na^{23}(p, \gamma_0)Mg^{24}$ reaction, which is the inverse of the $Mg^{24}(\gamma, p_0)Na^{23}$ reaction [57]. Measurements of the cross section for the $Na^{23}(p, \gamma_0)Mg^{24}$ reaction were made on a tandem in the proton energy range 3.8-11.6 MeV (the corresponding Mg^{24} excitation energies are 15.5-22.5 MeV), and the γ-rays were measured at 90° with a scintillation spectrometer (NaI(Tl) crystal 12.7 cm in diameter and 15.2 cm high). The energy of the protons was changed in 200

TABLE 6*

Theoretical data		Experimental data	
E_γ	Transition	Group	E_γ
		11.2 (48)	
		13.8 (7)	
14.8 (80)			
15.1 (55)			
18.2 (30) 18.6 (50)	18.5 (80)	17.1 (14) 18.9 (44)	18.4 (58)
20.7 (30) 21.1 (35) 22.0 (35)	21.5 (100)	20.4 (57) 22.9 (106)	22.0 (163)
25.0 (55)	25.0 (55)	24.9 (18)	24.9 (18)
27.0 (70) 28.0 (80)	27.5 (150)	27.0 (146)	27.0 (146)

*Integrated cross sections (MeV-mb) are shown in parentheses.

keV steps. The target thickness, which determines the resolution, was about 100 keV for 10 MeV protons, corresponding to a resolution of the order of 0.6% for 16 MeV excitation energy and 0.25% for 20-22 MeV. Considerably better resolution than that for the measurements of the nuclear absorption cross section and of photoneutron emission cross section σ_n, enabled Gove to detect a much greater number of peaks. This obviously indicates that the peaks observed in the nuclear absorption cross section really represent groups of narrow, overlapping peaks.

An analysis of the shape of the cross section for the $Na^{23}(p, \gamma_0)$ Mg^{24} reaction shows that the individual peaks in this cross section can be combined in groups which, with poorer resolution, are observed as single maxima (the location of the centers of gravity of these groups coincide approximately with the location of the maxima in the absorption cross section).

A similar structure was observed by Yamamura in the cross section for the $Mg^{24}(\gamma, p)Na^{23}$ reaction [39]. He measured the energy spectrum of photoprotons from magnesium for $E_\gamma^{max} = 21.5$ MeV by the emulsion method. The cross section for the $Mg^{24}(\gamma, p)Na^{23}$ reaction was computed from the photoproton spectra on the assumption that the final nucleus Na^{23} always remained in the ground state. In the cross section obtained in this manner, one can distinguish eight peaks whose location is given in column 4.

The structure of the cross section for the Mg^{24} $(\gamma, p)Na^{23}$ reaction, which was observed by Yamamura, is in good agreement with the data of Gove for the cross section for the inverse reaction $Na^{23}(p, \gamma_0)Mg^{24}$. The absence of several peaks in the photoproton spectrum is apparently associated with the poorer resolution of the emulsion method. The absence of additional peaks which might be caused by transitions into excited states of the Na^{23} nucleus may serve as a justification for the validity of the assumption that the probability of transitions to the ground state is relatively large for excitation energies not exceeding 20 MeV. As shown in recent measurements by Forkman and Stifler [58], the Na^{23} nucleus remains in an excited state with high probability at excitation energies above 20.5 MeV (in any case, in the 20.5-26.5 MeV range). This is evidence that the results of Yamamura may be distorted at excitation energies above 20 MeV.

The structure found in the nuclear absorption cross section is also qualitatively verified by the results of recently completed preliminary time of flight measurements of photoneutron spectra [59] at excitation energies of 18-24.3 MeV. However, a detailed analysis of these results is difficult because there is no way of knowing which peaks in the neutron spectrum correspond to transitions to the ground and excited states.

Although the experimental errors make it possible to pick out confidently the maxima in the nuclear absorption cross section given in Table 5 (except for the maximum at $E_\gamma = 13.8$ MeV), the existence of peaks in the cross sections for the partial reactions at these same excitation energies serve as additional verification

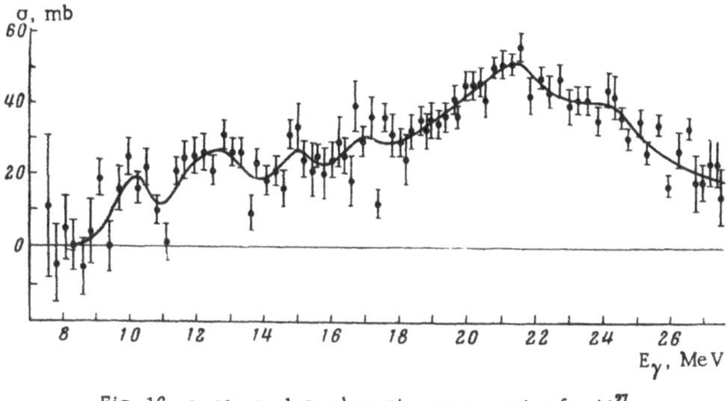

Fig. 16. γ-ray nuclear absorption cross section for Al^{27}

of the authenticity. The few experimental points in the region of the peak at E_γ = 13.8 MeV enable one to consider this maximum as only an indication of the existence of a transition (or group of transitions) there. The resonance at E_γ = 13.8 MeV is "held up" by a single point, in fact, the absence of which would eliminate the question of the existence of a maximum at that energy.

Theoretical consideration of E1 transitions during the absorption of γ-rays by magnesium nuclei was undertaken by Neudachin and Orlin [60], and also by Nilsson et al. [61]. Shell-model calculations were performed in both cases. Because the Mg^{24} nucleus is strongly deformed, level energies were computed with the Nilsson potentials. In [62], the residual interaction was treated phenomenologically, and the effects of configuration mixing were not considered. The calculations were made in the diagonal approximation. In [61], the residual interaction was considered systematically. In addition, the study was made with allowance for correlations in the ground state and for two type of nuclear forces – Ferrel-Visscher and Wigner. The results of these calculations are shown in Fig. 14.

As is clear from the figure, the transitions are clearly separated into two groups in all four types of calculations by Nilsson et al. [61]: one group of transitions (K = 0) are concentrated in the 16-18.5 MeV energy range and the other group (K = 1) in the 21.5-26.0 MeV range. The sum of the oscillator strengths of the groups with K = 0 is approximately half as much as the sum of the oscillator strengths of the K = 1 group. It is evident from Fig. 14 that there is a very serious discrepancy between the results obtained in [61] and the experimental data. It is impossible to explain this discrepancy by the contribution to the experimental cross section of transitions other than dipole because it would be necessary to assume in that case that the integrated nuclear absorption cross section in the 18.5-21.5 MeV range (amounting to almost 30% of the measured integrated cross section) resulted from transitions with multipolarity E2 and higher.

A similar picture is also observed in the comparison of experimental results with the calculations of Neudachin and Orlin [60]. The energy dependence of the nuclear absorption cross section obtained in this work gives a relatively large value for the integrated cross section in the 13-17 MeV range and a sharp reduction in the cross section at $E_\gamma > 23$ MeV, which completely disagrees with the shape of the experimental cruve.

In addition, all the transitions in [60] and [61] are at excitation energies below 27 MeV. At that point, every sum of oscillator strengths for electric dipole transitions is exhausted. At the same time, a comparison of the integrated cross section obtained from the sum rule with the experimental one (see below) indicates that the strength of transitions with energies above 30 MeV may amount to nearly 30% of the total sum of dipole transition strengths.

Considerably better agreement between experimental and theoretical data is observed in the consideration of differences in the binding energy of nucleons in filled and occupied shells of this nucleus. The important role of this effect was pointed out by Neudachin and Shevchenko [62] who found by analyzing spectroscopic data that this difference was about 1 MeV for a 1d-2s nucleus. Neudachin and Shevchenko calculated the absorption cross section for Mg^{24} in the diagonal approximation (similar to the calculation in [60]) with allowance for this effect. The results are shown in Fig. 15.

A comparison with experimental data (see Table 6) shows that consideration of the binding energy difference in filled and occupied shells considerably improves the agreement between experimental and theoretical data.

The main conclusion reached in [62] was that the difference in the binding energies of nucleons in filled and occupied shells leads to a separation of the absorption cross section into two groups, which correspond to transitions from unfilled (group A) and filled (group B) shells. Deformation of the nucleus only leads to some spreading of the groups, but there will be no mixing of the transitions in these groups as the result of residual interaction because their centers of gravity are considerably different in energy. However, considerable redistribution of intensity among the separate transitions may occur within each group because of the residual interaction.

Although the calculations of Neudachin and Shevchenko are not intended to be a quantitative estimate of the energy location and integrated cross section of the transitions, it is of interest to compare their results with the experimental data for transition energies and intensities. Such a comparison is made in Table 6, which gives the experimental and theoretical transition energies (MeV). It is clear from the table that the theoretical and experimental data for the location of the transitions agrees with satisfactory accuracy. If the transitions are collected into groups (as indicated by the braces), the agreement of the results becomes extremely good. In the latter case, one should also note the excellent agreement between the calculated and experimental values of the integrated cross sections.

The observed agreement of theoretical and experimental results testifies to the importance of considering the effect of nucleon binding energy differences in filled and occupied shells and to the reasonableness of the assumptions made, and of the computational methods used, in [62]. To prove this, it would be of interest to compare similar data for other nuclei also, silicon or sulfur, for example, but unfortunately detailed calculations have only been made for the case of Mg^{24}.

The integrated nuclear absorption cross section for Mg^{24} in the 11-30 MeV energy range where it was measured is 365^{+40}_{-20} MeV-mb, i.e., it is $72^{+8}_{-4}\%$ of the cross section which is obtained from the sum rule for electric dipole transitions (510 MeV-mb for an exchange force fraction $x = 0.5$). As already noted, this fact is evidence that about 30% of the sum of the transition strengths occurs in transitions with energies $E_\gamma \geq 30$ MeV. This is not contradicted by the results of [62] in which it was predicted that the contribution of transitions with $E_\gamma \sim 30$ MeV was about 15%.

A comparison of the measured integrated absorption cross section and that obtained by summation of the integrated cross sections for photoneutron ($\sigma_{\gamma,n} + \sigma_{\gamma,np} + 2\sigma_{\gamma,2n}$) [56] and photoproton ($\sigma_{\gamma,p} + \sigma_{\gamma,np}$) [58] yield within the same limits of integration (15 to 26 MeV), which are respectively 235 ± 20 MeV-mb and 190 ± 40 MeV-mb, show that they are in agreement within experimental error. Overestimate of the integrated cross section obtained by summation of the cross sections for the partial reactions because of twofold consideration of the (γ, np) and (γ, 2n) reactions will be insignificant because of the high thresholds of these reactions (24.1 and 26.0 MeV, respectively) and the low value of the cross sections. In addition, this contribution can be considered as some compensation for the neglected contributions from scattering and other processes.

Splitting of the giant resonance in magnesium because of the large internal quadrupole moment of the Mg^{24} nucleus was considered by King and McDonald [63]. The absorption curve that was obtained does not allow one to pick out objectively the predicted splitting for reasons similar to those discussed in Section 1 of this chapter; all the conclusions about the possibility of determining Q_0 from photoabsorption cross sections for light nuclei which were arrived at in the discussion of fluorine remain valid for the case of magnesium as well.

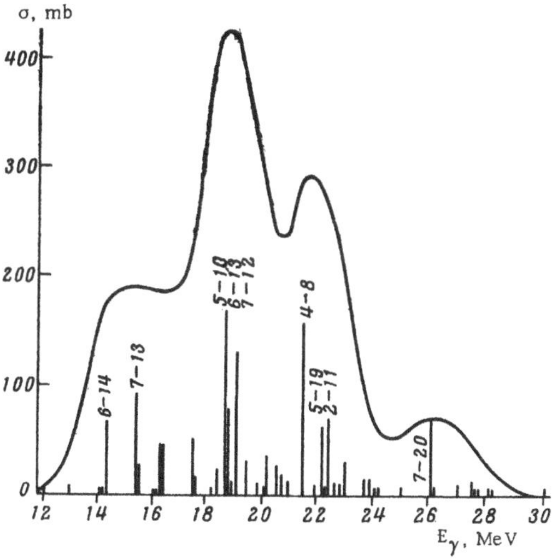

Fig. 17. Al^{27} γ-ray nuclear absorption cross section calculated in [72].

3. Aluminum Al^{27}

The nuclear absorption cross section of Al^{27} (Fig. 16, errors are standard deviations) does not have a clearly expressed structure resembling the structure in magnesium. The energy dependence of the Al^{27} nuclear absorption cross section exhibits a broad resonance ($\Gamma \simeq 10$ MeV) with a maximum at an excitation energy $E_\gamma = 21.5$ MeV. A number of irregularities in the curve through the experimental points at $E_\gamma = 10.5$ and 12.5 MeV, and possibly at 15.0, 17.0, and 26.5 MeV, can only be considered as indications of the existence of structure in the absorption cross section at the resolution used.

Such a sharp difference between the shape of the nuclear absorption cross section in aluminum and the shape of the absorption cross section in magnesium is evidence that a very significant rearrangement of the nuclear level scheme occurs in the transition from the Mg^{24} nucleus to the Al^{27} nucleus. If the dipole transitions are localized in relatively close-packed groups in the case of the Mg^{24} nucleus, they are distributed more uniformly within the 10-30 MeV excitation energy region in the case of Al^{27}, and their number is apparently considerably larger.

The γ-ray absorption cross section in Al^{27} was studied by similar methods in a number of experiments [13-16, 64-66]. However, a clearly expressed structure was observed in none of them. The Al^{27} nuclear absorption cross sections measured in [13, 14, 66]* agree only qualitatively with one another with regard to shape and absolute magnitude. Evidently, this is primarily associated with the choice of the values for the non-nuclear portion of the absorption cross section, the variation of which, even within narrow limits (~ 1%), can significantly distort both the shape and the magnitude of the nuclear absorption cross section, and also with the inadequate statistical accuracy of the measurements (the error in σ_n at the cross section maximum was about 20%) as was noted by the authors themselves. The accuracy of the present measurements is 2-3 times better than the accuracy in [13, 14, 66], and the error in the determination of the "zero" of the nuclear absorption cross section was considerably reduced by normalizing the non-nuclear portion of the absorption cross section. With respect to shape and absolute magnitude, the present measurements agree best with the results

*The results of [64] and [65] were reviewed by the authors of [14] and [13], and therefore will not be discussed here.

TABLE 7

E_p, MeV	σ_{max}, mb	Γ, MeV	∫σdE, MeV-mb
12.5	60	1.5	130
19.1	80	2.5	340
21.1	100	2.0	320
24.7	50	1.0	80
26.1	70	0.5	60

of Ziegler [14] who used as a detector a magnetic, pair, γ-spectrometer with a resolution of about 1% at E_γ = 20 MeV (roughly equal to the resolution of the spectrometer we used).

Results obtained in investigations of partial reactions also gave evidence of complexity in the structure of the aluminum absorption cross section. This is clearly seen, for example, in a comparison of the cross sections for the $Al^{27}(\gamma, n)Al^{26}$ reaction, which were measured in [5] and [67]. Both these results were obtained by the same method and with approximately the same statistical accuracy. The only difference was that the measurements in [67] were made in "steps" half as large (0.25 MeV) as those in [5]. Such an increase in accuracy considerably "enriched" the structure observed in the cross section for the $Al^{27}(\gamma, n)Al^{26}$ reaction; the number of maxima in the excitation energy range from threshold (~ 13 MeV) to ~ 22 MeV was almost doubled. The maxima observed in [67] were not the result of errors in the analysis of the yield curve because there are distinct breaks in the latter which are associated with peaks in the cross section.

A similar situation in also seen in comparisons of photoproton spectra measured with different energy resolutions. The locations of the maxima in spectra of protons from $Al^{27}(\gamma, p)$ [68] and $Al^{27}(e, e'p)$ [16] reactions are given below.

(γ, p) reaction . . .	3.5	4.3	5.1	6.0	6.8	8.0	9.0	10.3	11.8
(e, e'p) reaction . .		4.4		6.1			8.7	11.0	12.0

The measurements of proton spectra from the (e, e'p) reaction were made with somewhat poorer energy resolution (by ~ 30%) because the target used in these measurements was approximately 30% thicker than that used in [68] (20.6 and 15.3 mg/cm^2, respectively). This led to considerable "washing out" of structure. Unfortunately, the lack of information about transition probabilities to excited states of the residual nucleus does not permit calculation of the cross section from measured spectra.*

It is obvious that it is necessary to make measurements with considerably better energy resolution in order to reveal structure in the Al^{27} nuclear absorption cross section in a reliable manner. Some light would be thrown on this problem by measurements of the cross section for the inverse $Mg^{26}(p, \gamma_0)$ reaction, but they have only been made at low excitation energies [70] ($E_p \leq 3.0$ MeV).

In the paper of Ferrero et al. [71], the process of γ-ray absorption was treated in the framework of the Wilkinson [73] independent-particle model with a spherical potential. The Wilkinson model gives an underestimate of the energy value of the giant resonance. In order to get around this difficulty, Ferrero assigned the energy corresponding to the maximum of the cross section for the $Al^{27}(\gamma, n)Al^{26}$ reaction to the strongest single-particle transition $1d_{5/2} \rightarrow 1f_{7/2}$, which makes a 60% contribution to the integrated cross section. Such treatment is extremely qualitative in the case of the Al^{27} nucleus because this nucleus is strongly deformed (quadrupole moment $Q_0 \geq 0.35 \times 10^{-24}$ cm^2), and the use of a spherical potential is unjustified.

*Existing data on the structure of photoneutron spectra [69] are statistically unreliable (435 tracks in the entire spectrum), and duplicate measurements have not been made of the photoneutron spectrum from Al^{27}.

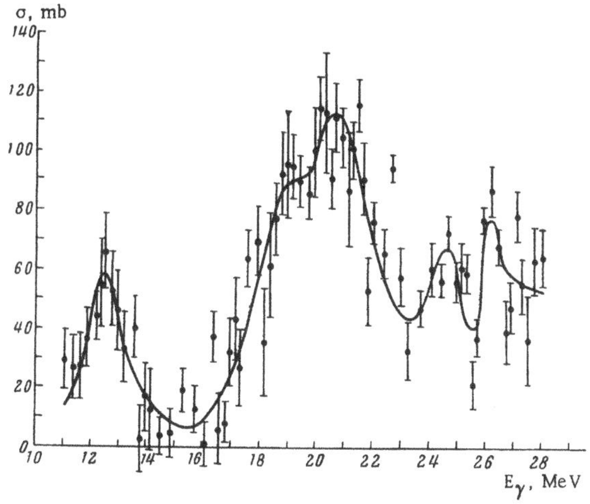

Fig. 18. γ-Ray nuclear absorption cross section for Ca^{40}.

Deformation of the Al^{27} nucleus was taken into account by Baglin et al. [72], who used the Nilsson level scheme for deformed nuclei [74]. It was further assumed that the Nilsson calculations were valid for any amount of deformation and that each absorption line had a triangular shape and 2 MeV width. In order to match the calculated location of the giant resonance to the experimental value, a "reduced nucleon mass," equal to half the mass of the free nucleon, was introduced (this was done in [75] for the Nilsson model). The results of these calculations are shown in Fig. 17.

It is easy to see that they do not agree with the experimental data (Fig. 16) either with respect to shape or with respect to absolute magnitude. More than that, the location of the maximum predicted by these calculations (~ 18.5 MeV) lies 3 MeV below the experimental maximum (21.5 MeV). The same sort of discrepancy is also observed in a comparison of these calculations with the experimental data for partial reactions ([71, 76], for example) which also give the location of the giant resonance at E_γ = 21-22 MeV. This fact is evidence of the unsoundness of using the Nilsson model in the form in which it was employed by Balgin et al., in the considering E1 absorption in aluminum. The discrepancy is evidently associated with the neglected particle-hole interaction, the importance of which was demonstrated by Elliott and Flowers [77], Brown and Bolsterli [78], and also Neudachin et al. [79].

Consideration of this interaction (the residual interaction) significantly changes the energies of the dipole transitions, leading to grouping in a relatively small range, and gives the correct value for the energy of the giant resonance. In this way, the need for the introduction of a "reduced nucleon mass" vanishes, and the agreement between theoretical and experimental data is considerably improved. In addition, consideration of the residual interaction leads to redistribution of transition strengths, which must change the shape of the absorption cross section significantly. However, because of the tremendous complexity and laboriousness of the computations, this type of calculation has not been performed for aluminum even with an electronic computer. Evidently, new methods which would simplify the calculations are needed to resolve these difficulties.

Since the Al^{27} nucleus is strongly deformed, two maxima corresponding to oscillations directed along, and perpendicular to, the nuclear symmetry axis should be observed in the Al^{27} absorption cross section according to the predictions of Spicer [53] (see Section 1 of this chapter). The resonance energies of these maxima are E_b = 19.7 MeV and E_a =15.1 MeV according to the estimate of Spicer based on data for the magnitude of the electric quadrupole moment [80] (Q_0 =0.44 × 10^{-24} cm²). However, as is clear from Fig. 16, such a splitting

TABLE 8

Absorption cross section	σ (γ, n)		Photoneutron spectrum [6]	σ (p, γ$_0$) [83]
	[44]	[82]		
12.5				12.37 *
	16.6			
	18.0			
			18.6	
				18.7
			18.8	
19.1	19.0	19.1	19.2	
			19.5	19.5
	20.25	19.9	20.3	20.0
21.1	21.5	21.4	20.9	21.1
24.4	24.5			
26.1				

*From [86].

is not observed in the Al^{27} photoabsorption cross section. This fact is an important verification of the conclusion reached during the discussion of the F^{19} absorption cross section that the existence of deformation in light nuclei need not necessarily lead to splitting of the giant resonance as observed in heavy nuclei because the transitions corresponding to different ΔK may overlap strongly.

The integrated cross section for γ-ray absorption predicted for aluminum by the dipole sum rule is 6.0 MeV-mb (for exchange force fraction x = 0.5), and it is in satisfactory agreement with the experimental value of 555 ± 50 MeV-mb. At the same time, the sum of the integrated cross sections for the partial reactions (γ, n) and (γ, p), which were measured to 23 MeV, is about 50% of the integrated absorption cross section in the same energy range (integrated absorption cross section, 415 MeV-mb; cross section for the (γ, n) reaction, 100 Mev-mb [81]; cross section for the (γ, p) reaction, 120 MeV-mb [76]). This may be because the contributions from (γ, np) and (γ, 2n) processes, whose energy thresholds are 19.2 and 24.4 MeV, respectively, were not taken into account in the summation of the integrated cross sections for the partial reactions. It is improbable that the integrated cross sections for these reactions could compensate for the existing discrepancy because, in that case, they would have to be of the same order of magnitude as the integrated cross section for the (γ, n) and (γ, p) cross sections. This discrepancy is evidently associated with either a large contribution from the scattering process, or, as is more likely, with errors in the measurements of partial reaction cross sections, or with both concurrently.

4. Calcium Ca^{40}

One can confidently pick out four maxima at excitation energies of 12.5, 20.5, 24.7, and 26.11 MeV in the Ca^{40} nuclear absorption cross section (Fig. 18, errors are standard deviations). The positions of the experimental points in the 18.5-20.0 MeV energy range and the asymmetric shape of the peak at E_γ = 20.5 MeV lead one to assume that this resonance consists of two maxima (or groups of unresolved maxima) with resonance energies of 19.1 and 21.1 MeV. The resonance parameters of the resonances observed in the nuclear absorption cross section of calcium are given in Table 7.

Measurements of the nuclear absorption of calcium by a similar method were made previously by Dular et al. [13], who used a magnetic, Compton γ-spectrometer (resolution 2%) as a detector. Only the principal maximum in the nuclear absorption cross section (E_γ = 20 MeV) was clearly visible in this measurements. The absence of a more detailed structure is evidently associated with insufficiently high accuracy, which was 2-3 times less than the accuracy of the present measurements, with large energy "steps" (~ 0.5 MeV), and with poor spectrometer resolution.

TABLE 9*

Experiment	Theory		
Absorption cross section	Brown et al. [88]		Shitikova and Yadrovskii [90]
	Wigner δ-force	Soper force	(Serber force)
			9.6 (1.5)
12.5 (14)			12.3 (1.0)
	13.6 (1.0)		
			14.3 (11.5)
			15.1 (5.0)
	16.6 (1.0)	16.8 (1)	
	17.6 (9.0)		
	18.5 (80.0)		
19.1 (37)		19.2 (55)	19.1 (79.0)
		20.6 (44)	
21.1 (35)	21.5 (9.0)		21.4 (2.0)
24.4 (8)			
26.1 (6)			

*Numbers in parentheses are the contribution, in percent, to the total integrated cross section of a given transition.

The structure of the giant resonance in calcium has been observed in recent years by a number of authors [6, 44, 82-86] investigating partial reactions. The results of their work is given in Table 8. The same table gives the location of the maxima observed in the nuclear absorption cross section (column 1). Columns 2 and 3 give the locations of the maxima in the cross section for the $Ca^{40}(\gamma, n)$ reaction which were found by the method of detecting the total number of neutrons from bremsstrahlung [44] and "monochromatic" [82] γ-ray spectra (the degree of monochromaticity in the latter case was ~2%). Column 4 gives the location of peaks in the cross section for the (γ, n) reaction calculated from photoneutron spectra measured by the time-of-flight method [6], and column 5 gives the locations of peaks observed in the cross section for the inverse reaction $K^{39}(p, \gamma_0)Ca^{40}$ [83].

Apparently, structure is also present in the cross section for the $Ca^{40}(\gamma, p)K^{39}$ reaction as is evidenced by the data on photoproton spectra from calcium obtained by Komar and Dragnev [84] who made measurements at a maximum bremsstrahlung spectrum energy of 80 MeV. However, calculation of the cross section for the $Ca^{40}(\gamma, p)K^{39}$ reaction from this spectrum is made difficult. In this case, it is impossible to use the assumption that the final nucleus K^{39} remains in the ground state because it was shown in [85-87] that a large part of the protons correspond to transitions into excited states of K^{39} even at excitation energies in the neighborhood of 20 MeV.

As can be seen from Table 8, the locations of the maxima in the nuclear absorption cross section agree satisfactorily with the results obtained in studies of partial reactions (allowing for resolution). The nuclear absorption cross section does not show the maxima at E_γ = 16.6 and 18.0 MeV, which were observed in the cross section for the (γ, n) reaction; it does not exhibit the maxima at E_γ = 18.7 and 19.5 MeV which were observed in the cross section for the (p, γ_0) reaction and in the photoneutron spectrum, nor does it show peaks at E_γ = 20.0-20.3 MeV which were observed in the cross sections for the (γ, n) and (p, γ_0) reactions and in the photoneutron spectrum. The absence of these maxima in the nuclear absorption cross section is explained by the fact that they cannot be separated from the background of the principal maximum in the nuclear absorption cross section at the accuracy and resolution achieved.

The maximum in the nuclear absorption cross section at E_γ = 12.5 MeV was observed for the first time. The absence of this maximum in the cross section for the (γ, p) reaction and in photoproton spectra apparently results from the fact that it is located near the effective threshold for the $Ca^{40}(\gamma, p)$ reaction.

TABLE 10

Nucleus	σ_0, MeV-mb	σ_{-1}, mb	σ_{-2}, mb/MeV	$\bar{E}$, MeV	E_H, MeV	E_M, MeV	E', MeV
F^{19}	335^{+80}_{-40}	19.8 ± 4.0	1.28 ± 0.30	18.26	16.9	16.2	15.5
Mg^{24}	365^{+40}_{-20}	17.9 ± 2.0	0.95 ± 0.15	21.6	20.4	19.3	18.9
Al^{27}	555 ± 50	31.2 ± 3.5	1.88 ± 0.28	19.7	17.8	17.1	16.6
Ca^{40}	920 ± 100	49.6 ± 4.5	2.60 ± 0.50	21.6	18.5	18.8	19.1

Theoretical consideration of E1 absorption in calcium was carried out in the framework of the shell model by Brown et al. [88], Balashov et al. [89], Shitikova and Yadrovskii [90], and Gillet and Sanderson [91]. The calculations of Brown were carried out with the δ-forces of Wigner and Soper, but those of Balashov used Serber forces. The residual interaction was taken into account in both papers. Because of the lack of experimental data, the locations of the $s_{1/2}$ and $d_{5/2}$ hole levels in Ca^{40} (K^{39} levels correspond to them) were incorrectly chosen in the calculations of Balashov et al. [89]. As shown by recently made measurements [92], the energies of the $s_{1/2}$ and $d_{5/2}$ states in the K^{39} nucleus are 2.6 ± 0.15 and 6.3 ± 0.15 MeV, respectively. Of the three alternatives used in the Balashov calculations, the ones corresponding best to these values were $E_{s1/2} = 3.2$ MeV and $E_{d5/2} = 6.5$ MeV. These errors in the selection of the zero approximation can significantly distort the results of the calculations. Therefore the Balashov calculations were recently repeated by Shitikova and Yadrovskii [90], who used the locations of the Ca^{40} hole states found in [92]. In addition, this work investigated the dependence of the position of the giant resonance and of the contributions of various transitions to the integrated absorption cross section on various types of exchange forces and pair interaction amplitudes.

The calculations of Gillet and Sanderson [91] were carried out in the framework of the particle-hole model including consideration of Coulomb effects, the energies of the hole states being taken from experiment as in [89, 90].

A comparison of experimental and theoretical data on the location and strength of individual transitions is presented in Table 9.

As is clear from the table, the various types of calculations lead one to conclude that the absorption cross section for calcium results mainly from the single transition at $E_\gamma = 17.0$-19.0 MeV, whose contribution to the integrated cross section is about 80%. Only in the case of Soper forces [88] is the energy of the principal transition increased to approximately 20 MeV and the transition broken into two at $E_\gamma = 19.2$ and 20.6 MeV giving roughly identical contributions to the integrated cross section, which is in accordance with experimental data. However, the Soper force calculations does not predict the existence of any kind of strong transition in regions of higher and lower excitation energies. In the region of low excitation energies, the experimental data agrees better with the results of the calculations by Shitikova and Yadrovskii [90], who predict the existence of a weak transition at $E_\gamma = 12.2$ MeV and a rather strong transition at $E_\gamma = 14.3$ MeV. As far as excitation at $E_\gamma > 22$ MeV is concerned, none of the variants shown in the table indicates the existence of transitions in this energy region. (A transition at $E_\gamma = 22.4$ MeV (4.7%) was predicted in the paper of Balashov et al.[89]; however, this paper is not being considered for the reasons discussed above.) The calculations of Gillet and Sanderson lead to results similar to those of Brown et al. [88] (Wigner force version) with the difference that Gillet and Sanderson predict only two maxima in the absorption cross section at $E_\gamma = 18.4$ MeV ($\sim$ 90%) and 21.4 MeV ($\sim$ 10%).

Such a situation does not enable one to make a unique choice of the various calculations presented in Table 9. The discrepancy between experimental and theoretical data is apparently associated with one of the following circumstances:

1. The locations of the $s_{1/2}$ and $d_{5/2}$ hole states in Ca^{40}, which were used by Shitikova and Yadrovskii on the basis of the data in [92], are erroneous. For example, their energies are 2.9 and 5.9 MeV, respectively, according to the data of Johansson and Forkman [87]. Variation in the energy of these states, even within narrow limits, can significantly change the results of computations.

TABLE 11

	Nucleus			
	F^{19}	Mg^{24}	Al^{27}	Ca^{40}
$\sigma_0 = \int \sigma dE$, MeV-mb	335	365	555	920
$60\frac{NZ}{A}$, MeV-mb	285	360	405	600
$\frac{\sigma_0}{60\frac{NZ}{A}}$	1.18	1.02	1.37	1.53

2. While only dipole transitions were considered in all the theoretical papers, transitions with multipolarities higher than E1 play an important role in the 10-15 MeV energy range. This is supported by data on inelastic electron scattering [93], in which there are indications that the multipolarities of the transitions with energies of 12.2 and 13.9 MeV are E3 and E2, respectively.

3. The theoretical studies do not take into account all the effects which play an important role in the mechanism of γ-ray absorption.

The integrated nuclear absorption cross section in the energy region investigated is 920 ± 100 MeV-mb, which agrees within limits of error with that predicted from the sum rule for dipole transitions (840 MeV-mb). About 60% of the integrated cross section is found in two closely-located maxima at E_γ = 19.1 and 21.1 MeV with a total width $\Gamma \simeq 5$ MeV. The integrated cross section which is obtained by summation of the integrated cross sections for the partial reactions is 630 ± 120 MeV-mb in the same energy range (~555 MeV-mb for the (γ, n) reaction [44]). This quantity is somewhat below the value obtained for measurement of the nuclear absorption cross section. However, the discrepancy is small and can be explained by incomplete consideration of the contribution of other processes to the integrated cross section and by experimental errors.

A comparison of the absorption cross sections of the nuclei studied shows that the shapes of these cross sections change abruptly from nucleus to nucleus. At the resolution achieved, a clearly expressed structure is observed in the case of even-even nuclei while it appears much more weakly, or not at all, in the case of odd-even nuclei. This is seen most clearly in a comparison of the shapes of the γ-ray absorption cross section curves for O_8^{16} [17], F_9^{19}, Mg_{12}^{24}, and Al_{13}^{27}. If such a regularity is valid, the absorption cross section of Ne^{20} should have a structure similar to that observed for Mg^{24}, and Na^{23} should have one similar to that for Al^{27}. A clearly expressed structure is indeed observed in the proton spectrum from the Ne^{23}(e, e'p) reaction [16], which is in good agreement with the indicated behavior. Unfortunately, similar measurements have not been made for Na^{23}. This relation can be understood qualitatively because the coupling of an unpaired nucleon with the nuclear core evidently should lead to an increase in the number of possible dipole transitions, to their "smearing" in energy, and possibly to significant redistribution of transition strengths. This effect will obviously be best expressed in the case of rather strong coupling between the unpaired nucleon and the nuclear core.

5. Integral Characteristics of the Absorption Cross Section

The measured nuclear absorption cross sections allow one to calculate not only the integrated cross sections $\sigma_0 = \int \sigma dE$, but also the moments $\sigma_{-1} = \int (\sigma/E)dE$, $\sigma_{-2} = \int (\sigma/E^2)dE$, and various average energies for dipole absorption. Table 10 gives values for σ_0, σ_{-1}, σ_{-2}, the arithmetic mean energy $\bar{E} = (1/\sigma_0)\int \sigma E dE$, the harmonic mean energy $E_H = \sigma_0/\sigma_{-1}$, and the mean energies $E_M = \sqrt{\sigma_0/\sigma_{-2}}$, and $E' = \sigma_{-1}/\sigma_{-2}$ for the four nuclei investigated.

1. Integrated Cross Section. The theoretical value of the integrated cross section for E1 absorption of γ-rays by nuclei can be computed by means of the sum rule. Such calculations have been made by a number of authors. As the detailed analysis made by Levinger [94] demonstrated, if one assumes that the nuclear Hamiltonian is velocity independent and that there are no exchange forces, the integrated cross section for any model is

$$\sigma_0 = \frac{2\pi^2 e\hbar}{Me} \frac{NZ}{A} = 60 \frac{NZ}{A} \text{ MeV-m}$$

Consideration of two-particle exchange forces or the use of a velocity-dependent potential considerably increases the magnitude of σ_0. As was shown by Levinger and Bethe [95]

$$\sigma_0 = 60 \frac{NZ}{A} (1 + \Delta) \text{ MeV-mb},$$

for various assumptions about the shape and parameters of the proton-neutron potential and for $\Delta = 0.8x$ (x is the fraction of exchange forces). According to their evaluations, $x = 0.5$. For a shell-model potential which depended on velocity [96], Rand obtained the value $\Delta = 0.9$ [97]. A comparison of the experimental values with the values determined by the classical sum rule allows one to determine the quantity Δ. Such a comparison was made for medium-heavy and heavy nuclei, for which the absorption cross section equals the cross section $\sigma_{\gamma,n} + \sigma_{\gamma,2n}$ with sufficient accuracy. Analysis of the experimental data showed that $\Delta = 0.3$-0.5 [98]. This is in good agreement with the Levinger-Bethe estimates. At the present time, $\Delta = 0.4$ is considered acceptable for nuclei with $A > 40$. In the case of light nuclei, the experimental values of the absorption cross section, which were obtained by summation of the cross sections for the (γ, n) and (γ, p) reactions, were less than the values predicted by the sum rule, generally speaking.

The measured integrated nuclear absorption cross sections allow one to determine the magnitude of Δ for nuclei with $A < 40$. Table 11 gives the experimental values of the integrated cross section and their ratio to the integrated cross sections given by the classical sum rule.

The average value of Δ obtained from Table 11 is 0.28 ($x = 0.35$). It should be considered as the lower limit of the exchange force contribution because the measurements being discussed were only made for energies up to $E_\gamma \simeq 30$ MeV, and the fraction of the integrated cross section at excitation energies $E_\gamma > 30$ MeV may be a significant quantity. For example, about 50% of the integrated absorption cross section in the case of O^{16} occurs in the excitation region $E_\gamma > 30$ MeV [18]. Allowing for this circumstance, one can consider that the resulting value $x = 0.35$ is in good agreement with the value $x = 0.5$ assumed for medium-heavy and heavy nuclei. Extension of the excitation energy region to 170 MeV, for example, as was done in the papers of Gorbunov et al. [7-9], considerably increases x. (The mean value of x is 0.77 from the data of Gorbunov alone.) From this point of view, it would be of interest to make systematic measurements of nuclear absorption cross sections over a broad region of A and to excitation energies of the order of 100 MeV.

It should be pointed out that the contribution to the integrated absorption cross section which is made by the excitation region $E_\gamma < 30$ MeV grows significantly with increasing A. This is clearly seen in Table 12 where, in addition to the results already discussed, data is presented for C^{12} and O^{16}, which was obtained earlier by means of similar techniques [18, 19], as well as the ratio of the experimental values of the integrated cross section to the theoretical values for an exchange force fraction $x = 0.5$.

The data in Table 12 indicate that the fraction of dipole transitions located in the region of excitation $E_\gamma > 30$ MeV grows with decrease in A. This emphasizes the need for making detailed studies in that energy

TABLE 12

	Nucleus					
	C^{12}	O^{16}	F^{19}	Mg^{24}	Al^{27}	Ca^{40}
σ_0, MeV-mb	84	150	335	365	555	920
$\frac{\sigma_0}{60\frac{NZ}{A}\cdot 1.4}$, %	34	45	80	75	95	100

TABLE 13

	F^{19}	Mg^{24}	Al^{27}	Ca^{40}
$R_{\sigma_{-1}}$, 10^{-13}cm	2.73±1.17	2.46±0.77	3.04±0.98	3.42±1.00
R_e, 10^{-13} cm	—	2.98	—	3.52
R_{equ} 10^{-13} cm	3.51±1.50	3.17±0.99	3.90±1.26	4.40±1.28
r_0, 10^{-13} cm	1.30±0.56	1.09±0.35	1.03±0.42	1.28±0.37

region. In addition, these results enable one to hope that the experimental values of the arithmetic mean energy (and also E_H, E_M, and E'), which corresponds to the "center of gravity" of the dipole oscillations, will increase with a reduction in A when measurements are made up to an energy of 100 MeV. In that case, one can expect agreement with the theoretical predictions of the proportionality of resonance energy to $A^{-1/3}$ for light nuclei also (see [94], for example).

2. Cross Section Weighted by the Bremsstrahlung Spectrum (σ_{-1}). In contrast to the integrated cross section, the magnitude of σ_{-1} is independent of exchange forces and it is completely determined by the wave function for the ground state alone. This makes it very suitable for comparison of experimental and theoretical data.

Theoretical treatment of the quantity σ_{-1} has been carried out by a number of authors [95, 99, 100] on the basis of the independent-particle model. It was shown [95] that σ_{-1} was determined only by the root-mean-square radius of the charge distribution in the nucleus in the case of E1 absorption when all correlation effects were neglected. Khokhlov [99], using the independent-particle model with a potential in the form of a rectangular well of finite depth and considering correlations resulting from the Pauli principle as well as Coulomb forces, showed that

$$\sigma_{-1} = \frac{4\pi^2}{3}\left(\frac{e^2}{\hbar c}\right)\frac{NZ}{A}(1-\Lambda)\langle R_c^2\rangle,$$

where $\Lambda = 0.84/(1+22/A)$ is a correction associated with the effects listed; $\langle R_c^2\rangle$ is the square of the root-mean-square radius of the charge distribution in the nucleus.

After consideration of the finiteness of the proton charge distribution, which was performed by Foldy [101], we obtain

$$\sigma_{-1} = \frac{4\pi^2}{3}\left(\frac{e^2}{\hbar c}\right)\frac{NZ}{A}(1-\Lambda)(\langle R_c^2\rangle - \langle R_p^2\rangle),$$

where $\langle R_p^2\rangle$ is the square of the root-mean-square radius for the proton charge distribution $\langle R_p^2\rangle^{1/2} = (0.805 \pm 0.011) \times 10^{-13}$ cm^2 [102].

Table 13 contains: values for the root-mean-square radius $R_{\sigma_{-1}}$ computed from measurements of the value of σ_{-1}; values for the root-mean-square radius R_e determined by Hofstadter [103] in electron scattering experiments; the radius R_{equ}, recalculated from $\langle R_c^2\rangle^{1/2}$ on the assumption that the charge is uniformly distributed over the the entire volume of the nucleus (equivalent uniform model); and the characteristic radius $r_0 = R/A^{-1/3}$. As is evident from the table, values for the radius which were calculated from measured values of σ_{-1} are in satisfactory agreement with the results of electron scattering experiments.

Experimental values of σ_{-1} for nuclei with A > 50 (open circles) and the results of Gorbunov et al. [7, 10] for He^4, N^{14}, O^{16}, and Ne^{20} (triangles) are given in Fig. 19. The solid line corresponds to the relation $\sigma_{-1} = 0.36\ A^{4/3}$ mb, which was obtained by Levinger [104] who used the shell model with a harmonic oscillator potential. Calculations based on the independent-particle model with a rectangular well of finite depth [99, 104] gave similar values for σ_{-1} (crosses) which are well described by the relation $\sigma_{-1} = 0.30\ A^{4/3}$ mb (dashed

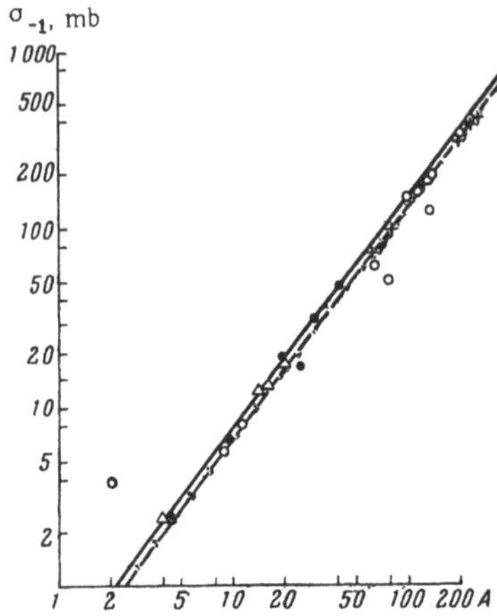

Fig. 19. Dependence of σ_{-1} on A.

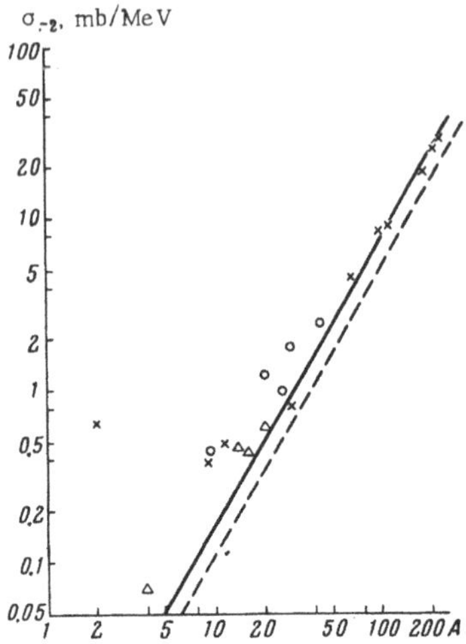

Fig. 20 Dependence of σ_{-2} on A.

line). The same figure shows the result of the present measurements (solid circles). As can be seen in Fig. 19, the experimental data are in good agreement with the theoretical values (with the exception of the data for Mg^{24}) not only in the case of nuclei with A > 50, but also in the region of light nuclei. This conclusion, arrived at previously by Gorbunov et al. [10] on the basis of an analysis of nuclei with $A \leq 20$, is confirmed by the expansion of the region to $A \leq 40$.*

3. Nuclear Polarizability (σ_{-2}). The static polarizability of the nucleus was treated by Migdal [105]. The dipole moment which is induced in the nucleus by a constant electric field was calculated on the assumption that the protons and neutrons can be considered as two interpenetrating fluids and that the symmetry energy $k(N - Z)^2/A$ in the Weizsacker formula is distributed uniformly over the entire nucleus. On the basis of these assumptions, Migdal showed that the nucleus polarizability is

$$\alpha = \frac{e^2 r_0^2 A}{40k} = \frac{\hbar c}{2\pi^2}\int \frac{\sigma(E)}{E^2}\,dE = \frac{\hbar c}{2\pi^2}\sigma_{-2}$$

or

$$\sigma_{-2} = \frac{\pi^2}{20}\left(\frac{e^2}{\hbar c}\right)\frac{r_0^2 A}{k}.$$

When $r = 1.2 \times 10^{-13}$ cm and k = 23 MeV, σ_{-2} is

$$\sigma_{-2} = 2.25\, A^{5/3} \text{ mb/MeV}.$$

It is obvious that the values of σ_{-2} will agree if one uses for their calculation different models yielding identical values for the symmetry energy $k(N - Z)^2/A$ in the Weizsacker formula. The quantity is very convenient for comparisons of experimental and theoretical data because it is insensitive to contribution to the cross sec-

*The small deviation of σ_{-1} from the theoretical values in the case of Mg^{24} is evidently connected with the fact that a considerable part of the E1 transitions of this nucleus are located where $E_\gamma > 30$ MeV. Analysis of the integrated cross section for this nucleus leads to the same conclusion (see Table 10).

tion at energies above the giant resonance. Such a comparison was made by Levinger [110] and also by Gorbunov et al. [10]. In Fig. 20 is shown the dependence of σ_{-2} on A obtained by Migdal (dashed line), experimental data for heavy nuclei (crosses), the results of Gorbunov et al. [10] (triangles), and the results of the present measurements (circles). It is clear from the figure that all the experimental values of σ_{-2} lie above the dashed line. Based on the results of the measurements for heavy nuclei, Levinger showed that these data are well fitted by the relation $\sigma_{-2} = 3.5\ A^{5/3}$ mb/MeV (solid line). However, the experimental values of σ_{-2} exceed the theoretical values for nuclei with $A < 40$, particularly for odd nuclei. As Levinger showed [94], this discrepancy can be almost completely eliminated if the values of r_0 used for these nuclei are those determined by Hofstadter [103], which for Be^9, Mg^{24}, and Ca^{40}, for example, are 1.89×10^{-13} cm, 1.33×10^{-13} cm, and 1.32×10^{-13} cm, respectively. In that way, agreement of experimental and theoretical data can also be obtained for σ_{-2} in the $A \leq 40$ region.

Because it is impossible to consider the nuclear absorption cross section of γ-rays at excitation energies $E_\gamma > 30$ MeV a negligibly small quantity (particularly in the case of light nuclei), it is impossible to make any kind of comparison of the experimental and theoretical values for the mean energies of dipole absorption. The theoretical values for these energies are close to $100\ A^{-1/3}$ and, as shown by Levinger [100], $E' \leq E_M \leq E_H \leq E$. The experimental data obtained do not contradict this relationship. One can hope that by carrying the measurements of the nuclear absorption cross section to energies of the order of 100 MeV, the theoretical and experimental values for the average energies of dipole absorption will be found to be in satisfactory agreement.

Conclusions

1. The nuclear γ-ray absorption cross sections of F^{19}, Mg^{24}, Al^{27}, and Ca^{40} have considerable structure, which is evidence of the existence of individual levels or groups of closely-spaced levels with the following excitation energy values:

F^{19}–(10.1), 11.4, 14.0, 16.8, 19.5, 24.0, and 28.0 MeV
Mg^{24}–11.2, (13.8), 17.1, 18.9, 20.4, 22.9, 24.9, and 27.0 MeV
Al^{27}–a large number of unresolved resonances in the 10-30 MeV excitation energy range
Ca^{40}–12.5, 19.1, 21.1, 24.4, and 26.1 MeV.

2. With a resolution of about 200 keV, structure is more clearly revealed in even-even nuclei than in odd-even nuclei.

3. Theoretical treatments of the nuclear γ-ray absorption cross section based on the shell model agree only qualitatively with experimental data. Consideration of the differences in binding energy for nucleons in filled and occupied shells considerably improves the agreement of experimental and theoretical data.

4. The integral characteristics of the photoeffect in light nuclei agree with the results of calculations for E1 absorption based on the independent-particle model (σ_0 and σ_{-1}) and on the hydrodynamic model (σ_{-2}). The resulting data confirm the earlier conclusion that the shape and magnitude of the γ-ray absorption cross section is determined by general properties of the nucleus which depend little on the model used.

5. The lower limit of the exchange force fraction in proton-neutron interactions, found by analysis of the integrated absorption cross sections of the nuclei investigated, is $x = 0.35$, which agrees satisfactorily with the data for medium-heavy and heavy nuclei.

6. The root-mean-square radii of charge distribution which were determined from the measured values are:

F^{19}–$(2.73 \pm 1.17) \cdot 10^{-13}$ cm
Mg^{24}–$(2.46 \pm 0.77) \cdot 10^{-13}$ cm
Al^{27}–$(3.04 \pm 0.98) \cdot 10^{-13}$ cm
Ca^{40}–$(3.42 \pm 1.00) \cdot 10^{-13}$ cm.

7. No clearly expressed separation of transitions resulting from deformation is observed in the nuclear absorption cross section of light nuclei. This prevents one from determining the internal electric quadrupole moment from data on the photodisintegration of light nuclei.

8. In order to establish a more complete structure for the nuclear γ-ray absorption cross section, particularly in the case of odd-even nuclei, it is necessary to make measurements with a resolution no worse than 100 keV. The region of excitation energies investigated should be extended to 50-100 MeV.

In conclusion, the author considers it his pleasant duty to express his gratitude to L. E. Lazareva, scientific director for candidates in the physical and mathematical sciences, for continuing interest and assistance, to B. S. Dolbilkin and V. I. Korin for help in carrying out this research, to V. A. Zapevalov for development of the electronic equipment, to N. S. Kozhevnikov for assisting in setting up the apparatus and in making measurements, and to B. A. Tulupov for discussions of the results. The author also thanks Prof. Cerenkov, corresponding member of the Academy of Sciences of the USSR, and A. N. Gorbunov, candidate in the physical and mathematical sciences, who made it possible to perform the measurements at the FIAN 260-MeV synchrotron, and the synchrotron operating group.

LITERATURE CITED

1. L. Katz, R. Haslam, et al., Phys. Rev., 95:464 (1954).
2. I. Goldemberg and L. Katz, Phys. Rev., 95:471 (1954).
3. A. Penfold and B. Spicer, Phys. Rev., 100:1377 (1955).
4. C. Tzara, J. Phys. Radiusm, 17:1001 (1956).
5. L. Bolen and W. Whitehead, Phys. Rev. Letters, 9:458 (1962).
6. F. Firk, Nucl Phys., 52:437 (1964).
7. A. N. Gorbunov, Tr. Fiz. Inst. Akad. Nauk SSSR, 13:174 (1960).
8. V. N. Fetisov, A. N. Gorbunov, and A. T. Vorfolomeev, FIAN Preprint A-68 (1964).
9. G. G. Taran and A. N. Gorbunov, Zh. Eks. i Teor. Fiz., 46:1492 (1964).
10. A. N. Gorbunov, V. D. Dubrovina, et al., FIAN Preprint A-27 (1962).
11. P. Fisher, D. Measday, F. Nikolaev, A. Kalmykov, and A. Clegg, Nucl. Phys., 45:113 (1963).
12. E. Fuller and E. Hayward, Phys. Rev., 101:692 (1956).
13. J. Dular, G. Kernel, et al., Nucl. Phys., 14:131 (1959).
14. B. Ziegler, Nucl. Phys., 17:238 (1960).
15. N. A. Burgov, G. V. Danilyan, et al., Izv. Akad. Nauk SSSR, Ser. Fiz., 27:867 (1963).
16. W. Dodge and W. Barber, Phys. Rev., 127:1746 (1962).
17. B. S. Dolbilkin, this volume, p. 17.
18. A. N. Burgov, G. V. Danilyan, et al., Zh. Eks. i Teor. Fiz., 43:70 (1962).
19. A. N. Burgov, G. V. Danilyan, et al., Zh. Eks. i Teor. Fiz., 45:1693 (1963).
20. G. V. Danilyan, Dissertation, ITEF (1965).
21. E. Keil and E. Zeitlor, Nucl. Instr. and Methods, 10:301 (1961).
22. R. Walker and B. McDaniel, Phys. Rev., 74:315 (1948).
23. Yu. D. Bayukov, M. S. Kozodaev, et al., Pribory i Tekhn. Eksp., No. 6, p. 23 (1958).
24. B. Kinsey and G. Bartholomew, Canad. J. Phys., 31:573 (1953).
25. G. Bartholomew, P. Campion, and K. Robinson, Canad. J. Phys., 38:194 (1960).
26. V. A. Zapevalov, Dissertation, FIAN (1964).
27. V. I. Gol'danskii, M. I. Podgoretskii, and A. V. Kutsenko, Counting Statistics in Nuclear Particle Detection. Fizmatgiz (1959).
28. G. White, NBS Report 1003 (1952).
29. G. White Grodstein, NBS Circular 583 (1957).
30. K. Siegbahn, Ed., Beta- and Gamma-Ray Spectroscopy. Interscience, New York (1955).
31. O. I. Leipunskii, B. V. Novozhilov, and V. N. Sakharov, Propagation of γ-Rays in Matter. Fizmatgiz (1960).
32. H. Greenberg, J. Taylor, and R. Haslam, Phys. Rev., 95:1540 (1954).
33. A. N. Gorbunov and V. A. Osipova, Zh. Eks. i Teor. Fiz., 43:40 (1962).
34. W. Dawson and D. Livesey, Canad. J. Phys., 34:241 (1956).
35. A. Penfold and E. Garwin, Phys. Rev., 116:120 (1959).

36. B. Spicer, F. Allum, et al., Austral. J. Phys., 11:273 (1958).
37. P. Yergin, Phys. Rev., 104:1340 (1956).
38. B. S. Ishkhanov, I. M. Kapitonov, et al., Phys. Letters, 9:162 (1964).
39. N. Yamamura, J. Phys. Soc. Japan, 18:11 (1963).
40. K. Shoda, K. Abe, et al., J. Phys. Soc. Japan, 17:735 (1962).
41. F. Bobard, G. Bouleque, and P. Chanson, Compt. Rend., 244:1761 (1957).
42. B. S. Ishkhanov, I. M. Kapitonov, et al., Zh. Eks. i Teor. Fiz., 46:1484 (1964).
43. B. S. Ratner, Zh. Eks. i Teor. Fiz., 46:1480 (1964).
44. K. Min, L. Bolen, and W. Whitehead, Phys. Rev., 132:749 (1963).
45. J. King, R. Haslam, and W. McDonald, Canad. J. Phys., 38:1069 (1960).
46. K. Murrary and W. Dendel, Phys. Rev., 132:1134 (1963).
47. B. Forkman and I. Wahlstrom, Arkiv. Fys., 18:339 (1960).
48. J. Taylor, L. Robinson, and R. Haslam, Canad. J. Phys., 32:238 (1954).
49. S. Fujii, Progr. Theoret. Phys., 21:511 (1959).
50. R. Sherr, C. Li, and R. Cheristy, Phys. Rev., 96:1258 (1954).
51. K. Okamoto, Phys. Rev., 110:143 (1958).
52. M. Danos, Nucl. Phys., 5:230 (1958).
53. B. Spicer, Austral. J. Phys., 11:490 (1958).
54. S. F. Semenko, Zh. Eks. i Teor. Fiz., 43:2188 (1962).
55. G. Ferguson, J. Halpern, R. Nathans, and P. Yergin, Phys. Rev., 95:776 (1954).
56. J. Miller, S. Schuhe, et al., CEN Preprint, Saclay, France.
57. H. Gove, Nucl. Phys., 49:270 (1963).
58. B. Forkman and W. Stifler, Nucl. Phys., 56:604 (1964).
59. F. Firk, Private Communication.
60. V. G. Neudachin and V. N. Orlin, Nucl. Phys. 31:338 (1962).
61. S. Nilsson, I. Sawicki, and N. Glendenning, Nucl. Phys., 33:239 (1962).
62. V. G.Neudachin and V. G. Shevchenko, Phys. Letters, 12:18 (1964).
63. J. King and W. McDonald, Nucl. Phys., 19:94 (1960).
64. B. Zeigler, Z. Phys., 152:566 (1958).
65. M. Mihailovic, G. Pregl, et al., Phys. Rev., 144:1621 (1959).
66. G. Tamas, I. Miller, et al., J. Phys. Radium, 21:532 (1960).
67. N. Mutsuro, K. Kageyama, et al., J. Phys. Soc. Japan, 17:1672 (1962).
68. K. Shoda, I. Ishizuka, et al., J. Phys. Soc. Japan, 17:1536 (1962).
69. C. Milone, Nucl. Phys., 47:607 (1963).
70. C. Van Der Leun and P. Endt, Physica, 29:990 (1963).
71. F. Ferrero, R. Malvano, et al., Nucl. Phys. 9:32 (1958).
72. J. Baglin, M. Thompson, and B. Spicer, Nucl. Phys., 22:216 (1961).
73. D. Wilkinson, Physica, 22:1039 (1956).
74. S. Nilsson, Mat. Fys. Medd. Kgl. Danske. Vid. Selskad., Vol. 29, No. 16 (1955).
75. W. Frahn and R. Lemmer, Nuovo Cimento, 6:664 (1957).
76. J. Halpern and A. Mann, Phys. Rev., 83:370 (1953).
77. J. Elliott and B. Flowers, Proc. Roy. Soc., A242:57 (1957).
78. G. Brown and L. Bolsterli, Phys. Rev. Letters, 3:472 (1959).
79. V. G. Neudachin, V. G. Shevchenko, and N. P. Yudin, Zh. Eks. i Teor. Fiz., 39:108 (1960).
80. R. Blin-Stoyle, Rev. Mod. Phys., 28:75 (1956).
81. R. Montalbetti, L. Katz, and J. Goldemberg, Phys. Rev., 91:659 (1953).
82. J. Miller, G. Schuhl, et al., Phys. Letters, 2:76 (1962).
83. N. Tanner, G. Thomas, and E. Earle, In: Proc. Rutherford Jubilee Internat. Conf., Manchester (1961), p. 297.
84. A. P. Komar and T. N. Dragnev, Dokl. Akad. Nauk SSSR, 126:1234 (1959).
85. T. N. Dragnev and B. P. Konstantinov, Zh. Eks. i Teor. Fiz., 42:344 (1962).

86. R. Helm, Thesis, Stanford University (1956).
87. S. Johansson and B. Forkman, Nucl. Phys., 36:141 (1962).
88. G. Brown, L. Castilleja, and J. Evans, Nucl. Phys., 22:1 (1962).
89. V. V. Balashov, V. G. Shevchenko, and N. P. Yudin, Nucl. Phys., 27:323 (1961).
90. K. V. Shitikova and E. L. Yadrovskii, Izv. Akad. Nauk SSSR, Ser. Fiz., 29:230 (1965).
91. V. Gillet and E. Sanderson, Nucl. Phys., 54: 472 (1964).
92. G. Kavaloski, G. Bassani, and N. Hintz, Phys. Rev., Letters, Vol. 11, No. 5, A11 (1963).
93. M. Hors, H. Nguen Ngoc, J. Perez, and I. Jorbe, Phys. Letters, 9:40 (1964).
94. J. Levinger, Nuclear Photo-disintegration. Oxford Univ. Press (1960).
95. J. Levinger and H. Bethe, Phys. Rev., 78:115 (1950).
96. N. Johnson and E. Teller, Phys. Rev., 98:783 (1955).
97. S. Rand, Phys. Rev., 107:208 (1957).
98. E. Fuller and M. Weiss, Phys. Rev., 112:560 (1958).
99. Yu. K. Khokhlov, Dokl. Akad. Nauk SSSR, 97:239(1954); Zh. Eks. i Teor. Fiz., 32:124(1957).
100. J. Levinger, Phys. Rev., 107:554 (1957).
101. L. Foldy, Phys. Rev., 107:1303 (1957).
102. L. Hand, D. Miller, and W. Wilson, Rev. Mod. Phys., 35:335 (1963).
103. R. Hofstadter, Rev. Mod. Phys., 28:214 (1956).
104. J. Levinger, Phys. Rev., 97:122 (1955).
105. A. Migdal, J. Phys. USSR, 8:331 (1949).

PHOTOPROTONS FROM MEDIUM WEIGHT NUCLEI ($50 \lesssim A \lesssim 125$) AND A MECHANISM FOR THE DECAY OF EXCITED NUCLEI IN THE GIANT DIPOLE RESONANCE REGION*

R. M. Osokina

Introduction

1. Photonuclear reactions have been intensively studied both experimentally and theoretically in the past 10-15 years. This interest was not accidental: γ-ray reactions as an instrument for the investigation of nuclear properties possess undoubted advantages when compared with nucleon reactions because the properties of electromagnetic radiation have been studied thoroughly.

One can distinguish two fields for such investigations. One of them is the analysis of the statistical moments $\sigma_k = \int E_\gamma^k \sigma_\gamma^k(E_\gamma) dE_\gamma$ ($k = 0, \pm 1; 2; 3$) of nuclear γ-ray absorption cross section curves by means of the sum rules for electromagnetic transitions. The sum rules for the nuclear photoeffect depend on the wave function for the nuclear ground state or its Hamiltonian (some of them depend on both), and do not depend on the mechanism for γ-ray absorption or on the structure of the excited states. A comparison of experimental data (more precise data than exists at present) with the sum rules obtained on the basis of these or other assumptions can prove to be extremely useful for the choice of a model of the nuclear ground state.

The second field is the analysis of the structure of photoabsorption cross section curves and of the cross sections of individual photonuclear reactions, and the analysis of the angular and energy distributions of their products. By such means, ideas about the mechanism involved in γ-ray interactions with nuclei and ideas about the dynamics of nuclear processes can be refined, and valuable information can also be obtained about highly-excited nuclear states.

It should be pointed out that the experimental investigation of photonuclear reactions is accompanied by considerable difficulty resulting from the smallness of the cross sections for these processes, on the one hand, and from the necessity of using the continuous bremsstrahlung radiation spectrum from electrons as a source of γ-rays, on the other. Despite this, however, a number of valuable results have been obtained through the study of photonuclear processes. For example, as far back as 1947 when the idea that nuclear reactions took place by "evaporation" of particles from a long-lived compound nucleus was prevalent, a mechanism different from evaporation was first observed [1]. This mechanism, the so-called direct photoeffect [2, 3], called into being by the concept that prompt nuclear processes were possible which resulted from the direct ejection of nucleons from the nucleus by incident particles. It was further shown that exchange forces played an important role in the nuclear Hamiltonian [4].

There is no doubt but that recently revealed prospects for the use of high-current electron accelerators and for the use of monochromatic γ-rays from positron annihilation [5] will lead to a considerable increase in information about the structure of the nucleus which can be obtained by means of photonuclear processes.

*Dissertation offered in satisfaction of the requirements for the scientific degree of candidate in the physical and mathematical sciences. Defended on December 14, 1964 at the P. N. Lebedev Institute of Physics, Academy of Sciences of the USSR.

2. The investigations described in this paper, which were published in part previously [6-13], are a continuation of work carried out on the 30-MeV FIAN synchrotron [14-16] to investigate the (γ, p) reaction in medium-weight nuclei at excitations in the giant dipole resonance region.

The spectra and angular distribution of photoprotons emitted from natural mixtures of copper and nickel isotopes exposed to bremsstrahlung radiation were measured in [14-16]. It was first observed that one could "probe" the internal nuclear shells even by means of a continuous γ-ray spectrum if one observed the changes in the shapes of the energy and angular distributions of fast photoprotons accompanying changes in the upper limit $E_{\gamma\ max}$ of the bremsstrahlung radiation spectrum.

In those same papers, there were indications that the role of processes other than evaporation in the emission of photoprotons by medium-weight nuclei was apparently considerably greater than was customarily thought. The data obtained indicated, in particular, that such processes, in all probability, made a contribution not only to the high-energy portion of the photoproton spectrum, which amounted to 5-10% of the yield, but also to the soft portion of the spectrum to a considerable extent.

However, the results of [14-16] were not sufficiently precise, and the conclusions drawn from them required verification. In order to arrive at more definite conclusions with respect to the role of various mechanisms for the emission of photoprotons and with respect to the possibility of using photonuclear reactions for investigating nuclear shell structure (particularly for determining the binding energy of nucleons in deeper levels), we undertook measurements of the yields, and angular and energy distributions, of photoprotons from a number of medium-weight nuclei in the range $50 \leqslant A \leqslant 125$.

A large part of the measurements were made with targets enriched in particular isotopes for the purpose of obtaining more easily interpreted data. Studies were made of nuclei in the neighborhood of closed proton shells with Z = 28 (Ni^{58}, Ni^{60}, Ni^{64}, Cu^{65}, and a natural mixture of Zn isotopes), Z = 50 (Cd^{114}, Sn^{224}, Sn^{124}, Sb^{121}, and Sb^{123}), and also niobium having the intermediate value Z = 41 and 100% natural content of the isotope Nb^{93}. The protons were detected with nuclear emulsions. Bremsstrahlung radiation from the FIAN synchrotron producing accelerated electron energies up to 30 MeV served as the γ-ray source.

3. The first chaper of this paper is in the nature of a review. It presents ideas about the nature of the levels out of which the giant dipole resonance is formed and about nuclear reaction mechanisms in that excitation energy region. It is pointed out that the (γ, p) and (γ, n) reactions, together with the (n, d) and (p, d) pickup reactions and direct processes like (p, 2p) and (e, e'p), can be used in principle to determine the binding energy of nucleons in internal shells if direct decay of particle-hole configurations (from which the dipole state is formed according to the shell model of the giant resonance) or a nonresonance direct photoeffect have a large role in photoprocesses.

Further consideration is given to work dealing with the as yet incompletely resolved problem of the description of the width of the giant resonance in doubly magic nuclei (and the width of each of the peaks into which this resonance is split in strongly deformed nuclei). It is emphasized that the problem of describing the width of the giant resonance is intimately bound up with the description of the decay of a nucleus excited by the absorption of γ-rays. In this connection, the discovery of the relative contributions of the various mechanisms for photonuclear reactions in the giant resonance region is of considerable interest.

Particular attention is devoted to a comparison of photonucleon spectra and of experimental data on partial cross sections and on yields of various photonuclear reactions with statistical theory. It is shown that it is evidently impossible to bring the existing set of data into accord with the prevalent idea that evaporation of photonucleons is the dominant process.

The second chapter gives a description of the experiment and discusses the problems associated with background measurements and with the analysis of the experimental data. The third chapter contains a summary of the experimental data. The fourth chapter presents a discussion of the results.

First and foremost, an attempt was made to determine the relative contribution of the evaporation process to the observed yields of photoprotons from nuclei with Z = 20-60 in the giant resonance region. For this purpose, calculations were made of the relative probability of photoproton emission according to the statistical model, and the results of these calculations were compared with the experimental values obtained from photoproton yields for $E_{\gamma\ max}$=24 MeV in this work and in that of others. The calculations made use of parameters by means of which the (n, p) reaction in similar nuclei was well described for the case of 14 MeV neutrons. In particular, consideration was given to the effect of pair correlations on the level density in the final nucleus.

As a rule, it was found that the observed photoproton yield significantly exceeded that expected on the basis of statistical theory.

A study was made of the observed photoproton spectra and angular distributions from the point of view of existing models of photonuclear processes in order to determine the nature of a mechanism different from evaporation.

In the case of Cu^{65}, Nb^{93}, and Cd^{114} (where measurements were made at several values of $E_{\gamma\ max}$, and it proved possible to construct by difference techniques the spectra of photoprotons produced by comparatively monochromatic γ-rays), an attempt was made to estimate roughly the binding energy of protons in inner shells.

The final section gives a brief resume of the results, and presents the conclusions that can be reached on the basis of the analyses that were made.

CHAPTER I

Photonuclear Reactions in Medium and Heavy Nuclei (A > 40) in the Giant Resonance Region (A Review)

1. The Nuclear Photoabsorption Cross Section $\sigma_\gamma(E_\gamma)$

In studying photonuclear processes, particular attention is given to reactions in the region of the giant dipole resonance which is observed in the $E_\gamma \approx$ 10-30 MeV region of the nuclear γ-ray absorption cross section. In the case of medium and heavy nuclei, the giant resonance is a peak the energy E_d of which falls smoothly from approximately 20 to 14 MeV in the region from calcium to lead.

The experimental dependence of E_d on mass number A can be approximated by the power function

$$E_d = aA^{-b}, \tag{1.1}$$

where a =40.7 ± 1.8 and b= 0.20 ± 0.01 [17], or by the function

$$E_d = cA^{-1/3} + c', \tag{1.2}$$

where c = 40 ± 6 MeV and c' =7.5 ± 1.5 MeV [18].

Frequently presented in various reviews (for example, [19, 20]), is a dependence of the type

$$E_d = cA^{-1/3}, \tag{1.3}$$

where c = 82 MeV, which was based on quite inaccurate data of Fuller and Hayward for the elastic scattering of photons by nuclei [21], and which fits the experimental data rather poorly (Fig. 1).

The width of the giant resonance at half height $\Gamma(\sigma_0)$ is about 4 MeV in the case of nuclei with doubly closed shells, and is larger for aspherical nuclei.

As Okamoto [23] has shown, there is a clear correlation between the width of the giant resonance $\Gamma(\sigma_0)$ and nuclear eccentricity. A splitting of the giant resonance into two components is observed in strongly deformed nuclei [24].

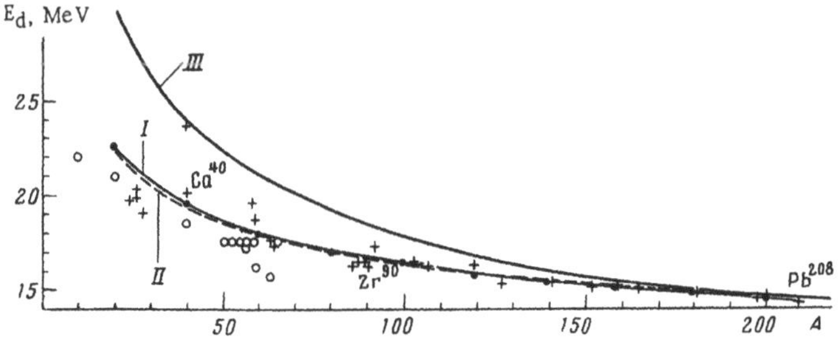

Fig. 1. Position of E_d, the maximum of the giant dipole resonance, as a function of atomic weight A. (I) $E_d = 40.7\,A^{-0.2}$ [17]; (II) $E_d = 7.5 + 40\,A^{-1/3}$ [18]; (III) $E_d = 82\,A^{-1/3}$ [21]. The crosses are experimental values obtained from the cross section for photonuclear absorption $\sigma_\gamma(E_\gamma)$ [17]; the circles are experimental values obtained from the cross section for inelastic scattering of high-energy protons [22].

The value of the cross section $\sigma_\gamma(E_\gamma)$ integrated over the giant resonance region is close to the value which is given by the sum rules for dipole transitions [4]:

$$\sigma_{\text{int}} = \sigma_0 = \int \sigma_\gamma (E_\gamma)\, dE_\gamma = \frac{2\pi^2 \hbar e^2}{Mc} \cdot \frac{NZ}{A}(1 + 0.8x) \approx 60 \frac{NZ}{A}(1 + 0.8x) \tag{1.4}$$

(x is the exchange force fraction), and it is several orders of magnitude greater than that calculated for other multipolarities.

The chief problem in the initial stages of theoretical studies of the giant resonance was the explanation of the nature of its levels and of the possibilities for its formation on the basis of existing nuclear models. Two types of models were proposed which appeared to be mutually exclusive at first glance. In collective models [25-27], the giant resonance was treated as a consequence of the excitation of dipole oscillations of all the protons with respect to all the neutrons while the shell model considered that it was formed of single-particle (more precisely, particle-hole) states [28]. Both kinds of models explained more or less successfully such characteristics of the nuclear photoabsorption cross section curve $\sigma_\gamma(E_\gamma)$ as the location E_d of the giant resonance, the broadening of the $\sigma_\gamma(E_\gamma)$ curve associated with static nuclear asphericity [23, 29, 30], and the magnitude of the cross section $\sigma_\gamma(E_\gamma)$ integrated over the giant resonance region.

The question as to which of the pictures of nuclear dipole excitation was the more correct was the subject for lively discussion even recently. However, Brink [31] showed in 1957 that the same wave function described an excited nuclear dipole state both for the case of the shell model with an isotropic harmonic oscillator potential and for the case of two rigid proton and neutron spheres oscillating with respect to one another. Although it was clear that so complete an agreement did not exist for a more realistic shell model potential, Brink's paper played an important role in demonstrating that the dilemma as to whether collective or single-particle states formed the dipole resonance apparently did not exist at all.

At the present time, there is almost no doubt but that the chief contributors to dipole excitation of the nucleus are states in which particle-hole configurations are mixed. Many nucleons participate in these states and, in a sense, the collective and shell model aspects of nuclear excitation by γ-rays are synthesized; as the result of dipole absorption of γ-rays, collective nuclear states are excited which, from the microscopic point of view, are a coherent mixture of particle-hole states.

2. Shell Model of the Giant Resonance

According to the shell model, the giant resonance results from dipole transitions of nucleons from filled shells to neighboring, higher shells. This model has received widespread acceptance since 1956 when Wilkinson showed [28] that among the vast number of E1 transitions possible at arbitrary energies, generally speaking, according to the shell model, there were several very strong transitions concentrated in a narrow energy region and nearly exhausting the dipole sum rule. In an oscillator potential model, these transitions correspond to a change in the principal quantum number $\Delta N = 1$, i.e., to transitions of the type $1l \rightarrow 1(l \pm 1)$. Dipole transitions in the region of higher oscillator levels, $1l \rightarrow 2(l \pm 1)$ for example, proved to be 200-300 times weaker.

Wilkinson further assumed that the quasistationary states produced decayed by direct emission of an excited nucleon into the continuum (resonance quasidirect photoeffect), or that the excitation of numerous overlapping levels of the compound nucleus (in the giant resonance region) occurred through the interaction of this nucleon with core nucleons and the levels decayed according to the laws of statistical theory.

Thus, in contrast to the collective model, the shell model proved to be quite attractive because, in addition to γ-ray absorption, it could describe the decay of excited nuclei along different channels, i.e., it could predict partial cross sections as well as the angular and energy distributions of the products of photonuclear reactions.

Initially, Wilkinson developed the model on the basis of the simplest single-particle type of shell model. The basic assumption of the shell model in this approximation was that all pairing interactions between nucleons reduced to a self-consistent, spherically symmetric, central potential. The spectrum of excited levels of an odd nucleus was determined by the spectrum of excitations of the outer odd nucleon, and the rest of the nucleons formed an inert core. The spacing between $1l$ and $1(l \pm 1)$ levels, the transitions between which produced the giant resonance, were 6-8 MeV in this model, i.e., they were 1.5-2 times less than the energy E_d at the maximum of the giant resonance.

Wilkinson tried to get around this difficulty by introducing an effective mass m^* equal to approximately half the true nucleon mass M. However, from the very beginning, the introduction of a small effective mass was not particularly justified, and it has been experimentally established at the present time that the spacing between single-particle $1l$ and $1(l \pm 1)$ levels in medium and heavy nuclei is indeed 6-8 MeV [32].

The difficulties created by the need to explain the location of the giant resonance were resolved by means of a modern, multiparticle type of shell model where consideration is given to "residual" pairing interactions not reducible, in principle, to central interactions in addition to the interactions described by a central potential. This interaction significantly changes the nuclear excitation spectrum in comparison with the independent particle model, and leads to nucleon correlations of a different kind.

A description of the giant resonance on the basis of this model is the subject of papers by Elliott and Flowers [33], Brown and associates [34-36], Balashov, Neudachin, Yudin, et al. [37-40]. Their results can be summarized in the following manner. Three types of residual interactions have an effect on the magnitude of the energy of E1 transitions from closed shells which make the principal contribution to the dipole sum:

1. Residual interactions in the ground state. This interaction between nucleons in a closed shell is of the pairing kind which leads to significant change in nucleon binding energy in proportion to filling of the shell in comparison with that given by the single-particle shell model.

2. The diagonal part of the residual interaction of excited particle-hole states. This interaction corresponds to a correction of the change in the value of the potential when a single nucleon is removed from a core with closed shells.

3. The off-diagonal part of the interaction of excited particle-hole states, which leads to mixing of pure shell model configurations.

In light nuclei, even the zeroth approximation (consideration of residual interactions of the first kind), based on empirical data for nuclear levels, leads to an increase in the energy at the maximum of the giant resonance by a factor of 1.5-2 in comparison with the energy in the single-particle model. The diagonal part of the particle-hole interaction raises the excitation energy by an additional 2-3 MeV. The off-diagonal part of the interaction has no significant effect on the magnitude of dipole transition energies and displaces the cross section maximum by 1-1.5 MeV at the most.

In heavy nuclei, consideration of only the zero and diagonal parts of the residual interaction leads to no significant change in the location of the giant resonance in comparison with the Wilkinson model. In the Pb^{208} nucleus, for example, the dipole γ-ray absorption cross section curve, calculated in the diagonal approximation, has a maximum in the 6-8 MeV region while the experimental value for the maximum is 13.5-14 MeV. Consideration of just the off-diagonal part of the interaction shifts the calculated value of the maximum by almost a factor of two and leads to agreement with experiment [39].

The work of Brown and Bolsterli [34], in which it was shown that the off-diagonal part of the interaction leads to coherence effects that significantly increase the energy of dipole states in heavy nuclei, played an important role in the construction of a theory for the giant resonance in heavy nuclei. In this case, a high-lying isolated state is formed which exhausts almost the entire dipole sum.

An important effect of the off-diagonal part of the residual interaction is that it leads to mixing of pure shell model states, i.e., to the formation of collective states, in a certain sense, in which a large number of nucleons participate. In light nuclei, the role of the off-diagonal part is comparatively minor, and the dipole shell model states resemble pure single-particle states. In heavy nuclei, configuration mixing is considerable, and the dipole states forming the giant resonance are essentially collective.

Because each of the dipole states is a mixture of shell model configurations (of the particle-hole type), additional decay channels are opened which are of great significance for predicting partial cross sections for the emission of protons and neutrons and for predicting spectra and angular distributions in both heavy and light nuclei.

Recent investigations have shown that the modern shell model correctly predicts the magnitude of the integrated cross section for nuclear photoabsorption in the giant resonance region, and the location of the maximum, over the entire range of atomic weights A, and that it is able to describe quantitatively the splitting of the giant resonance in deformed nuclei [30]. Obviously, this can be regarded as conclusive proof that particle-hole configurations are really the main contributors to dipole excitation of the nucleus in the giant resonance region.

However, using this model, no one has managed to explain by simple variation of parameters such important characteristics as the fine structure of the photoabsorption cross section curve in light nuclei and the width of the giant resonance in medium and heavy doubly magic nuclei [$\Gamma_{theor}(\sigma_\gamma) < \Gamma_{exp}(\sigma_\gamma)$]. This is because the model as it presently exists predicts a considerably weaker spectrum of nuclear dipole excitation than is actually the case.* For this reason, apparently, no one has succeeded in explaining the spectrum of emitted nucleons [41].

An important deficiency of the model being discussed is the fact that calculations based on this model are extremely laborious, require the diagonalization of high-order energy matrices, and can only be accomplished with high-speed computers in the case of comparatively heavy nuclei. In addition, rigorous calculations are only possible for doubly magic nuclei (O^{16}, Ca^{40}, Pb^{208}). Such calculations are accompanied by even greater technical difficulties in the case of nonmagic nuclei.

*In this connection, it is of interest to note the method for describing nuclear dipole photoabsorption based on the Fermi-fluid model with the help of Green's functions recently developed by Migdal, Zaretskii, and Lushnikov [42, 43]. In this approach, dipole transitions into higher shells are taken into account along with the usually considered transitions between neighboring shells. As a result, the spectrum of nuclear dipole excitations is considerably enriched, and one can explain the width of the nuclear photoabsorption curve successfully.

In conclusion, it should be pointed out that the decay process for particle-hole configurations should play a definite role in photonuclear reactions on the basis of the shell model of the giant resonance. If the role of such processes is large, (γ, p) and (γ, n) reactions may prove to be an important source of information about the binding energy of nucleons in inner shells of the nucleus.

Indeed, one-hole excited states of the nucleus are produced by the decay of the configurations mentioned, by (p, d) and (n, d) pickup reactions, and by direct process of the type (p, 2p), (e, e'p), etc. As a result, one ought to expect the appearance of peaks in spectra of photonucleons produced by monochromatic γ-rays at energies ε_i equal to the difference between the γ-ray energy E_γ and the binding energy B_i of a nucleon in the i-th shell. It is clear that the peak energy ε_i, which corresponds to the excitation of the i-th hole will shift in proportion to the change in γ-ray energy in such a way that the energy difference $E_\gamma - \varepsilon_i$ remains constant while the magnitude of the peak, which is determined by the excitation cross section of the i-th hole in the final nucleus, may undergo considerable change.

3. The Problem of the Giant Resonance Width

As already noted, the experimental width of the giant resonance $\Gamma(\sigma_\gamma)$ is about 4 MeV in nuclei with A > 40 for spherical nuclei with doubly closed shells, and has a larger value where the shell structure is not complete. In the case of strongly deformed nuclei, the giant resonance is split into two components.

A considerable amount of work has been devoted to the theoretical description of the width of the giant resonance. In the greater part, the effect of static or dynamic nuclear deformations (β and γ-oscillations) and of anharmonicity in nuclear motion on the width of the giant resonance far from a nucleus with doubly closed shells is discussed phenomenologically or on the basis of the collective model [30, 44-52]. It was established in these papers that the magnitude of the spread in the giant resonance curve in such nuclei is characterized by the amount of nuclear level splitting, which arises as the result of the removal of degeneracies associated with spherical symmetry of the nuclear potential through deviation of nuclear shape from the spherical in either ground or excited states.

However, there is as yet no satisfactory explanation of the reason for the nearly 4 MeV width of the giant resonance in spherical, doubly magic nuclei (and for the width of each of the peaks into which the giant resonance is split in strongly deformed nuclei as well). Work is discussed below where an attempt has been made to describe the width of the giant resonance in spherical nuclei.

It is completely reasonable to assume that the width of the giant resonance results from two causes: a) a spread ΔE_d in the transition energies which make the main contribution to the dipole sum; b) the natural width of the excited levels from which the giant resonance is formed, $\Gamma = \hbar/\tau$, where τ is the lifetime of the dipole state. Hence,

$$\Gamma(\sigma_\gamma) = \Delta E_d + \Gamma. \tag{1.5}$$

It is evident that there exists an unique frequency of dipole motion, i.e., $\Delta E_d = 0$ and $\Gamma(\sigma_\gamma) = \Gamma$, in spherical nuclei in the case of collective models.

Wuldermuth and Wittern [53] evaluated the width Γ by means of the collective hydrodynamic model [26, 27] on the assumption that dipole vibrational motion damps as the result of collisions between nucleons over the entire volume of the nucleus. They found that

$$\Gamma = \frac{\hbar}{\tau} = 5.55\left(\frac{200}{A}\right)^{4/3}\text{MeV}. \tag{1.6}$$

It is easy to see that the width Γ is always greater than 5.5 MeV according to this formula, and that the calculated widths are in the range from 16 to 7.5 MeV for nuclei with A = 50-100, which considerably exceeds the experimental value $\Gamma(\sigma_\gamma) \approx 4$ MeV.

In a very brief note [54], Danos pointed out that the calculation of Wuldermuth and Wittern gave an overestimate because it was carried out on the assumption of a strongly heated nucleus. An estimate made by Danos with more rigorous consideration of the Pauli principle showed that from the time expended in heating the nucleus by collisions over its entire volume, one can determine that

$$\Gamma \approx A^{-4/3}. \quad (1.7)$$

For A = 200, for example, this results in the unacceptably low value $\Gamma \approx 1$ MeV. Because of this, Danos expressed the opinion (without giving any numerical estimates) that damping of dipole oscillations was produced, in all probability, not by internal friction but by surface effects, the latter not being associated with the excitation of surface waves but with irreversible direct processes of particle emission.

Businaro and Gallone [55] estimated the width Γ on the assumption that the collective oscillation of two rigid proton and neutron spheres [26] damped as the result of the excitation of single-particle degrees of freedom. They obtained a width $\Gamma \approx 4$ MeV for a spectrum of single-particle states in an infinite rectangular well and an oscillator potential. A similar idea that the excitation of the nucleus in the giant resonance bore a collective nature but that the decay width was single-particle was put forth by Fujii and Takagi [56].

However, despite the fact that widths close to the observed one were obtained in [55, 56], the estimates were based on extremely crude assumptions, and it is impossible to consider them as a quantitative description of the width of the giant resonance.

Attempts to describe the width of the giant resonance in the framework of the shell model are sounder from the physical point of view, but these attempts have not as yet led to a conclusive and unambiguous explanation of the magnitude of $\Gamma(\sigma_\gamma)$. According to Wilkinson [28], the width $\Gamma(\sigma_\gamma)$ can be represented in the form:

$$\Gamma(\sigma_\gamma) = \Delta E_d + \Gamma_{part} + \Delta\Gamma, \quad (1.8)$$

where ΔE_d is the energy spread of shell model dipole transitions (particle-hole type) because of anharmonicity of the nuclear potential or because of zero-order oscillations of the nuclear surface (it is obvious that $\Delta E_d = 0$ for an isotropic harmonic oscillator potential; Γ_p is the width associated with quasidirect decay of particle-hole configurations leading to the emission of excited nucleons into the continuum; $\Delta\Gamma$ is the broadening of the giant resonance resulting from the coupling of shell model dipole particle-hole states with other degrees of freedom of nuclear motion.

In the papers of the various authors, completely different ideas are expressed with regard to how much significance ought to be attached to the quantity Γ and to what the relative roles of ΔE_d, Γ_p, and $\Delta\Gamma$ are in respect to the width of $\Gamma(\sigma_\gamma)$.

Using the independent particle model, Wilkinson [28] found that the energy spread ΔE_d of the dipole transitions forming the giant resonance was 1.5-2 MeV in the case of a more or less realistic nuclear potential (rectangular well with spin-orbit interaction).

In the single-particle model, the quantity Γ_p is determined by

$$\Gamma_{part} = \frac{\hbar^2}{2MR^2} 2kP_l, \quad (1.9)$$

where M, k, and P_l are the mass, wave number, and barrier penetrability of the emitted nucleon, respectively; R is the nuclear radius. Estimates have shown that the width of Γ_p is no more than several hundred kilovolts.

In order to describe quantitatively the magnitude of $\Gamma(\sigma_\gamma)$, which connects the initially excited single-particle states with overlapping levels of the compound nucleus that decay in accordance with the laws of statistical theory, Wilkinson introduced an imaginary part $W = \Delta\Gamma = 2$-3 MeV into the shell model potential. The value of W was gotten from experiments on nucleon scattering. As Levinger pointed out [19], however, the imaginary part of the optical potential, W = 2-3 MeV, was obtained in experiments involving nucleons with small orbital momentum l. In the giant resonance, nucleons states with large l have the greatest weight. Since there is no basis for assuming that W is independent of l, the agreement of the theoretical value for $\Gamma(\sigma_\gamma)$ with experiment which was obtained by Wilkinson cannot be considered convincing.

The quantity $\Delta E_d \to 0$ in the modern shell model where allowance is made for residual interactions as a consequence of the Brown-Bolsterli effect [34] (separation of practically a single state in the giant resonance region which exhausts the greater part of the dipole sum). On the other hand, the width of Γ_p increases in comparison with that given by the single-particle model because configuration mixing increases the number of decay channels:

$$\Gamma_{\text{part}} = \sum_i a_i \Gamma_i, \tag{1.10}$$

where α_i is the weight assigned to the i-th particle-hole configuration in the dipole state; Γ_i is the partial width for decay through the i-th channel as defined by relation (1.9).

As calculations have shown, the quantity Γ_p makes a significant contribution to the width of the photoabsorption curve in light nuclei; in the case of Ca^{40} [40], $\Gamma_p \approx 2$ MeV, while the halfwidth of the cross section curve $\Gamma(\sigma_\gamma)$ is about 4 MeV as measured by Dolbilkin et al. [57]. At the same time, the calculated width Γ_p is considerably reduced in heavier nuclei because of a reduction in particle penetration through the barrier and a reduction in the energy of the giant resonance. Thus Γ_p is 200-300 keV at most in Zr^{90} according to [41].

Therefore, it proves to be difficult to explain $\Gamma(\sigma_\gamma)$ through consideration of residual interactions within the framework of particle-hole excitations, just as it did in the case of the single-particle model.

Balashov and Chernov [58] phenomenologically obtained a quantitative description of the giant resonance width in Pb^{208} by the introduction of coupling between dipole motion and one-phonon excitation of the nuclear surface.

If we keep to the microscopic point of view, one-phonon excitation corresponds to direct excitation of two particle-hole pairs by the absorption of γ-rays;* on the other hand, the presence of slow oscillations of the nuclear surface leads to removal of the degeneracy of particle-hole states associated with spherical symmetry in close similarity to what occurs with the Nilsson deformed potential [45]. Both these factors increase ΔE_d. The imaginary part of the potential introduced by Wilkinson leads to dissipation of the initially excited configurations over states of more complex microstructure as the result of subsequent collisions of the excited nucleons with core nucleons, i.e., it introduces into the quantity Γ a contribution resulting from the natural width of the dipole levels.

In a recently published paper, Danos and Greiner [60] made an attempt to explain quantitatively the width of the giant resonance in heavy nuclei by the transfer of energy from collective Brown-Bolsterli particle-hole states to two particle-hole pair states. Hundreds of states of that kind exist in heavy nuclei for excitations in the giant resonance region. Among them are many for which the excited nucleons are in the positive energy region. The authors consider that summation of the small decay widths of two particle-hole pair states leads to the large total width of the giant resonance, which can be represented in the form:

$$\Gamma = 2\pi \,|\langle \psi_u^{(2)} | V | \psi_v^{(1)} \rangle|^2_{\text{av}} \cdot \rho(E), \tag{1.11}$$

*One should point out in this connection the interesting work of Mihailovic and Rosina [59] in which it is shown that a certain number of states, each of which takes up part of the dipole strength of the Brown-Bolsterli state, are generated in the nonmagic nuclei C^{12} and Ne^{20} as the result of considering such configurations.

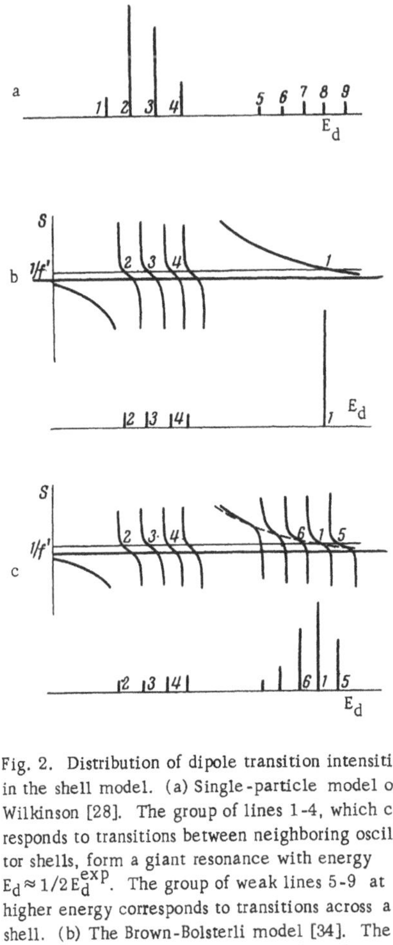

Fig. 2. Distribution of dipole transition intensities in the shell model. (a) Single-particle model of Wilkinson [28]. The group of lines 1-4, which corresponds to transitions between neighboring oscillator shells, form a giant resonance with energy $E_d \approx 1/2 E_d^{exp}$. The group of weak lines 5-9 at higher energy corresponds to transitions across a shell. (b) The Brown-Bolsterli model [34]. The transitions 5-9 are neglected here. Consideration of the off-diagonal part of the residual reaction leads to one of the levels 1-4 being shifted over to the right and exhausting almost the entire dipole sum (S is a function the poles of which determine the location of the dipole levels E_d, and the residues, the transition intensities). (c) The Migdal, Zaretskii, Lushnikov model [42]. Both groups of poles (1-4 and 5-9) are considered. Inclusion of the interactions leads to a shift of pole 1 into the region of poles 5-9. As a result, the residues at points 5 and 6 are increased. This leads to the creation of a spread ΔE_d.

where $|\langle \psi_u^{(2)} | V | \psi_v^{(1)} \rangle|^2_{av}$ is the average value of the matrix element for transitions from particle-hole states described by the wave function $\psi_v^{(1)}$ to two particle-hole pair states with the wave function $\psi_u^{(2)}$; $\rho(E)$ is the density of states of the latter kind. Both factors were calculated for a simplified model with a rectangular well.

It was found that 1 MeV $\leq \Gamma \leq$ 5 MeV. The authors ascribed the uncertainty in the resulting value of Γ to inexact knowledge of the internucleon interaction potential V in the nucleus.

From the above discussion of [28, 34, 58, 60], it follows that there is no possibility of explaining the width of the giant resonance $\Gamma(\sigma_\gamma)$ by the sum of the energy spread ΔE_d of the particle-hole states and the decay width Γ_p of these states; it is necessary to bring in $\Delta\Gamma$, which describes the coupling of dipole states with other degrees of freedom of intranuclear motion. However, Migdal, Zaretskii, and Lushnikov [42, 43], solving the problem of nuclear photoabsorption on the basis of the Fermi-fluid theory for finite systems, came to the opposite conclusion. Estimates presented in [42] showed that the width of single-particle states associated with their decay into more complex multiparticle states was of the order of $\varepsilon_0 A^{-2/3}$, where ε_0 is the Fermi energy and A is the atomic weight of the nucleus. It follows from this that $\Delta\Gamma$ is no more than 1 MeV. Therefore, the transient nature of single-particle states connected with nucleon interaction cannot be a fundamental cause of the giant resonance width $\Gamma(\sigma_\gamma)$. The authors of [42] further asserted that, in principle, there is no possibility for the existence of the isolated degrees of freedom of nuclear dipole excitation which arise in the hydrodynamic interpretation of dipole oscillations and in the Brown-Bolsterli model.

Using the Green's function method to solve the problem of nuclear dipole excitation in the manner of a problem involving the behavior of interacting quasiparticles in an electromagnetic field, the authors [42] obtained the width of the giant resonance as the result of the spread ΔE_d of particle-hole excitations. In Fig. 2, which is adapted from [42], it is shown that the absence of the spread ΔE_d in the Brown-Bolsterli model [34] is associated with the fact that dipole transitions to higher oscillator shells are neglected in the usual shell model calculations.

The results obtained in [42, 43] are extremely interesting. However, it should be pointed out that the appearance of the spread ΔE_d in this model is principally associated with the fact that the "displaced" Brown-Bolsterli state happens to fall into a region where the next group (in comparison with the ones usually considered in the giant resonance) of single-particle states are concentrated. Whether this is so in actual nuclei remains obscure.

Thus it can be pointed out that there are at the present time very diverse concepts with respect to the chief reasons for the width $\Gamma(\sigma_\gamma)$ of the giant resonance in doubly magic nuclei:

1. The basic factor is the spread in particle-hole excitations ΔE_d (Zaretskii, Lushnikov).
2. It is impossible to explain the width of the giant resonance on the basis of particle-hole excitations; the main factor is the spread in energy of the dipole states resulting from the excitation of one-phonon oscillations of the nuclear surface, i.e., the excitation of two particle-hole pair configurations play an important role in the giant resonance region (Balashov, Chernov).
3. ΔE_d is small; one of the main factors is the "natural" width of the dipole states resulting from damping of the initial particle-hole states as the result of excitation of states with more complex microstructure (Wilkinson, Danos, and Greiner).

4. Photonuclear Reaction Mechanisms and Giant Resonance Width

The solution of the problem of giant resonance width $\Gamma(\sigma_\gamma)$ is connected with the explanation of the role played by states of varying microstructure in the giant resonance region, and it is closely bound up with the description of the decay of nuclei excited by γ-ray absorption. In this respect, the clarification of the relative roles of various mechanisms for nuclear process in photoreactions is extremely useful for understanding the true reasons underlying the spread in the nuclear photoabsorption cross section curve.

As a function of their time course, one can divide nuclear reactions into the following three classes:

I. Prompt reactions, or nonresonance direct processes, which proceed without the formation of a compound nucleus. In photoprocesses, this corresponds to the direct photoeffect proposed by Courant [2], which involves the ejection into the continuum of one of the nucleons in the nucleus by an incident γ-ray. Numerical estimates indicate that the role of this mechanism in the total nuclear photoabsorption cross section in the giant resonance region is quite small [2, 61, 62].

II. Reactions of an intermediate type in which the initial ("input") nucleon configuration produced by the absorption of an incident particle decays by the emission of an excited nucleon into the continuum without interaction with the remaining nucleons. Among such processes, in particular, is the process of emission into the continuum of an excited nucleon in a dipole particle-hole configuration located in the positive energy region, a process called the quasidirect photoeffect by Wilkinson [28].

III. Statistical processes, which pass through the stage of a long-lived compound nucleus in the region where a large number of its levels overlap. In such a nucleus, an important role is played by multiparticle configurations produced as the result of multiple collisions between nucleons. The lifetime of such a nucleus is very large in comparison with nuclear times, and is sufficient for the establishment of "thermal" equilibrium between the various degrees of freedom of intranuclear motion. In this case, the decay of the nucleus resembles the evaporation of particles from a heated liquid drop, and it is described by the statistical theory of nuclear reactions.

Also possible are II' processes which are intermediate between II and III. However, their role is unclear at the present time, and they are not usually considered.

Each of the processes II, II', and III passes through the dipole resonance state and makes its contribution to the natural width Γ of the state

$$\Gamma = \Gamma_B + \Gamma_{B'} + \Gamma_{\text{stat}}. \quad (1.12)$$

Assuming that $\Gamma_{B'} \ll \Gamma_B$ and Γ_{stat}, i.e., the decay of the configurations initially produced as the result of interactions with other nucleons necessarily leads to a compound nucleus, where decay proceeds by way of evaporation, we obtain

$$\Gamma = \Gamma_B + \Gamma_{stat} \tag{1.13}$$

In the particle-hole model of the giant resonance, $\Gamma_B = \Gamma_p$ and the absolute value of Γ_p varies roughly from 2 MeV to 200 keV in the region from Ca to Zr. At the moment, it is impossible to compute the value of Γ_{stat}. As was pointed out, Wilkinson completely arbitrarily assigned to it the imaginary part of the potential $W \approx 2$-3 MeV. However, this arbitrariness can be eliminated if one determines experimentally the fraction of evaporation ΔY_{evap} in the total yield of photoprocesses (for example, by comparing the observed relative contributions from (γ, p), (γ, n), and (γ, 2n) processes with those predicted by statistical theory). It is clear that this fraction is related to the widths Γ_p and Γ_{stat} by the relation

$$\Delta Y_{evap} = \frac{\Gamma_{stat}}{\Gamma_{part} + \Gamma_{stat}} = \frac{\Gamma_{stat}}{\Gamma}. \tag{1.14}$$

Hence, knowing the quantity Γ_p (from calculation) and ΔY_{evap} (from experiment), one can determine the numerical value of Γ_{stat} and, therefore, the total width Γ associated with damping of dipole motion.

From relation (1.14), it is easy to derive

$$\Gamma = \frac{\Gamma_{part}}{1 - \Delta Y_{evap}}. \tag{1.15}$$

Because of the smallness of Γ_p, it is clear from this that one can expect to explain the width of the giant resonance by damping of initially excited dipole motion only through the dominant role of the evaporation process ($\Delta Y_{evap} \approx 0.9$; $\Gamma_{stat} \gg \Gamma_p$). For equal contributions from evaporation and from the quasidirect photoeffect, the width Γ remains small in comparison with $\Gamma(\sigma_\gamma)$ in the case of fairly heavy nuclei.

Thus information about the relative contribution of the evaporation process to photonuclear reactions is very important for an understanding of the causes responsible for the width of $\Gamma(\sigma_\gamma)$. A review of the work from which such information can be obtained is given in the following section.

5. Statistical Theory and Photonuclear Reactions in the Giant Resonance Region

The Statistical Theory of Nuclear Reactions (or the Evaporation Model) [63, 64]. This theory is used to describe nuclear reactions in the region of fairly high excitations fulfilling the condition

$$\Gamma_c \gg D_c, \tag{1.16}$$

i.e., the widths of the levels in the compound nucleus are much greater than the spacing between them.

Two principles are fundamental to the theory: 1) the hypothesis of independence, i.e., the concept of Bohr that the decay of a compound nucleus does not depend on the method of its formation; 2) averaging of cross sections over a large number of final states. The latter is possible if one of two conditions is fulfilled: the widths of the levels of the final nucleus are considerably greater than the spacing between them:

$$\Gamma_R \gg D_R \tag{1.17}$$

or the experimental resolution of the final states is much poorer than the average spacing between levels of the final nucleus:

$$\Delta E_{R\,exp} \gg D_R. \tag{1.18}$$

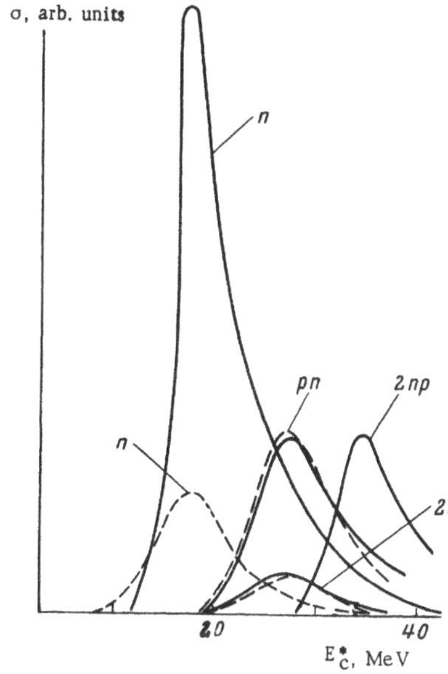

Fig. 3. A comparison of the relative probabilities for the (x, n), (x, 2n), and (x, p) processes going by way of the compound nucleus Zn^{64*} produced in various ways. Solid line is the data from Sagane, the dashed line is the data from Goshal.

The hypothesis of independence in association with the principle of detailed balance (the large number of degrees of freedom of the final system makes it possible to use statistical methods) offers an opportunity to express the cross section for any nuclear reaction through the cross section for compound nucleus formation in the inverse process and the phase space of the final system.

The basic consequences of statistical theory can be formulated briefly in the following manner:

1. Both integral and differential cross sections for any reaction are smooth functions of the incident particle energy.

2. Emitted particle spectra are described by the relation:

$$n(\varepsilon)\,d\varepsilon \sim \varepsilon\sigma_c(\varepsilon)\cdot\rho_R(\varepsilon_{max}-\varepsilon)\,d\varepsilon, \quad (1.19)$$

where $\sigma_c(\varepsilon)$ is the cross section for compound nucleus formation in the inverse process; $\rho_R(\varepsilon_{max}-\varepsilon)$ is the level density in the final nucleus; $\varepsilon_{max} = E_c^* - B$ is the maximum possible energy of the emitted particles where E_c^* is the excitation energy of the compound nucleus and B is the binding energy in that nucleus for the emitted particle.

3. Angular distributions are always symmetric with respect to 90°. The degree of anisotropicity is small (no more than a few percent [65, 66]).

It should be emphasized once again that statistical theory yields cross sections averaged over a large number of states; the fulfillment of the conditions $\Gamma_R \gg D_R$ or $\Delta E_R \gg D_R$ in the final nucleus is a necessary requirement for the validity of consequences 1-3. Otherwise, one should expect deviations from these predictions which by no means result from the existence of any special reaction mechanisms not associated with compound nucleus formation. The question of cross section fluctuations predicted by statistical theory was first formulated and discussed in detail by Ericson [67, 68].

In recent years, considerable progress has been made in substantiation of the statistical model and in the understanding of its region of application [67]. The model parameters characterizing the level densities in excited nuclei and the inverse reaction cross sections [69-72] have been significantly refined.

Analyses made by a number of authors [73-75] have shown that the statistical theory successfully describes a large portion (not less than 90%) of the total yield for nucleon reactions at excitations in the giant resonance region (in particular, for reactions involving 14-MeV neutrons). This conclusion is deserving of confidence because the range of permissible variation of the parameters used in evaporation model calculations has been considerably reduced, and the predictions of this model become somewhat more definite, from the quantitative point of view, than they were several years ago.

A Comparison of Data from Photonuclear Reactions with the Predictions of Statistical Theory. The giant resonance in the case of fairly heavy nuclei (A > 40) lies in a region where the density of excited levels in the compound nucleus is extremely high. It, therefore, appears completely reasonable that a large number of compound nucleus states which decay in accordance with statistical theory should be excited as the result of a coupling of dipole states (regardless of their nature) with these levels.

The experimental data gathered in the 1950's, at first glance, confirmed this picture completely. Thus, Weinstock and Halpern [76], comparing the relative probability of photoproton emission measured at $E_{\gamma\,max}$ = 24 MeV with that predicted by statistical theory, showed that the photoproton yields from medium nuclei could be explained by the evaporation model.

On the other hand, photoproton spectra observed for approximately the same energy $E_{\gamma\,max}$ (for example, from Cu [77], Co [78], and Mo^{92} [79]), as well as photoneutron spectra (Cu [77] and Bi [80]), proved to be low-energy spectra, and noticeable deviation from an evaporation spectrum was observed only in the high-energy region. Significant difference from evaporation spectra were found only in the cases of those nuclei where the photoproton emission cross section, which considerably exceeded that expected on the basis of the evaporation model, was nevertheless an insignificant part of the total γ-ray absorption cross section (Mo^{100} [79], In, Ce, Bi, Ta, and Pb [81]).

On the basis of these facts, the idea that the statistical theory explained the fundamental portion of cases of decay in medium-heavy and heavy nuclei, and that only 5-10% of the total yield of photonuclear reactions in the giant resonance region resulted from a mechanism other than evaporation, has received widespread acceptance and is actually with us at the present time (for example, see the monograph of Levinger [19] and the review by Wilkinson [20]).

However, it is difficult to reconcile with such a picture a whole set of experimental data, both previously available and recently obtained, which has not received proper evaluation. For example, consider the following data.

Experiment of Sagane [82]. A fundamental hypothesis of statistical theory is the one stating that the decay of a compound nucleus is independent of the method of formation. One of the few experiments to check this hypothesis is the well known experiment by Goshal [83], who measured the dependence of the cross sections for the (x, n), (x, np), and (x, 2n) processes on the excitation energy E_C^* of the compound nucleus Zn^{64*} produced in the reactions

$$\alpha + Ni^{60} = Zn^{64*};$$

$$p + Cu^{63} = Zn^{64*}.$$

Sagane measured the same excitation functions for the γ-ray reaction

$$Zn^{64} + \gamma = Zn^{64*}.$$

One can see from Fig. 3, which was adapted from [83], that the data of Goshal favors the independence hypothesis in the case of the decay of Zn^{64*} which is produced by heavy particles, but the behavior of the cross section curve for the (γ, n) reaction contradicts that hypothesis.

Contradictory Results Involving the (γ, p) Reaction in Copper. The yield, spectral,and angular distributions of photoprotons from copper for $E_{\gamma\,max}$= 24 MeV (with the exception of a small group of fast particles amounting to 5-10% of the yield) are in agreement with the predictions of statistical theory according to the data in [76, 77]. However, in the work of Leikin, Ratner, and the author [14], where measurements were made of the angular and energy distributions from copper for several values of $E_{\gamma\,max}$, it was found that the photoproton spectra resembled an evaporation spectrum. At the same time, however, it turned out that:

1. A significant anisotropic part was observed in the angular distribution of photoprotons for $E_{\gamma\,max}$ = 19 and 24 MeV amounting to approximately 40% of the yield and which could in no way be the result of evaporation.

2. The behavior of the yield curve as a function of $E_{\gamma\,max}$ contradicts the statistical theory.

3. If it is assumed that 40% of the yield for E_{γ} max = 19 MeV results from a mechanism other than evaporation, then from a comparison of the dependence of photoproton yield on $E_{\gamma\,max}$ predicted by statistical theory with that observed, there follows the conclusion that the fraction resulting from this mechanism increases with

increase in E_γ max, and reaches 50-70% for E_γ max = 28 MeV. This estimate, which is based on the anisotropic part of the yield, determines the lower limit of the contribution from processes other than evaporation because these processes can also contribute to the isotropic part.

The Fraction of (γ, 2n) Processes in the Total Cross Section for Neutron Emission. Consideration of the cross sections for the (γ, n) and (γ, 2n) reactions in several heavy nuclei has shown that it is necessary to assume that 30-50% of $\sigma_{\gamma\,neut}$ results from a mechanism other than evaporation [84] in order to explain the observed fraction of (γ, 2n) reactions in the total cross section for the emission of neutrons ($\sigma_{\gamma\,neut} = \sigma_{\gamma,n} = \sigma_{\gamma,2n}$).

Photoneutron Spectra. Measurements of photoneutron spectra made after 1960, which are more accurate than earlier measurements, have shown that there is a peak in the region of neutron energies $\varepsilon_n > 4$ MeV which amounts to several tens of percent of the neutron yield and which cannot be explained by the evaporation model [85, 86]. In addition, the behavior of the spectrum in the low-energy region evidently cannot be fitted by means of acceptable values for the nuclear temperature [86-88].

Estimates made in [86] show that it is necessary to assume about half the photoneutron yield results from a mechanism other than evaporation in order to explain the photoneutron spectrum from In^{115} which is observed for $E_{\gamma\,max} = 28$ MeV.

Summing up, one can point out that a large amount of data indicates the role of processes other than evaporation is evidently important in photoreactions. On the other hand, there is the idea that evaporation plays a fundamental part. However, the latter conclusion follows from estimates made about 10 years ago based on quite imprecisely established model parameters.

The establishment of the true role of the evaporation process in photonuclear reactions is of considerable interest. In this connection, it is extremely useful to carry out a detailed comparison of the data from photonuclear reactions with the predictions of statistical theory, using up-to-date model parameters for the calculations. Such a comparison is made in Chapter IV for (γ, p) reactions in medium-weight nuclei.

CHAPTER II

Description of the Experiment

1. Photoproton Detection Technique

In this work, photoprotons were detected by means of thick-layer emulsions. Despite the labor involved, this method played an important role, until very recently, in the study of the angular and energy distributions of photonuclear reaction products, and particularly of photoprotons. The widespread use of thick-layer emulsions for the study of photonuclear processes is chiefly the result of the fact that emulsion assures 100% discrimination of protons from electrons whereas investigators working at electron accelerators and using any other method are faced with tremendous problems in distinguishing proton pulses from pulses produced by pileup of a large number of electron background pulses. The effect of instrumental background from heavy charged particles is also successfully reduced to a minimum by working with emulsion because only those tracks in the emulsion are visually selected which are located in an angular range corresponding to particle emission from the target while in counter work, the solid angle for the background is always greater than the solid angle for the signal (except in the case of a 4π-geometry).

This method is convenient and economical from the point of view of expenditure of accelerator operating time because one can obtain proton angular and energy distributions simultaneously, the latter being determined with sufficient accuracy if the emulsion thickness and the experimental geometry permit one to avoid tracks going completely through the emulsion. Another favorable feature of the thick-layer emulsion technique is the separation of irradiation time and measurement time; there is also the simplicity in experimental setup.

This technique is rather effective for the study of (γ, p) reactions in those nuclei where the photoproton yield is comparatively large. In the case of low yields of the particles under study, it is difficult to obtain

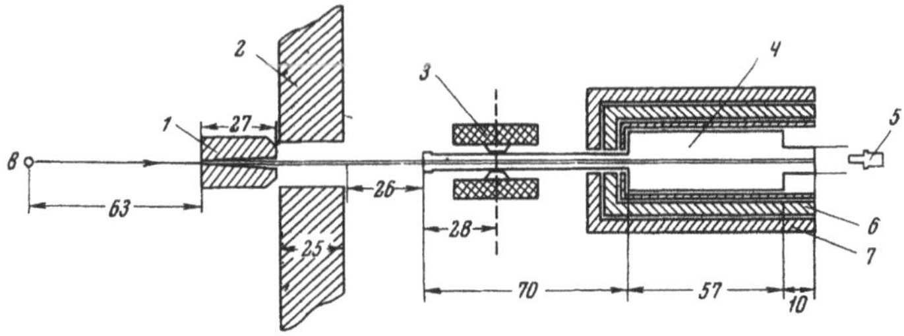

Fig. 4. Diagram of the arrangement of the apparatus (dimensions shown in centimeters). (1) Lead collimator; (2) lead shield wall; (3) clearing magnet; (4) plate exposure chamber; (5) monitor; (6) paraffin; (7) lead; (8) synchrotron target.

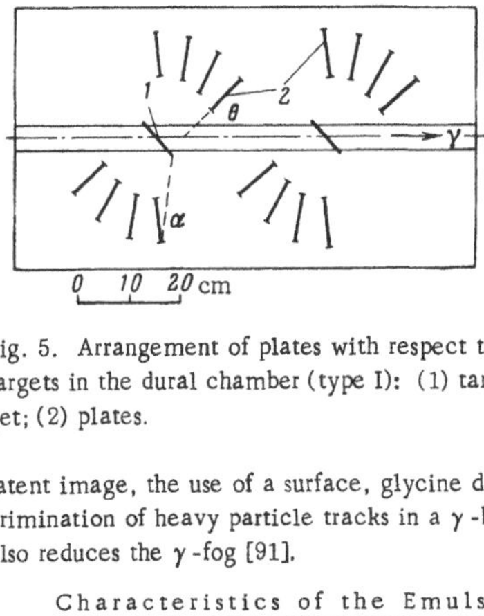

Fig. 5. Arrangement of plates with respect to targets in the dural chamber (type I): (1) target; (2) plates.

high statistical accuracy because the plates must be arranged to subtend small solid angles at the target in order to avoid the appearance of straight-through tracks, and it is impossible to use very large irradiation doses because they produce a dense, granular fog from scattered γ-rays and electrons which hinders scanning.

The following methods are used to reduce the fogging: removal of electron admixture from the γ-radiation; the use of emulsions insensitive to the γ-background; various methods for discriminating fog and heavy particle tracks based on the fact that the nature of the latent image in emulsion grains ionized by light and heavy particles is, in all probability, different [89]. Because light particles form a deep latent image and heavy particles a surface latent image, the use of a surface, glycine developer [90], or simply underdevelopment [89], improves the discrimination of heavy particle tracks in a γ-background fog. There are indications that the Herschel effect also reduces the γ-fog [91].

Characteristics of the Emulsions Used. Locally-made NIKFI Ya-2 and NIKFI T-3 emulsions 300 and 400 μ thick were used in this work. For good-quality proton tracks, these emulsions are less sensitive to γ-background than the Ilford C-2 emulsion. We performed control experiments which showed that the NIKFI Ya-2 emulsion allowed 2-3 times greater exposure for a given fog density. The NIKFI T-3 emulsion was even less sensitive to γ-background. In addition, it is recommended for its ability to undergo storage in a vacuum (10^{-1} mm Hg) for several days without peeling. The composition of the NIKFI Ya-2 emulsion we used resembled that of the Ilford C-2 emulsion. The NIKFI T-3 emulsion had a lower density – 3.50 g/cm^3. The AgBr content was 79%. The corresponding values for dry Ilford C-2 emulsions are 3.92 g/cm^3 and 86% [92].

Emulsion Development. The emulsions were developed in amidol developer using the dry heat development method recommended by NIKFI. A unique departure was that we slightly underdeveloped the emulsion in order to get better discrimination of protons from the grain background by reducing the temperature of the heated stage from 27 to 22-23°C; in the case of particularly long exposures, we used in amidol developer devised at FIAN by T. A. Romanova which has the following composition: amidol – 4.5 g, boric acid – 35 g, anhydrous sulfite – 18 g, potassium bromide – 0.8 g, water to make 1 liter. This developer has better capability than the standard amidol developer for discrimination of proton tracks from fog.

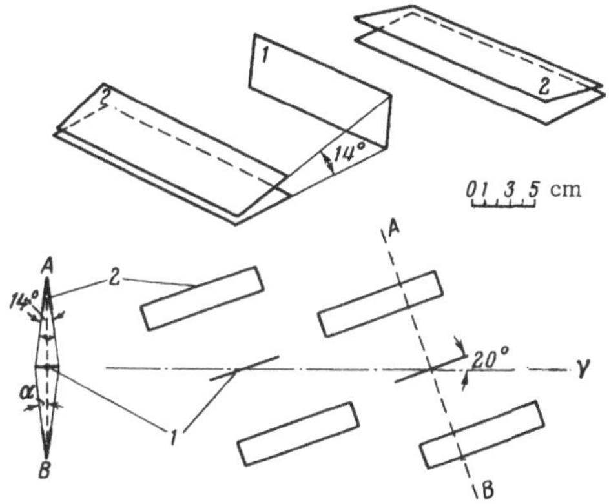

Fig. 6. Plate arrangement with respect to targets in the polystyrene-graphite chamber (type II): (1) target; (2) plates.

2. Arrangement of the Apparatus

A diagram of the arrangement of the equipment is shown in Fig. 4. The γ-ray beam from the synchrotron target was defined by a mobile lead collimator 27 cm thick having a conical opening with an entrance diameter of 8 mm. The angular spread of the beam defined by the collimator was 0.013 radians or 0.73°. In order to avoid γ-ray scattering at the collimator exit, the divergence of the conical opening of the collimator was somewhat greater than the divergence of the beam. The collimator was located in front of a solid lead shield 25 cm thick.

The chamber where the targets and plates were located was immediately behind the collimator. The chamber was fastened on a movable stand, which allowed displacements in all the necessary degrees of freedom. Preliminary establishment of the alignment of γ-beam, collimator, and chamber was made visually with the help of special equipment. Precise setting was accomplished by means of beam images on x-ray film located directly outside the collimator and at the target position. Checking of alignment was done several times during each series of irradiations.

In order to reduce the background from electrons mixed in with the γ-ray beam, the front fitting of the chamber was located between the poles of an electromagnet with a field strength of approximately 3000 Oe.

For additional protection against background from scattered radiation, the chamber was surrounded on all sides by layers of lead (5 cm) and paraffin (10 cm).

A monitor was located on the beam axis beyond the chamber; it was a thin-walled, integrating, ionization chamber for the determination of γ-ray dose. The monitor was calibrated by means of the current in an absolute, thick-walled aluminum chamber. Its operational stability was checked by means of a standard Co^{60} source.

3. Emulsion Exposure Chamber and Arrangement of Plates with Respect to the Target

The chamber was a hollow, duraluminum cylinder with a wall thickness of 0.5 cm, a length of 50 cm, and an internal diameter of 20 cm. In order to avoid contamination of the plates by particles from photonuclear reactions in air nuclei and to eliminate additional proton energy loss, the chamber air was pumped down to a pressure of 0.1 mm Hg. A vacuum gauge and valved tubing which connected the chamber interior with a forepump were located on the removable front and back faces of the cylinder. The γ-beam passed

TABLE 1. Target Characteristics

Enriched isotope	Isotopic computation, %	Impurities, %	t_0, mg/cm²	S, cm²	ΔR_{equ}, μ
Ni_{28}^{58}	Ni^{58}—98.6 Ni^{60}— 1.0 Ni^{61}— 0.2 Ni^{62}— 0.1 Ni^{64}— 0.1	Traces of Fe, Cu	4.6	12	4.6
Ni_{28}^{60}	Ni^{58}—21.2 Ni^{60}—77.9 Ni^{61}— 0.4 Ni^{62}— 0.5 Ni^{64}— 0.1	Fe — 0.001 Cu $<$ 0.001 Mg — 0.01 Sn — 0.01	3.1	12	3.0
Ni_{28}^{64}	Ni^{58}— 9.8 Ni^{60}— 6.9 Ni^{61}— 0.5 Ni^{62}— 5.0 Ni^{64}—77.8	Light nuclei: K, Na, Mg, Ca, Al, S — 0.075; Medium nuclei: Cr, Mn, Fe. Co; Cu — 0.046; Sn $<$ 0.001; Pb $<$ 0.03	3.5	12	3.2
Cu_{29}^{65}	Cu^{63}— 6.3 Cu^{65}—93.7	$<$0.01	27.2	1.6 (I) * 11.1 (II) *	25
Natural mixture of isotopes Zn_{30}	Zn^{64}—48.89 Zn^{66}—27.81 Zn^{67}— 4.11 Zn^{68}—18.56 Zn^{70}— 0.62	Chemically pure	25	1.6	22
Nb_{41}^{93} – Natural mixture of isotopes	Nb^{93}—100	Chemically pure	27.7	12	21
Cd_{48}^{114}	Cd^{110}— 0.3 Cd^{111}— 0.4 Cd^{112}— 1.2 Cd^{113}— 4.3 Cd^{114}—93.0 Cd^{116}— 0.8	Fe, Cu $<$ 0.05	29.7	1.6 (I) 12 (II)	20
Sn_{50}^{114}	Sn^{112}— 0.6 Sn^{114}—57.2 Sn^{115}— 3.3 Sn^{116}—19.6 Sn^{117}—10.8 Sn^{118}— 6.8 Sn^{119}— 0.4 Sn^{120}— 1.3 Sn^{122}— 0.1 Sn^{124}— 0.1	Fe, Cu $<$ 0.05	19.9	9.0	14

* Number I and II in parentheses indicate type of experimental geometry.

TABLE 1 (Continued)

Enriched isotope	Isotopic computation, %	Impurities, %	t_0, mg/cm²	S, cm²	ΔR_{equ}, μ
Sn^{124}_{50}	$Sn^{112,\,114-117}$—0.1 Sn^{118}— 0.8 Sn^{119}— 0.4 Sn^{120}— 1.5 Sn^{122}— 1.0 Su^{124}—96.3	Cu — 0.06 Fe — 0.004	30	12	19
Sb^{121}_{51}	Sb^{121}—18.5 Sb^{123}—81.5	Traces of iron, polystyrene, $(CH_2)_n$	Sb — 15.8; $(CH_2)_n$ — 9.4	12	29
Sb^{123}_{51}	Sb^{121}—18.5 Sb^{123}—81.5	Traces of iron, polystyrene, $(CH_2)_n$	Sb — 13.6; $(CH_2)_n$ — 8.5	12	26

through entrance and exit apertures covered by Al foil 100 μ thick which were located rather far from the central portion of the chamber by means of long fittings in order to avoid the appearance of a proton background from reactions in the window material.

The photographic plates and targets were kept in position by means of special holders on movable bases. This made it possible to change targets and plates without moving the chamber and without disturbing the alignment of the equipment. Two targets could be irradiated simultaneously in the chamber.

Two different types of construction were used for the movable parts of the chamber. In type I (Fig. 5), the entire internal portion of the chamber, including target and plate holders, was made of duraluminum. The target, in a special frame, was fixed at an angle of 45° to the beam. In order that the frame not be struck by the beam, the dimensions of the target had to be considerably greater than the dimensions of the beam; angular distribution was measured by means of several photographic plates 3 × 4 cm in size arranged at angles of 30°, 50°, 70°, 90°, 110°, 130°, and 150° with respect to the direction of the beam.

Subsequently, the internal portion of the chamber (type II) was made of polystyrene (CH_2) and graphite, and the relative positions of targets and plates were changed (Fig. 6). The 2 × 6 cm targets were held by nylon threads at an angle of 20° to the beam in such a way that they were located entirely within the beam.

Angular distribution was determined by means of two plates 2.5 × 11.5 cm in size symmetrically located with respect to the targets so that one of them could measure proton yields in the angular range from 10° to 120°, and the other in the range from 60° to 170°.

When compared with type I, type II exhibited a number of advantages:

1. The reduction in the amount of material in the immediate vicinity of the beam and the use of polystyrene and graphite having a high (γ, p) reaction threshold (B_p = 16 MeV for C^{12} and 16.2 MeV for C^{13}) as construction material made it possible to reduce the magnitude of the proton background by almost an order of magnitude. This circumstance significantly improved the signal-to-noise ratio for (γ, p) reaction studies in nuclei where the proton yield is comparatively small (Ni^{64}, Nb^{93}, and nuclei in the neighborhood of Z = 50).

2. The geometry of type II ensured the determination of angular distributions with great accuracy. One can show that the principal inaccuracy in the determination of solid angle in both case I and case II is associated with the inaccuracy of the angle α (see Figs. 5 and 6). An estimate shows that the error in solid angle may amount to 10-20% because of an uncertainty Δα = ± 1°.

In the case of type I, there are two reasons for an error in the angle α: inaccurate centering of the beam, which leads to a displacement of the irradiated portion of the target because the dimensions of the target are larger than those of the beam; inaccurate positioning of the plate holders leading to rotation with respect to point 0.

The first cause leads to the production of a systematic error in the relative solid angles of the plates located above and below the targer, as a result of which the shape of forward-backward angular distribution may be distorted; the second cause leads to uncontrolled spread of the points corresponding to proton yields at different angles not associated with statistical errors.

In the case of type II, the error associated with inaccurate centering of the beam is generally absent because the dimension of the target is less than that of the beam. An error in the angle α because of inaccurate positioning of the holder leads to an identical relative error in the magnitude of the solid angle over the entire range of the angle θ covered by a given plate. Therefore, an additional spread in the points, unconnected with statistics, is not produced in the angular distribution. A systematic error in the forward-backward angular distribution which arises because the angle α for plates located in opposite directions from the target is not the same can easily be evaluated and eliminated by a comparison of the proton yields in the $\theta = 70\text{-}110°$ range on both plates.

3. The relative location of targets and plates in the case of type II is also more suitable from the point of view of corrections for proton energy loss in a target because the effective half-thickness of the target over a wide range of angles θ is close to half the true thickness of the target $t/2$. In the case of type I, the effective half-thickness is considerably greater than $t/2$ for a large fraction of the angles, and it is particularly large for plates located at the angle $\theta = 90°$.

4. Target Preparation and Characteristics

A large part of the data obtained in this work was obtained by irradiation of targets enriched in particular isotopes. The targets were thin, rectangular samples with areas of 2 × 12 cm or 2 × 6 cm, depending on the construction of the exposure chamber (type I or type II, respectively).

In the case of sufficiently soft materials, the targets were prepared by rolling with special rollers. Foils enriched in the isotopes Cd^{114}, Sn^{114}, Sn^{124}, were obtained in this manner as were foils of Zn (natural mixture of isotopes) and Nb^{93}. The minimum thickness of the foils was limited by the construction of the rollers and was 25-30 μ.

Targets enriched in the nickel isotopes Ni^{58}, Ni^{60}, and Ni^{64} were prepared by electrolytic deposition of nickel on a thoroughly cleaned and polished stainless steel cathode according to the recommendations made in [93]. However, because we required rectangular samples, the construction of the electrolytic bath was changed from that described in [93].

Apparently the shape of the electrodes has an important effect on the achievable thickness of the deposited layer. Thus we did not succeed in obtaining rectangular samples 2 × 6 cm in size which were thicker than 5 μ, while the authors [93] obtained samples in the form of discs with thicknesses up to 20 μ. This is possible because the stresses leading to rupture of the deposited layer with increase in thickness may be considerably greater in the case of rectangular electrodes.

It was impossible to use either rolling or electrolytic deposition to obtain targets enriched in antimony isotopes (Sb^{121} and Sb^{123}). The method of vacuum coating on a thin backing, normally used for the preparation of samples of such easily volatilized materials as antimony, was also not suitable because there was the risk of losing the small amount of material at our disposal. It was, therefore, decided to prepare the targets from antimony powder, using polystyrene $(CH_2)_n$ as a binder.

The presence of foreign matter in a target is undesirable. However, the threshold for proton emission from carbon is large ($B_p \sim 16$ MeV) in comparison with the proton threshold for antimony ($B_p \approx 6$ MeV). One can, therefore, anticipate that the observed effect will not be perturbed by a photoproton background from carbon for γ-ray energies that are not too high. The minimum amount of polystyrene needed to obtain durable targets was determined in preliminary experiments with a natural mixture of antimony isotopes. It turned out to be two parts of polystyrene by weight to three parts of antimony powder.

The technique for preparing the targets was extremely simple. The antimony powder was mixed with the required amount of a 10% solution of polystyrene in dichlorethane. The resulting mixture was poured into a horizontally arranged form made of photographic film base, which is insoluble in dichlorethane. The form was placed in a fume hood. After the dichlorethane evaporated, the target could be easily removed from the form.

The characteristics of the targets irradiated, namely their isotopic content, chemical purity, thickness t_0, irradiated area S, proton range in emulsion ΔR_{equ} equivalent to the range in a target half-thickness, etc., are given in Table 1.

5. Emulsion Scanning and Selection and Measurement of Tracks

The emulsion was examined under MBI-2 binocular microscopes having 60 × immersion objectives and 5 × oculars. For scanning the 2.5 × 11.5 cm plates used in type II, the standard microscope stages was replaced by a stage permitting the scanning of large areas. In the scanning process, tracks starting at the surface of the emulsion were selected.

For track measurement, we used a 7 × ocular supplied with a scale for measuring the length of the track projection in the plane of the emulsion and with a goniometer for measuring the angle θ_{meas}, which characterizes the direction of the track in the plane mentioned. The angle θ_{meas} was determined from the tangent to the beginning of the track. The angle of inclination φ of the track in the emulsion was determined, in the case of short tracks (with projected length $l < 100\ \mu$), from the total track depth h in the developed emulsion by means of the relation

$$\varphi = \arctan \frac{Dh}{l}, \tag{2.1}$$

where D is the coefficient for emulsion shrinkage during development; it is approximately 2.3.

In the case of long tracks ($l > 100\ \mu$), the angle φ was determined not from the total depth h but from the depth of the track at the point where its projection equaled 100 μ in order to reduce the effect of multiple scattering on the accuracy of the subsequent selection process

$$\varphi = \arctan \frac{Dh_\varphi}{100}. \tag{2.2}$$

The depth h, or h_φ, was determined by means of a linear displacement gage fastened to the microscope tube. In this way, we managed to avoid errors in the determination of h caused by play in the micrometer screw which moved the microscope tube in the vertical direction.

The residual range R_{res} of the proton in the emulsion was determined from the length l of its projection and the angle φ;

$$R_{res} = \frac{l}{\cos\varphi}. \tag{2.3}$$

Tracks with angles θ_{meas} and φ corresponding to emission from an irradiated region of the target and with a residual range greater than 40 μ were selected for ultimate analysis.

6. Determination of Proton Energy ε_p

Total Proton Range $R(\varepsilon_p)$. The total range R is equal to the sum of the measured residual range R_{res} in the emulsion and some additional range $\Delta R(t, \theta_{meas}, \varphi)$ equivalent to the proton range in the target (t is the depth at which the proton was produced).

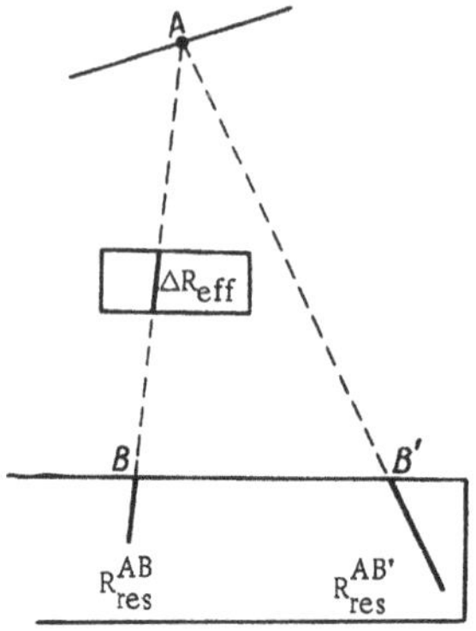

Fig. 7. Effect of target on background tracks.

Assuming that all the protons were produced in an infinitely thin layer at the half-thickness of the target, it is easy to obtain the following expression for the effective proton range in the target:

$$t_{\mathrm{eff}} = \frac{t_0}{2} \frac{1}{\sin\theta_{\mathrm{meas}} \cos\varphi \sin\beta + \sin\varphi \cos\beta}, \tag{2.4}$$

where t_0 is the target thickness, and β is the angle between the plane of the emulsion and the plane of the target.

Conversion of the effective range in the target to range in the emulsion was carried out in accordance with well-known relations for ionization loss. As a result, we found (see Appendix)

$$\Delta R(\theta, \varphi) = c \frac{Z^{1/2}}{A} t(\theta, \varphi), \tag{2.5}$$

where Z and A are the atomic number and weight of the target material and the constant C = 20.9 or 22.1, respectively, for the NIKFI Ya-2 or T-3 emulsions we used if the range R is expressed in microns and t is in milligrams per square centimeter.

Numerical values for the equivalent ranges corresponding to the half-thicknesses of the irradiated targets are given in Table 1.

Range – Energy Curves for the NIKFI Ya-2 and T-3 Emulsions. The proton energy $\varepsilon_p(R)$ was determined from range-energy curves.

At the time when this work was being done, there was no published data for the dependence of range on energy in the NIKFI Ya-2 and T-3 emulsions. The range-energy curve for the Ilford C-2 emulsion was, therefore, used for the NIKFI Ya-2, emulsion, the density of which was close to that of the Ilford emulsion. Such a curve for dry Ilford C-2 emulsion was measured by El Bedewi [94] for proton energies up to $\varepsilon_p = 5$ MeV, by Rotblat [95] for energies up to $\varepsilon_p = 15$ MeV, and by Catala and Gibson [96] up to 16 MeV. The accuracy of range determination in all these papers was no worse than 1%. There are also data by Bradner et al. [97], who measured the range-energy curve up to 40 MeV. However, the latter data are inaccurate because of deficiencies in the control of moisture content in the emulsion. The results of Gibson, Prowse, and Rotblat [98], which were obtained for an Ilford C-2 emulsion with a density of 3.94 g/cm^3, are distinguished by the greatest accuracy (to 0.3%). They determined the range-energy dependence up to a proton energy $\varepsilon_p = 21$ MeV. We used the Rotblat data [95, 98].

It was impossible to use the range-energy curve for Ilford C-2 emulsion in the case of the T-3 emulsion. The T-3 emulsion was calibrated with recoil protons from 14-MeV neutrons in order to determine the necessary correction. It was found that the range of protons of this energy in NIKFI T-3 emulsion was greater than the corresponding range in Ilford C-2 emulsion by 6 ± 3%.* In plotting the range-energy dependence, it was measured that the correction was constant over the entire proton energy range, and that the range should be increased by 6% for the NIKFI T-3 emulsion.

*This result agrees within experimental error with the data of Anashkina for the determination of the range of recoil protons with an energy of 14.2 ± 0.1 MeV [99], and with estimates made on the basis of a comparison of the densities of the Ilford C-2 and NIKFI T-3 emulsions. The latter indicated that the range in T-3 emulsion should be approximately 10% greater than in C-2 emulsion.

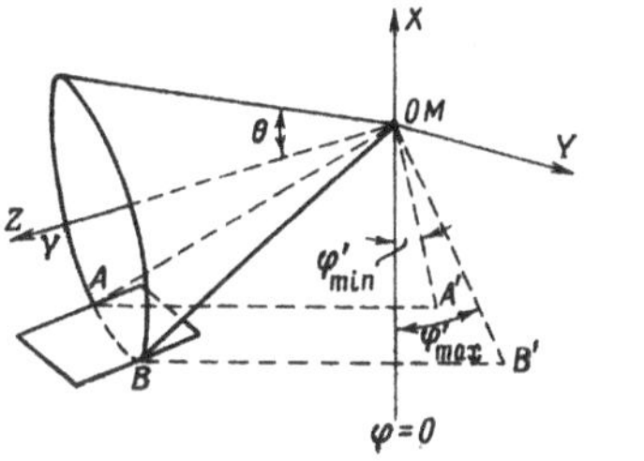

Fig. 8. Illustrating the calculation of the solid angle $\Omega(\theta)$.

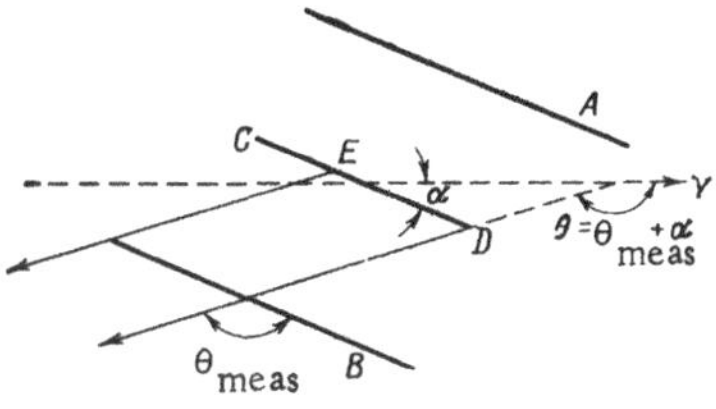

Fig. 9. Illustrating the effectively used target area $S(\theta)$. $ED = S(\theta)$; $CD = S$.

7. Background

Sources of Background. Background tracks of charged particles observed in the emulsion were produced for a number of reasons.

First, if discrimination of proton tracks from tracks of other charged particles is not made, the background arises from: (1) (γ, d) and (γ, α) reactions in target nuclei (background A); and (2) contamination by natural α activity (background B).

Second, there is the background produced by scattered γ-rays. This background is produced by: (1) (γ, p), (γ, α), and (γ, d) reactions in the walls of the exposure chamber, plate holders, etc. In this case, the particles are incident upon the plate from the outside (background C); and (2) the same reactions in emulsion nuclei: the particles produced within the emulsion escape outward in the required range of angles. In this case, the direction of the track can be determined for fast protons from the ionization density, while it becomes difficult to tell whether the beginning or end of a track is located in the emulsion in the case of slow protons (background D).

Finally, a third source of background is the neutron flux accompanying the γ-beam. It gives rise to: (1) (n, p), (n, d), and (n, α) reactions in target nuclei (background E); (2) the same reactions in the chamber walls, etc. (similar to case C; background F); and (3) the same reactions plus recoil protons in the emulsion itself (similar to case D; background G).

Determination of Background. The results given below were corrected for the background $\Phi_1 = B + C + D + F + G$, which is not associated with reactions in the target. This background can be measured rather simply by one of two methods. The first method consists in the carrying out of special irradiations without targets. The second method is to consider tracks having arbitrary direction in scanning the experimental plates and to plot the dependence of track number on the angle θ_{meas} of the track projection on the plane of the emulsion and on the slope angle φ with respect to this plane over the entire angular range $\theta_{meas} = 0°\text{-}360°$ and $\varphi = 0°\text{-}90°$. Then an extrapolation is made of the background which is observed outside the limits of the region of angles θ_{meas} and φ corresponding to a direction from the target into this very region.

It should be pointed out that the background Φ_1 is noticeably reduced in comparison with that found without a target when a comparatively thick target is used for the measurements. The effect of a target on background tracks is illustrated in Fig. 7.

In order to strike the plate within the acceptable angular range, background particles must necessarily pass through the target, thereby losing a part of their energy. As a result, a particle with range $R_{\Phi}(\varepsilon)$ is detected as a particle having a residual range $R_{res} = R_{\Phi}(\varepsilon) - 2\Delta R_{equ}(\theta, \varphi)$, where $\Delta R_{equ}(\theta, \varphi)$ is the range in emulsion defined above as equivalent to the effective half-thickness of the target. Hence it follows that: 1) background tracks from a region where $R_{\Phi}(\varepsilon) \leq 2\Delta R_{equ}$ ($R_{res} = R_{\Phi} - 2\Delta R_{equ} \leq 0$ for them) will be completely stopped in the target and will generally not reach the experimental plates; 2) one should introduce a correction to the number of acceptable tracks with residual range R_{res} to allow for background tracks with initial range $R_{\Phi} = R_{res} + 2\Delta R_{equ}(\theta, \varphi)$.

TABLE 2. The Correction $G(\theta_1, \theta_2)$ for Type II Geometry (Fig. 6)

Angular range $\theta_1 - \theta_2$, deg		$G(\theta_1, \theta_2)\cdot 10^3$	Angular range $\theta_1 - \theta_2$, deg		$G(\theta_1, \theta_2)\cdot 10^3$
Plate A	Plate B		Plate A	Plate B	
70—80	110—100	1.97	120—130	60—50	9.37
80—90	100—90	4.39	130—140	50—40	8.77
90—100	90—80	6.75	140—150	40—30	6.92
100—110	80—70	8.88	150—160	30—20	3.66
110—120	70—60	9.68			

The determination of the background from the accompanying reactions in the target (backgrounds A and E) is a rather complicated problem, and corrections were not made for this background. However, the error introduced by this background is evidently small as follows from the discussion below.

Generally speaking, one can determine the contribution from neutron reactions in the target (background E) by using data on the cross sections for these reactions and by measuring the flux of neutrons through the target and their mean energy in one way or another. However, the problem of measuring neutron fluxes and mean energies in an intense γ-ray beam is rather complicated. We, therefore, used an indirect method for estimating the background E; special irradiations were carried out beyond the lead wall without an opening in the collimator, both with and without a nickel target 50 μ thick. Under these conditions the γ-beam was absent, and the neutron flux, chiefly produced by (γ, n) reactions in the lead shield, was essentially unchanged. It is easy to see that the background $\Phi_2 = B + E + F + G$ was determined with the nickel target in position, and the background $\Phi_3 = B + F + G$ was determined without it. Therefore, the desired background E could be equated to the difference in backgrounds $\Phi_2 - \Phi_3$. Measurements showed that Φ_2 and Φ_3 were considerably less than the background Φ_1 with the γ-beam present, and the difference $\Phi_2 - \Phi_3 = E$ could generally be neglected in comparison with Φ_1. Byerly and Stephens [77] also arrived at the same conclusion concerning the insignificance of the background from neutron reactions in the target.

The contribution from (γ, α) and (γ, d) reactions (background A) was also negligibly small for the proton energies under consideration. According to data in [100-102], the yields of (γ, α) reactions are no more than 1-2% of the photoproton yield. The data on photodeuteron yield are contradictory: for example, Byerly and Stephens [77] found that the photodeuteron yield from copper was 30% of the photoproton yield for $E_{\gamma\,max} = 24$ MeV while the photodeuteron yield proved to be negligibly small in the case of the neighboring Co^{59} nucleus [78]. A noticeable photodeuteron yield from copper was also observed in [103, 104]. However, Forkman found not very long ago [105] that the photodeuteron yield from copper, as determined by magnetic analysis, was practically nonexistent for $E_{\gamma\,max} = 31$ MeV. It was 0.00 ± 0.03 of the photoproton yield. The results in [77, 103, 104] are in error because the authors of these papers identified protons and deuterons by the grain-count method, which is unsuitable for low-energy particles.

8. Determination of Solid Angles

In the case of type I geometry, the solid angles subtended by the scanned areas of the plates were calculated on the assumption of a point source located at the center of the irradiated portion of the target. To check on this, the relative solid angles of the various plates was calibrated by means of a polonium α source located at the target position and having the shape of the irradiated portion of the target. The results agreed within the calibration accuracy of ± 5%.

The type II geometry presented the case of extended and closely located detector and source. The following problems arise in the calculation of the geometry factor for this case:

1. It was necessary to transform from the measured parameters of track direction θ_{meas} and φ with respect to the plate to its azimuthal direction θ with respect to the γ-beam.

2. It was necessary to calculate the solid angle subtended by the detector in some angular range $\theta_1 < \theta < \theta_2$ under consideration, taking each elemental target area dS_M as a point source:

$$\Omega(\theta_1, \theta_2, dS_M) = \int_{\theta_1}^{\theta_2} \int_{\varphi'_{min}(\theta)}^{\varphi'_{max}(\theta)} \sin\theta \, d\theta \, d\varphi', \tag{2.6}$$

where θ and φ' are the polar coordinates of the detector in a system the origin of which coincides with the position of the elemental target area dS_M in which the Z-axis is directed along the beam. It is evident that the range of angles $\Delta\varphi'(\theta) = \varphi'_{max}(\theta) - \varphi'_{min}(\theta)$, which is subtended by the detector in a cone with half-angle θ and with its vertex in the area element dS_M, depends on the shape and dimensions of the detector and on its position with respect to dS_M (Fig. 8).

3. One had to calculate the fraction of the target area which is effectively used in the angular range $\theta_1 < \theta < \theta_2$,

$$K(\theta_1, \theta_2) = \frac{S(\theta_1, \theta_2)}{S_M}, \tag{2.7}$$

where $S(\theta_1, \theta_2)$ is the portion of the target from which particles emitted in the angular range $\theta_1 < \theta < \theta_2$ can be recorded by the detector (S_M is the total area of the irradiated target).

From Fig. 9, where the case of plane-parallel detector and radiator is shown for simplicity, it is clear that of the particles having a direction θ, only those emitted by the portion $S(\theta)$ of the target reach the detector.

As a result, we obtain the geometry factor $G(\theta_1, \theta_2)$ by means of which the photoproton yields the directions $\theta_1 < \theta < \theta_2$ can be normalized to the same solid angle or to the same number of target nuclei:

$$G(\theta_1, \theta_2) = K(\theta_1, \theta_2)\, \overline{\Omega(\theta_1, \theta_2)}, \tag{2.8}$$

where $K(\theta_1, \theta_2)$ is determined by formula (2.7) and $\overline{\Omega(\theta_1, \theta_2)}$ is the value of the solid angle averaged over the area $S(\theta_1, \theta_2)$; it is given by

$$\overline{\Omega(\theta_1, \theta_2)} = \int_{S(\theta_1, \theta_2)} \frac{\Omega(\theta_1, \theta_2, dS_M)\, dS_M}{S(\theta_1, \theta_2)}. \tag{2.9}$$

Computation of $G(\theta_1, \theta_2)$ for our geometry was performed with the Ural-2 computer. The results of the calculation are given in Table 2.

The accuracy of the relative values of $G(\theta_1, \theta_2)$ were at least 1%. The absolute values of G were computed with an accuracy of ±(7-8)%.

9. Use of the Ural-2 Computer for Analysis of Experimental Data

A fundamental drawback of the photographic plate technique is the labor involved in emulsion scanning, track measurement, and subsequent analysis of the data, which requires the performance of a great number of operations for each track. Experience has shown that it requires six to seven weeks for a single individual to analyze 10,000 proton tracks even with maximum use of every sort of nomogram to accelerate significantly the process of analysis.

We used the Ural-2 computer to speed up the analysis of experimental data. The programs written made it possible to automate the entire data analysis, starting with track selection on the basis of direction and ending with classification into energy and angular intervals. As a result, the time involved in analysis (including that required for manual punching of a large amount of input data on cards) was reduced by a factor of 5-6.

CHAPTER III

Results

1. Photoproton Angular and Energy Distributions

Table 3 contains information about target irradiation conditions during measurements of photoproton angular and energy distributions. It gives the type and emulsion thickness in microns, the type of exposure chamber (geometry type), the upper limit of the bremsstrahlung spectrum $E_{\gamma\ max}$, the dose M in terms of monitor counts, the total scanned area S_{scan}, the number of selected tracks satisfying the geometric selection criteria n_{meas}, n_{Φ} the background Φ normalized to radiation dose and area scanned, and, finally, $N_p = n_{meas} - n_{\Phi}$ the number of tracks corresponding to the effect produced by the target.

Comparing the ratio of the desired effect N_p to the background n_{Φ} in the cases of irradiations 5 and 6 for Cu^{65} and irradiations 15 and 16 for Cd^{114}, which were performed at neighboring values of the energy $E_{\gamma\ max}$ but in different chambers, one can see that, other conditions being equal, the signal-to-background ratio was considerably better for the polystyrene-graphite chamber (type II geometry) than for the dural chamber (type I geometry).

Photoproton Spectra. Photoproton spectra summed over all plates and angles are shown in Figs. 10-20.

The energy distribution was constructed in the following manner. After the total range $R = \Delta R_{res} + R_{meas}$ was determined for every track arriving from the irradiated portion of the target, the dependence of track number on R was plotted. Then the range-energy curve was drawn on the range scale with intervals $\Delta\varepsilon_p$ marked off, and the number of tracks in each interval were counted, and the energy dependence was determined. The size of the interval $\Delta\varepsilon_p$ was 0.2, 0.5, or 1 MeV depending on the statistics.

TABLE 3. Irradiation Conditions

Irradiation No.	Target	NIKFI emulsion		Type of geometry	$E_{\gamma\ max}$, MeV	Dose, M	S_{scan}, cm^2	No. of tracks n_{meas}	Background n_{Φ},	Effect $N_p = n_{meas} - n_{\Phi}$
		Type	Thickness, μ							
1	Ni^{58} (enriched)	T-3	300	II	24.0	4 580	4.0	462	9	453
2	Ni^{60} »	T-3	300	II	24.0	13 340	4.0	436	17	419
3	Ni^{64} »	T-3	300	II	24.0	17 920	20.0	843	93	750
4	Cu^{65} »	Ya-2	400	I	17.9	43 300	34.2	1 183	43	1 140
5	» »	Ya-2	300	I	19.5	6 000	36,6	2 480	65	2 415
6	» »	T-3	370	II	20,3	10 590	20.0	8 106	41	8 065
7	» »	Ya-2	400	I	24.5	6 310	50,9	9 377	359	9 018
8	» »	Ya-2	400	I	25.1	8 000	26.8	5 492	250	5 242
9	» »	Ya-2	400	I	28.5	9 490	66.2	11 082	579	10 503
10	Zn (natural mixture of isotopes)	Ya-2	430	I	20,8	5 520	26.5	5 415	185	5 230
11	Zn (natural mixture of isotopes)	Ya-2	430	I	28.6	3 700	12,2	7 096	176	6 920
12	Nb^{93}	T-3	300	II	19.5	12 830	16,0	2 234	13	2 221
13	»	T-3	300	II	23.5	7 060	16,0	2 635	30	2 605
14	»	T-3	300	II	27.5	10 400	8,0	2 975	36	2 937
15	Cd^{114} (enriched)									
16	» »	T-3	370	II	20,3	10 590	78.0	1 526	104	1 422
17	» »	Ya-2	400	I	24,5	6 310	50,6	3 663	755	2 908
18	» »	Ya-2	400	I	28,5	5 780	43.6	4 000	971	3 029
19	Sn^{114} (enriched)	T-3	370	II	23,5	5 250	40	852	56	796
20	Sn^{124} »	Ya-2	370	II	23,5	12 310	68	1 112	188	824
21	Sb^{121} »	Ya-2	370	II	19.5	13 012		1 994	353	1 600
22	Sb^{123} »	Ya-2	370	II	19.5	13 012	80	1 470	394	1 076

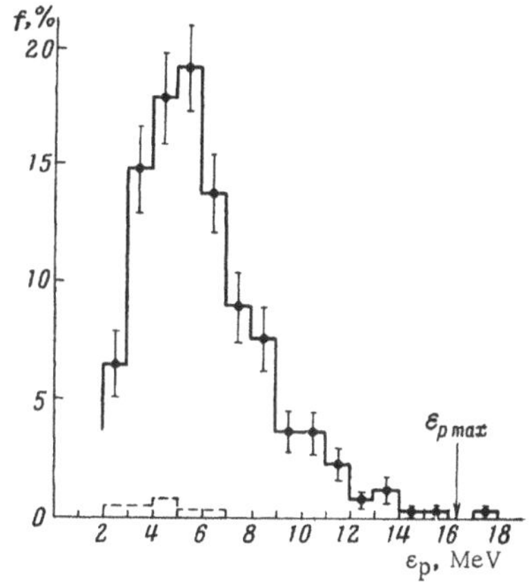

Fig. 10. Photoproton spectrum from Ni^{58} for $E_{\gamma\,max}$ = 24 MeV, N_p=453, $n_ф$ = 9. Dashed line is proton background spectrum.

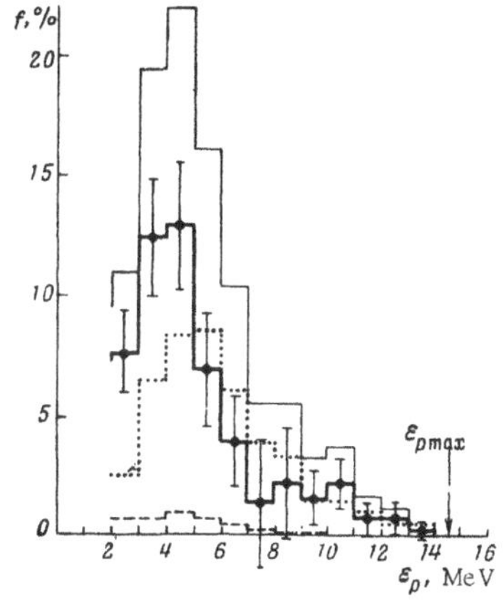

Fig. 11. Photoproton spectrum from Ni^{60} for $E_{\gamma\,max}$ = 24 MeV (heavy line); observed energy distribution for photoprotons from a target enriched in Ni^{60} (light line); proton background spectrum (dashed line); spectrum of photoprotons from Ni^{58} present in the target (dotted line).

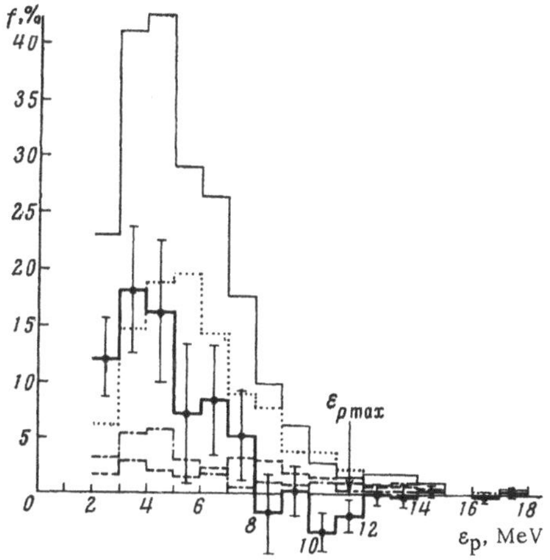

Fig. 12. Photoproton spectrum from Ni^{64} for $E_{\gamma\,max}$ = 24 MeV (heavy line); observed energy distribution for photoprotons from a target enriched in Ni^{64} (light line); proton background spectrum (dashed line); spectrum of photoprotons from Ni^{58} present in the target (dotted line); and from Ni^{60} (dot-dashed line).

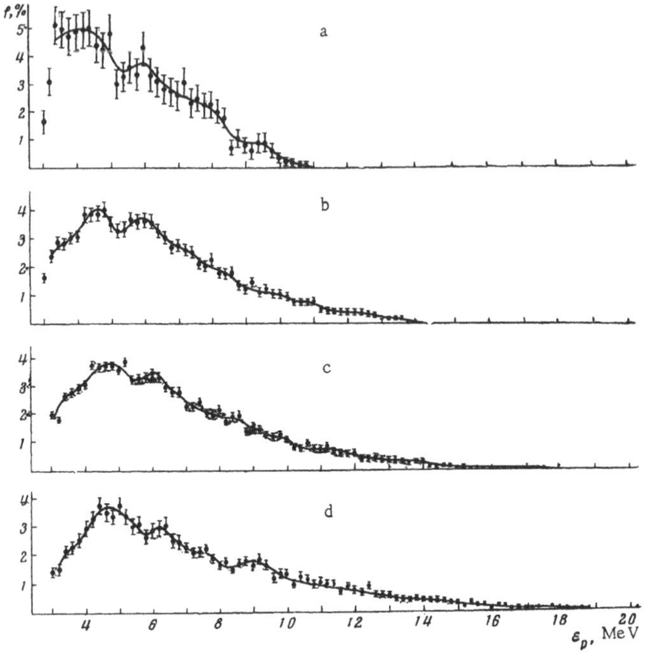

Fig. 13. Photoproton spectra from a target enriched in Cu^{65} for $E_{\gamma\,max} = 17.9$ MeV (a), 20 MeV (b), 24.5 MeV (c), and 28.5 MeV (d).

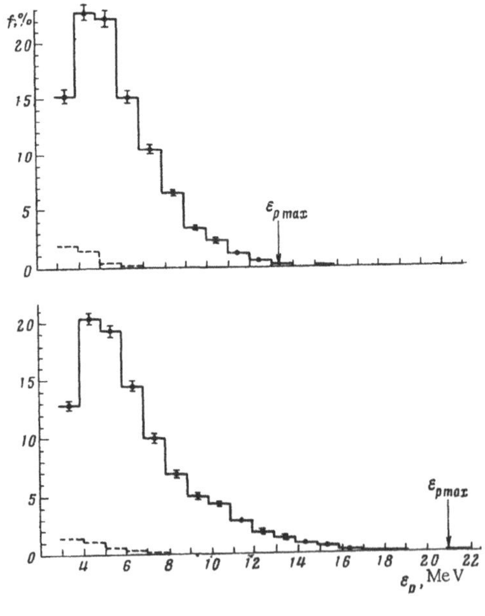

Fig. 14. Photoproton spectra from Zn (natural mixture of isotopes) for $E_{\gamma\,max} = 20.8$ MeV (above) and 28.6 MeV (below); proton background spectrum (dashed line).

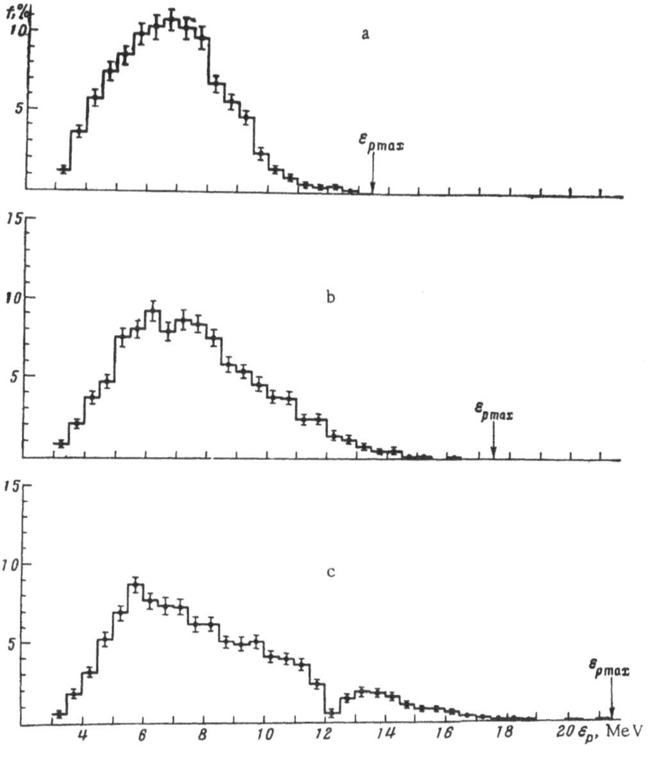

Fig. 15. Photoproton spectra from Nb^{93} for $E_{\gamma\ max}$ = 19.5 MeV (a), 23.5 MeV (b), and 27.5 MeV (c).

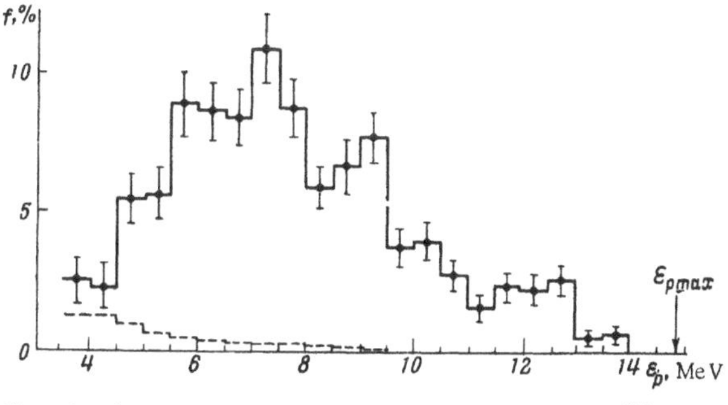

Fig. 16. Photoproton spectrum from a target enriched in Sn^{114} for E_γ max = 23.5 MeV (solid line); proton background spectrum (dashed line).

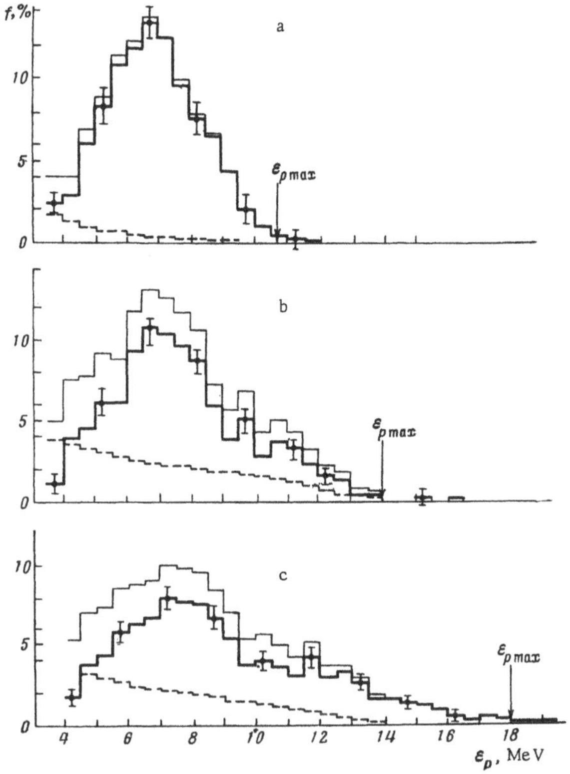

Fig. 17. Photoproton spectra from a target enriched in Cd^{114} (heavy line) for $E_{\gamma\ max}$ = 20.3 MeV (a), 24.5 MeV (b), and 28.5 MeV (c); observed energy distribution (light line); proton background spectra (dashed line).

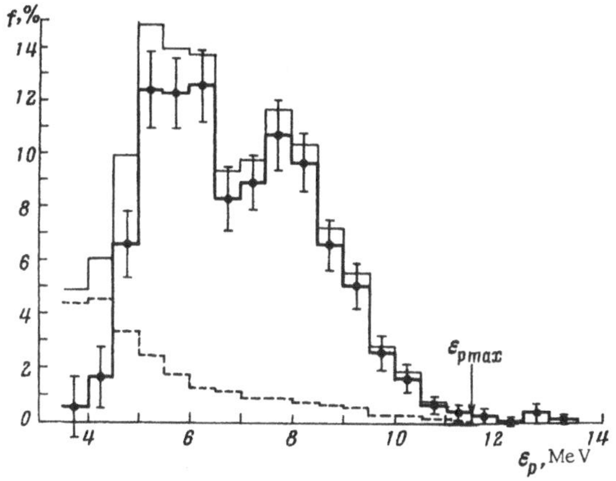

Fig. 18. Photoproton spectrum from a target enriched in Sn^{124} for $E_{\gamma\ max}$ = 23.5 MeV (heavy line); observed energy distribution (light line); proton background spectrum (dashed line).

TABLE 4.* Parameters for the Angular Distributions $f_1(\theta)$ and $f_2(\theta)$

Target	E_{γ} max, MeV	ε_p, MeV	b/a	c/b
Ni^{58}	24	≥2	0.10±0.20	1.30
		2–8	0.15±0.22	1.02
		≥8	−0.14±0.34	−0.07
Ni^{60}	24	≥2	1.3±1.0	0.95
Cu^{65}	17.9	≥4	0.81±0.40	—
		4–7	0.50±0.25	—
		7–9	1.60±1.0	—
		≥9	83±50	—
	20.3	≥4	0.68±0.30	—
		4–7	0.27±0.20	—
		7–9	0.65±0.25	—
		≥9	10.0±3.0	—
	24.5	≥4	0.76±0.22	0.79
		4–7	0.40±0.20	0.83
		7–9	1.20±0.42	0.69
		≥9	1.90±0.93	0.87
	28.5	≥4	0.52±0.21	0.86
		4–7	0.22±0.25	0.19
		7–9	0.47±0.30	0.42
		≥9	1.23±0.96	0.91

*Dashes in the last column indicate that the function $f_1(\theta)$ was chosen to approximate the observed distribution.

TABLE 4 (Continued)

Target	E_γ max, MeV	ε_p, MeV	b/a	c/b
Zn (natural mixture of isotopes)	20.8	⩾3 3—6 6—9 ⩾9	0.42±0.21 0.22±0.18 0.71±0.41 3.3±1.7	— — — —
	28.6	⩾3 3—6 6—9 ⩾9	—0.02±0.15 —0.10±0.20 —0.14±0.21 0.8±0.4	— — — 0.47
Nb^{93}	19.5	⩾3 3—6 6—10 ⩾10	0.75±0.18 0 32±0.20 1.25±0.34 —0.19±0.53	— — — —
	23.5	⩾3 3—6 6—10 ⩾10	—0.54±0.15 0.14±0.13 0.84±0.25 0.49±0.33	— — — —
	27.5	⩾3 3—6 6—10 10—12 ⩾12	0.28±0.11 —0.12±0.14 0.36±0.17 0.72±0.39 0.94±0.58	— — — — —
Cd^{114}	20.3	⩾3.5 3.5—6.5 ⩾6.5	0.14±0.14 —0.15±0.12 0.70±0.33	— — —
Cd^{114}	24.5	⩾3.5 3.5—6.5 6.5—9.5 ⩾9.5	1.33±0.35 0.13±0.23 1.14±0.42 12.5±11.1	— — — —
	28.5	⩾ 3.5 3.5— 6.5 6.5— 9.5 9.5—12.5 ⩾12.5	0.88±0.21 0.21±0.13 2.16±0.93 1.60±0.90 2.9±2.0	— — — — —
Sn^{114}	23.5	⩾3.5 3.5—8 8—11 ⩾11	0.10±0.17 —0.11±0.37 0.64±0.44 0.42±0.65	— — — —
Sn^{124}	23.5	⩾3.5 3.5—7 ⩾7	0.61±0.28 0.06±0.25 2.25±1.57	— — —
Sb^{121}	19.5	⩾3.5 3.5—6.5 6.5—9.5 ⩾9.5	0.49±0.16 0.12±0.22 0.76±0.34 1.6±1.0	— — — —
Sb^{123}	19.5	⩾3.5 3.5—5.5 ⩾5.5	1.1±0.6 0.45±0.43 5.0±3.5	— — —

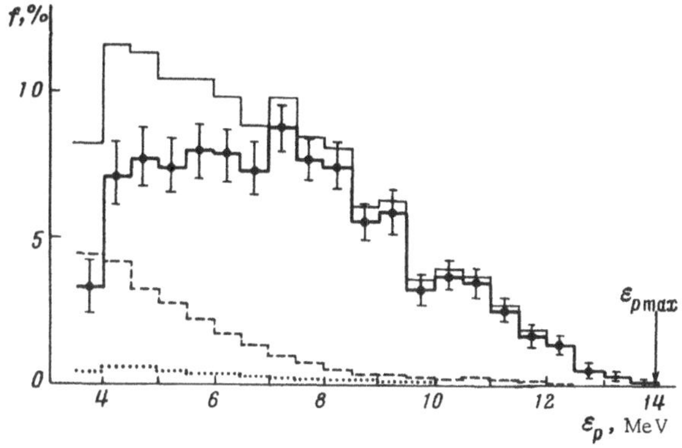

Fig. 19. Photoproton spectrum from Sb^{121} for $E_{\gamma\ max} = 19.5$ MeV (heavy line); observed energy distribution for photoprotons from a target enriched in Sb^{121} (light line); proton background spectrum (dashed line); spectrum of photoprotons from Sb^{123} present in the target (dotted line).

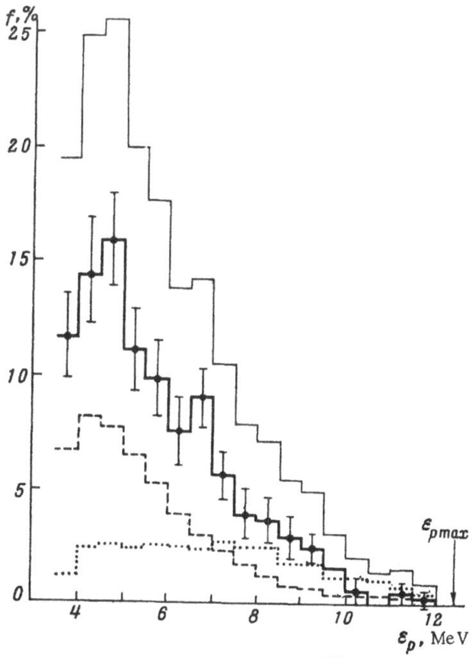

Fig. 20. Photoproton spectrum from Sb^{123} for $E_{\gamma\ max} = 19.5$ MeV (heavy line); observed energy distribution for photoprotons from a target enriched in Sb^{123} (light line); proton background spectrum (dashed line); spectrum of photoprotons from Sb^{121} present in the target (dotted line).

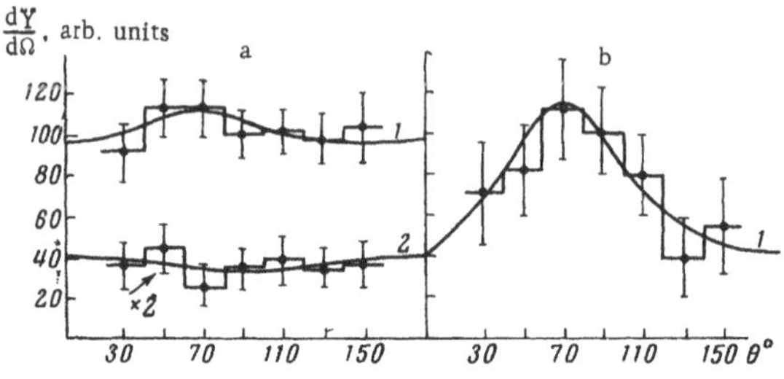

Fig. 21. Angular distribution of photoprotons from (a) Ni^{58}, (b) and Ni^{60} for $E_{\gamma\ max} = 24$ MeV. (a) $\varepsilon_p > 2$ MeV; (2) $\varepsilon_p > 8$ MeV.

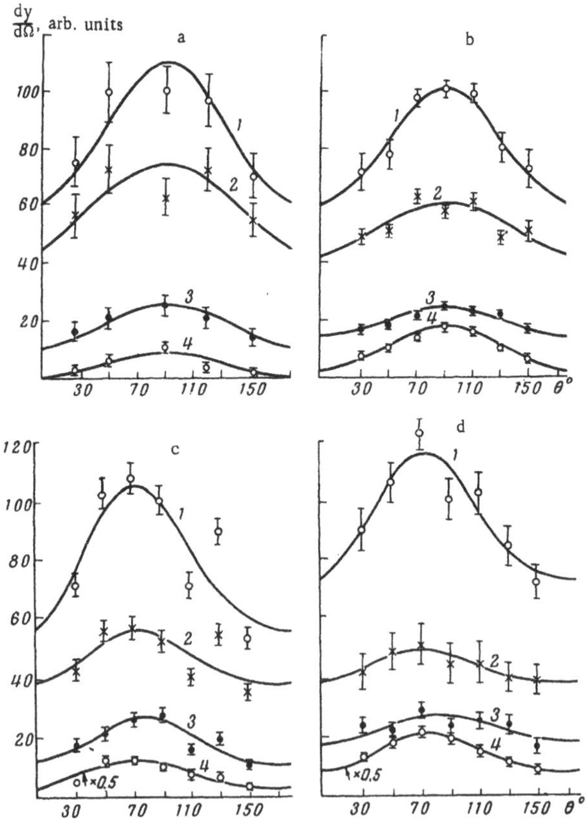

Fig. 22. Angular distribution of photoprotons in various energy groups from a target enriched in Cu^{65} and measured at (a) $E_{\gamma\ max} = 17.9$ MeV, (b) 20.4 MeV, (c) 24.5 MeV, and (d) 28.5 MeV. (1) $\varepsilon_p > 4$ MeV; (2) 4-7 MeV; (3) 7-9 MeV; (4) above 9 MeV.

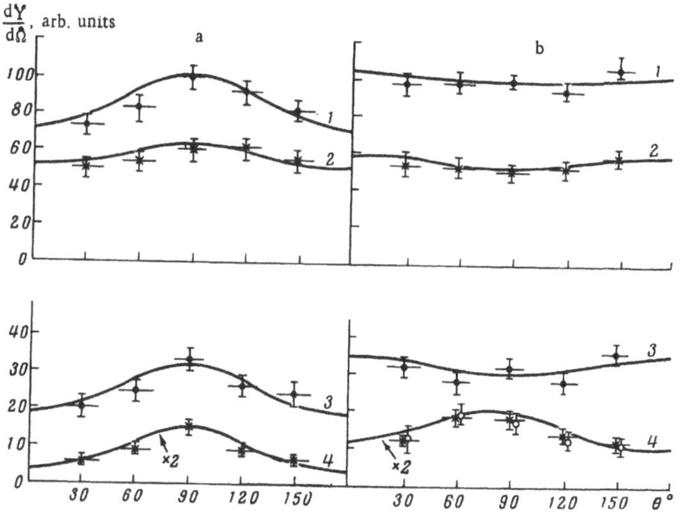

Fig. 23. Angular distribution of photoprotons in various energy groups from Zn for (a) $E_{\gamma\ max}$ = 20.8 MeV and (b) 28.6 MeV. (1) ε_p > 3 MeV; (2) 3-6 MeV; (3) 6-9 MeV; (4) above 9 MeV.

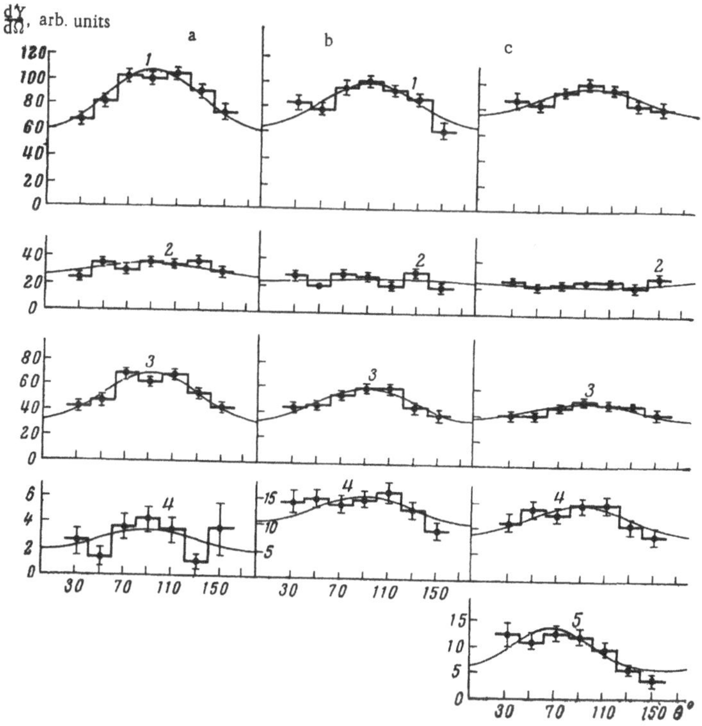

Fig. 24. Angular distribution of photoprotons in various energy groups from Nb^{93} for (a) $E_{\gamma\ max}$ = 19.5 MeV, (b) 23.5 MeV, and (c) 27.5 MeV. (1) $\varepsilon_p \geq 3$ MeV; (2) 3-6 MeV; (3) 6-10 MeV; (4) 10-12 MeV; (5) above 12 MeV.

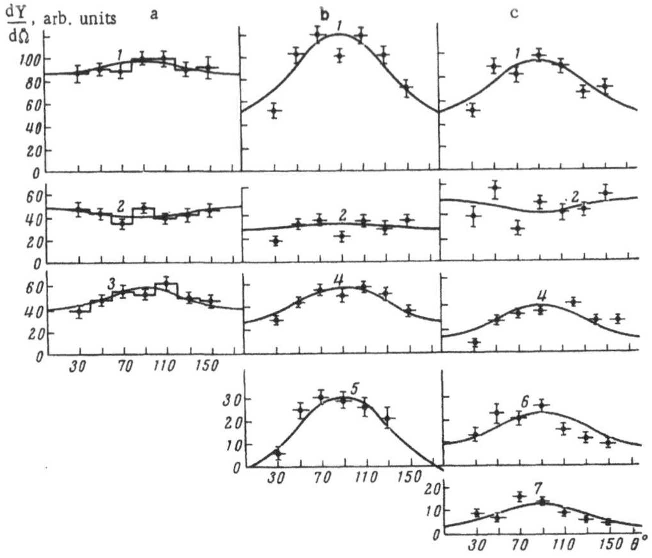

Fig. 25. Angular distribution of photoprotons in various energy groups from a target enriched in Cd^{114} for (a) $E_{\gamma\ max} = 20.3$ MeV, (b) 24.5 MeV, and (c) 28.5 MeV. (1) $\varepsilon_p > 3.5$ MeV; (2) 3.5-6.5 MeV; (3) above 6.5 MeV; (4) 6.5-9.5 MeV; (5) above 9.5 MeV; (6) 9.5-12.5 MeV; (7) above 12.5 MeV.

In those cases where the background was a few percent (≤5%), the distributions are shown in the figures with background subtracted. If the background was appreciable, the figures also show the directly observed proton distribution n_{meas} and the assumed spectral contribution from background tracks n_{Φ}.

Where possible (in the case of targets enriched in nickel and antimony isotopes) a separation of the photoproton spectrum N_p into the contributions from individual isotopes was made. In the remaining cases, spectra are given for targets with the isotopic compositions given in Table 1.

The spectra are given in relative units. The quantity $f(\varepsilon_p)$ plotted along the ordinate corresponds to the ratio of the number of protons with energies in the range $\varepsilon_p - \Delta\varepsilon_p/2$ to $\Delta\varepsilon_p/2$ to the total number of protons N_p expressed in percent.

Statistical errors and the value of $\varepsilon_{p\,max} = E_{\gamma\ max} - B_p$ (B_p is the proton binding energy), which corresponds to the maximum proton energy, are also indicated.

Angular Distribution of the Photoprotons. The angular distributions of the photoprotons in the various energy groups are shown in Figs. 21-29. Along the ordinate is plotted the proton yield $Y(\theta)$ per unit solid angle averaged over the angular range from $\theta = -10°$ to $\theta = +10°$ (θ is the azimuthal angle between the beam direction and the direction of proton emission):

$$Y(\theta) = \frac{N_p(\theta)}{G(\theta)}, \tag{3.1}$$

where $N_p(\theta)$ is the number of protons detected in the specified angular range; and $G(\theta)$ is the geometry factor defined above.

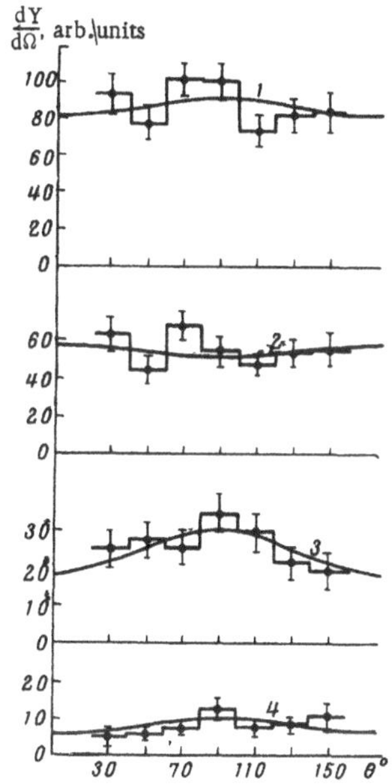

Fig. 26. Angular distribution of photoprotons in various energy groups from a target enriched in Sn^{114} for $E_{\gamma\,max} = 23.5$ MeV. (1) $\varepsilon_p > 3.5$ MeV; (2) 3.5-8 MeV; (3) 8-11 MeV; (4) above 11 MeV.

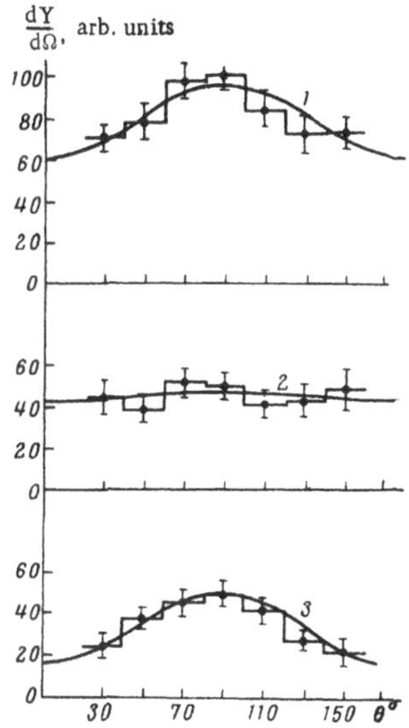

Fig. 27. Angular distribution of photoprotons in various energy groups from a target enriched in Sn^{124} for $E_{\gamma\,max} = 23.5$ MeV. (1) $\varepsilon_p > 3.5$ MeV; (2) 3.5-7 MeV; (3) above 7 MeV.

TABLE 5. Yield of Photoprotons with Energies $\varepsilon_p \geq \varepsilon_{p\,min}$ Normalized to Equal Ionization in the Thick-Walled, Aluminum Chamber and to the Yield from the Natural Mixture of Copper Isotopes for $E_{\gamma\,max} = 24$ MeV

Target	$\varepsilon_{p\,min}$, MeV	$E_{\gamma\,max}$, MeV	$\frac{Y_p}{Y_{Cu}}$
Ni^{58}	2	24.0	2.20±7%
Ni^{60}	2	24.0	0.70±10%
Ni^{64}	2	24.0	0.14±25%
Ni (natural mixture of isotopes)	3	24.5	1.76±20%
Cu (natural mixture of isotopes)	3	24.5	1.00
Cu^{65}	3	24.5	0.61±5%
Zn (natural mixture of isotopes)	3	24.5	1.43±20%
Nb^{93}	3	19.5	0.41±15%
	3	23.5	0.65±10%
	3	27.5	0.92±10%
Cd^{114}*	3.5	20.3	0.018±15%
	3.5	24.5	0.067±10%
	3.5	28.5	0.103±10%
Sn^{114}*	3.5	23.5	0.19±10%
Sn^{124}*	3.5	23.5	0.033±10%

*Isotopic composition of the targets are given in Table 1.

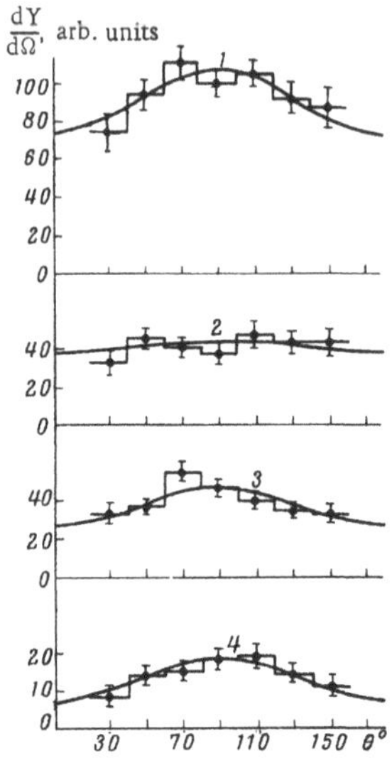

Fig. 28. Angular distribution of photoprotons in various energy groups from Sb^{121} for $E_{\gamma\,max}$ = 19.5 MeV. Correction was made for the contribution from Sb^{123}. (1) $\varepsilon_p > 3.5$ MeV; (2) 3.5-6.5 MeV; (3) 6.5-9.5 MeV; (4) above 9.5 MeV.

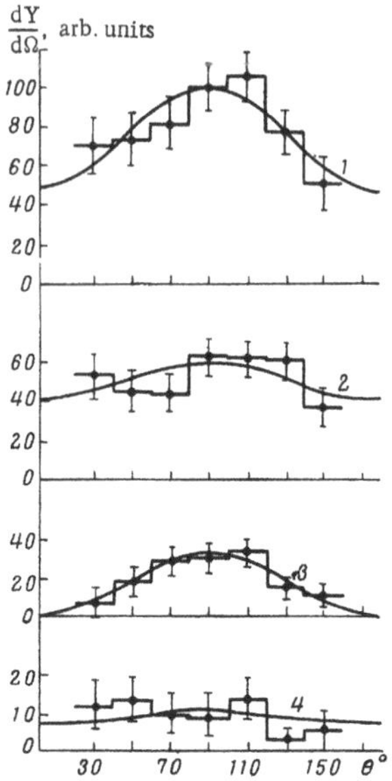

Fig. 29. Angular distribution of photoprotons in various energy groups from Sb^{123} for $E_{\gamma\,max}$ = 19.5 MeV. Correction was made for the contribution from Sb^{121}. (1) $\varepsilon_p \geq 3.5$ MeV; (2) 3.5-5.5 MeV; (3) 5.5-7.5 MeV; (4) above 7.5 MeV.

The yields $Y(\theta)$ are given in relative units. Furthermore, the total yield of protons with energies $\varepsilon_p \gtrsim 3$ MeV in the angular range $\Delta\theta$ = 80-100° was assumed to be 100 for each of the irradiations. The experimental values for $Y(\theta)$ are given in the form of a histogram (type II geometry), or are indicated by crosses (type I geometry). The distributions are shown with background subtracted. In the case of Ni^{58}, Ni^{69}, Sb^{121}, and Sb^{123}, the distributions shown were corrected for contributions from other isotopes.

Statistical errors are indicated. The photoproton distribution from Ni^{64} is not shown because of very large statistical errors.

The smooth curves correspond to an approximation to the angular distributions by a function of the form

$$f_1(\theta) = a + b\sin^2\theta, \tag{3.2}$$

if no asymmetry with respect to θ = 90° was observed, or by the function

$$f_2(\theta) = a + b\sin^2\theta + c\sin^2\theta\cos\theta, \tag{3.3}$$

if such asymmetry was present. The parameters a, b, and c were determined by the method of least squares. The ratio b/a, which characterizes the degree of anisotropy in the angular distributions, and the quantity $p = c/b$, which reflects the degree of asymmetry with respect to 90°, are given in Table 4.

2. Photoproton Yields

The integrated photoproton yields we obtained, normalized to the same amount of ionization in the absolute aluminum chamber, are shown in Table 5, and are given by

$$Y_p(E_{\gamma\,max}) = \int_0^{2\pi} Y(\theta)\sin\theta\, d\theta, \tag{3.4}$$

where $Y(\theta)$ is the yield observed at an azimuthal angle θ with respect to the beam. Since the yields $Y(\theta)$ are approximated by the functions $f_1(\theta)$ or $f_2(\theta)$, it is easy to show that the integrated yield can be expressed by the relation

$$Y_p(E_{\gamma\,max}) = A\left(1 + \frac{4}{3}\frac{b}{a}\right), \tag{3.5}$$

where the ratio b/a was determined by the least squares method; A is a normalization factor which depends on the choice of units for the measurement of yield.

The cross section integrated over angle for a reaction, weighted by the bremsstrahlung spectrum, can serve as a convenient objective characteristic and unit of yield for that reaction:

$$\overline{\sigma(E_{\gamma\,max})} = \int_0^{E_{\gamma\,max}} \sigma(E_\gamma)\,\varphi(E_\gamma, E_{\gamma\,max})\, dE_\gamma, \tag{3.6}$$

where $\sigma(E_\gamma)$ is the cross section integrated over angle for the energy E_γ; $\varphi(E_\gamma, E_{\gamma\,max})$ is the bremsstrahlung spectrum normalized so that

$$\int_0^{E_{\gamma\,max}} \varphi(E_\gamma, E_{\gamma\,max})\, dE_\gamma = 1.$$

If the angular distribution of the detected particles is isotropic, the measured value of $\overline{\sigma(E_{\gamma\,max})}$ is

$$\overline{\sigma(E_{\gamma\,max})} = \frac{4\pi N}{N_N N_\gamma S_M \Omega}, \tag{3.7}$$

where N is the total number of particles recorded by the detector; N_N is the number of target nuclei per cm^2; N_γ is the number of γ-quanta per cm^2 penetrating the target during irradiation; S_M is the irradiated area of the target; and Ω is the solid angle subtended by the detector.

However, the dose in terms of N_γ, the number of γ-quanta per cm^2, is not measured directly. The observed measure of radiation intensity is the flux of beam energy $W(E_{\gamma\,max})$, which is determined from the ionization in the absolute chamber. In this connection, the observed yields are sometimes referred to the energy flux $W(E_{\gamma\,max}) = N_\gamma \overline{E_\gamma(E_{\gamma\,max})}$, where $\overline{E_\gamma(E_{\gamma\,max})}$ is the γ-ray energy averaged over the bremsstrahlung spectrum as determined by the relation

$$\overline{E_\gamma(E_{\gamma\,max})} = \frac{\int_0^{E_{\gamma\,max}} E_\gamma \varphi(E_\gamma, E_{\gamma\,max})\, dE_\gamma}{\int_0^{E_{\gamma\,max}} \varphi(E_\gamma, E_{\gamma\,max})\, dE_\gamma} = \int_0^{E_{\gamma\,max}} E_\gamma \varphi(E_\gamma, E_{\gamma\,max})\, dE_\gamma. \tag{3.8}$$

In this way, the so-called yield per "effective" quantum is obtained:

$$Y_W(E_{\gamma\,\max}) = \frac{4\pi N}{N_N S_M W(E_{\gamma\,\max})\,\Omega} = \frac{\overline{\sigma(E_{\gamma\,\max})}}{E_\gamma(E_{\gamma\,\max})}, \tag{3.9}$$

which can also serve as an objective characteristic of a process.

Very often, however, yields are referred to a mole of target material per cm^2 and to unit charge released by ionization in the absolute chamber. The total ionization charge q is related to the energy flux W in the beam by the expression

$$q = \overline{K(E_{\gamma\,\max})} \cdot W(E_{\gamma\,\max})\,s, \tag{3.10}$$

where $\overline{K(E_{\gamma\,\max})}$ is the chamber sensitivity averaged over the bremsstrahlung spectrum; and s is the cross section of its sensitive volume. It is easy to show that the yield to unit ionization charge is

$$Y_q(E_{\gamma\,\max}) = \frac{N_{Av}\cdot 4\pi N}{N_N S_M \Omega q} = \frac{N_{Av}\cdot\overline{\sigma(E_{\gamma\,\max})}}{E_\gamma(E_{\gamma\,\max})\cdot\overline{S}(E_{\gamma\,\max})\cdot s}, \tag{3.11}$$

where N_{Av} is the Avogadro number. It is obvious that the quantity $Y_q(E_{\gamma\,\max})$, which depends on the sensitivity of the chamber $\overline{S(E_{\gamma\,\max})}$ and on its cross sections, can assume different values as a function of $E_{\gamma\,\max}$ when different chambers are used, and it is, therefore, not an objective characteristic of a process. In this connection, the yields in Table 5 are given in relative units. The yield of photoprotons with energies $\varepsilon_p \geq 3$ MeV from a natural mixture of copper isotopes for $E_{\gamma\,\max} = 24$ MeV was taken as a unit. The normalization of yields adopted is also convenient for comparisons with the data of other authors.

Furthermore, the accuracy of relative measurements is always greater than that of absolute measurements because one avoids the errors connected with inaccuracy in the calibration of the monitor with respect to the absolute chamber, and with uncertainty in the knowledge of its sensitivity, as well as errors in dose measurements, which are unavoidable in the case of irradiations separated by long intervals of time.

CHAPTER IV

Discussion of Results

1. Contribution of Evaporation to Photoproton Yields from Medium-Weight Nuclei

As was noted in Chapter I, the role of evaporation in the yields of photonuclear reactions in the giant resonance region is of great interest. However, an analysis of the experimental data revealed that the information about the role of the evaporation process in these reactions was contradictory, particularly with respect to the observed yields of photoprotons from medium-weight nuclei. Because of this, there is cause to doubt the widely accepted idea that the statistical theory explains the principal part of the photoproton yield in medium-weight nuclei. Estimates of photoproton yields according to the statistical theory, which were obtained previously by the authors of [76-78], were made using model parameters that were rather incorrect from the present-day point of view (in particular, they did not give proper consideration to the even-odd variation in the level densities of the final nucleus produced as the result of proton and neutron emission which is quite important for the prediction of the relative probability of proton emission).

It was, therefore, decided to carry out once again a detailed comparison of experimental data on photoproton yields (obtained by ourselves and others) with the predictions of the statistical model.

The contribution of evaporation to photoproton yield in the dipole resonance region was estimated by a comparison of the experimentally observed relative probability for proton emission, η_p^{exp}, for $E_{\gamma\,\max} = 24$ MeV*

*Most of the photoproton yield measurements were made in the neighborhood of $E_{\gamma\,\max} = 24$ MeV.

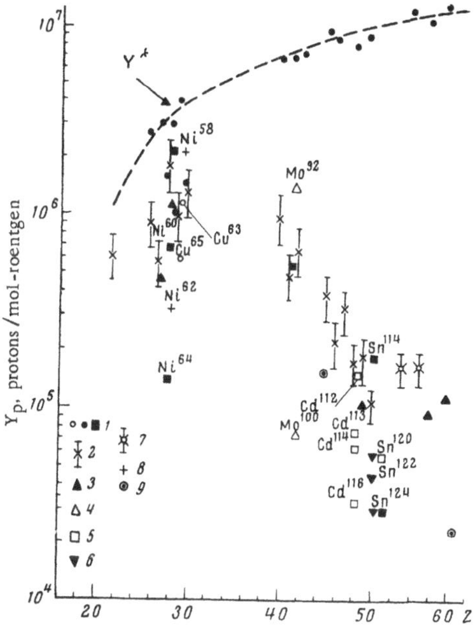

Fig. 30. Photoproton yields Y_p and $Y^* = Y_p + Y_n$ for $E_{\gamma\,max} = 24$ MeV. (1) Data from this work; (2) [76]; (3) [77, 78, 81]; (4) [79]; (5) [107]; (6) [108]; (7) [109]; (8) [110]; (9) [111].

with the average value of this probability weighted over the bremsstrahlung spectrum predicted by statistical theory, η_p^{theor}. The value of η_p^{exp} was assumed to be

$$\eta_p^{exp} = \frac{Y_p}{Y^*}, \tag{4.1}$$

where Y_p is the observed yield; Y^* is the total photoreaction yield (or total number of excited nuclei) for $E_{\gamma\,max} = 24$ MeV.

The yields of photoprotons from nuclei of medium weight (Z = 20-60) for $E_{\gamma\,max} = 24 \pm 1.5$ MeV, measured by ourselves and others [76-79, 81, 107-111], are given in Fig. 30. Also shown there is the experimental dependence of Y^* on Z. The small contribution from (γ, d), (γ, α), and (γ, γ') reactions was not taken into account. It was considered that the quantity Y^* equaled the sum of the yield Y_p of photoprotons and the yield Y_n of photoneutrons. Data on the yield of photoneutrons was taken from [112-114, 25-27]. The value of Y^* was estimated from data for neighboring nuclei in cases where experimental data for Y_n was lacking. Furthermore, use was made of the fact that the integrated nuclear γ-ray absorption cross section in the giant resonance region is proportional to NZ/A [4] according to the dipole sum rules.

It is easy to show that the relative contribution of evaporation in the observed photoproton yield

$$\Delta Y_p^{evap} = \frac{Y_p^{evap}}{Y_p} \tag{4.2}$$

can be determined with the help of the relation

$$\Delta Y_p^{evap} = \frac{\eta_p^{theor}}{\eta_p^{exp}}. \tag{4.3}$$

Indeed, according to statistical theory, one should expect a photoproton yield equaling

$$Y_p^{evap} = Y \cdot \eta_p^{theor}. \tag{4.4}$$

Combining (4.1) and (4.4), we obtain (4.3).

The quantity η_p^{theor} was found by numerical integration of the relation

$$\eta_p^{theor} = \frac{\int_0^{24} \sigma_\gamma(E_\gamma)\,\varphi(E_\gamma)\,\eta_p(E_\gamma)\,dE_\gamma}{\int_0^{24} \sigma_\gamma(E_\gamma)\,\varphi(E_\gamma)\,dE_\gamma}. \tag{4.5}$$

In the calculation, the Schiff bremsstrahlung spectrum [115] $\varphi(E_\gamma)$ with upper limit $E_{\gamma\,max}$=24 MeV and the experimental nuclear γ-ray absorption cross sections $\sigma_\gamma(E_\gamma)$ were used.

The cross sections $\sigma_\gamma(E_\gamma)$ are far from known for all the nuclei for which photoproton yields were determined. As can be seen from relation (4.5), however, the shape of the $\sigma_\gamma(E_\gamma)$ curve, and not its absolute magnitude, is important and it changes comparatively little from nucleus to nucleus in the case of sufficiently heavy nuclei. It is completely valid to regard it as the same for nuclei over some not-too-large range of A. We assumed that the shape of the cross section $\sigma_\gamma(E_\gamma)$ was identical for each of the nuclear groups Ti – Zn, Zr – Mo, Rh – Sn, and Sb – Pr. The cross section sum $\sigma_{\gamma,p} + \sigma_{\gamma,n} + \sigma_{\gamma,np} + \sigma_{\gamma,2n}$ for Ni^{58}, which was measured in [110], was used as $\sigma_\gamma(E_\gamma)$ for the first group. For the other groups of nuclei, where the proton yield was small, it was assumed that $\sigma_\gamma(E_\gamma) \approx \sigma_{\gamma,n}(E_\gamma)$; averaged data from [113, 114, 116, 117, 26, 27, 31, 32] was used for the Zr – Mo group, that from [113, 114, 117-120] for the Rh – Sn group, and that from [84, 121, 122] for the Sb – Pr group.

The relative probability $\eta_p(E_\gamma)$ for the evaporation of photoprotons as the result of the action of monochromatic γ-rays which appears in (4.5) was calculated on the assumption that the total decay width of compound nucleus levels equaled the sum of the total proton and neutron widths Γ_p and Γ_n, respectively, and that all the remaining widths were negligibly small, i.e.,

$$\eta_p(E_\gamma) = \frac{F_p(E_\gamma)}{F_p(E_\gamma) + F_n(E_\gamma)}, \tag{4.6}$$

where F is a function which is proportional to the appropriate total width usually used in statistical theory calculations. For particle emission, F is determined by the well-known relation

$$F = \frac{2M}{\hbar^2} \int_0^{\varepsilon_{max}} \varepsilon \sigma_c(\varepsilon)\, \rho_R(\varepsilon_{max} - \varepsilon)\, d\varepsilon, \tag{4.7}$$

where M is the mass of the particle emitted; ε is its energy; $\sigma_c(\varepsilon)$ is the cross section for compound nucleus formation in the inverse process; ρ_R is the level density in the final nucleus; and ε_{max} is the maximum possible energy of the emitted particle, equal to $E_c^* - B$ (E_c^* is the excitation energy of the compound nucleus, equal to E_γ in our case; B is the binding energy of the particle being considered in this nucleus).

TABLE 6. Relative Probabilities for Photoproton Emission η_p and the Contribution of Evaporation to Observed Photoproton Yields ΔY^{evap} Equal to $\eta_p^{theor}/\eta_p^{exp}$

Element (natural mixture of isotopes)	η_p^{exp}	η_p^{theor}	ΔY^{evap}	Isotope	η_p^{exp}	η_p^{theor}	ΔY^{evap}
Ti	0.484	0.116	0.240	Ni^{58} *	0.672	0.653	≈1
Fe	0.320	0,080	0.250	Ni^{60}	0.218	0,0283	0.13
Co (Co^{59}—100%)	0,164	0.0295	0.180	Ni^{62}	0.112	$3{,}66\cdot10^{-3}$	$3{,}3\cdot10^{-2}$
Ni *	0.606	0.505	0.833	Ni^{64}	0.0463	$2.64\cdot10^{-4}$	$5.7\cdot10^{-3}$
Cu	0.229	0.0895	0.290	Cu^{63}	0.379	0.124	0.33
Zn *	0.333	0.263	0.791	Cu^{65}	0,206	$1{,}23\cdot10^{-2}$	0.06
Zr	0.158	0.0186	0.117	Mo^{92} *	0.279	0.285	≈1
Nb (Nb^{93}—100%)	$7{,}23\cdot10^{-2}$	$7.4\cdot10^{-3}$	0,104	Mo^{100}	$1.05\cdot10^{-2}$	$3{,}84\cdot10^{-6}$	$3.81\cdot10^{-4}$
Mo *	$8.85\cdot10^{-2}$	$6.16\cdot10^{-2}$	0.700	Cd^{112}	$1.39\cdot10^{-2}$	$3.98\cdot10^{-6}$	$2.92\cdot10^{-3}$
Rh (Rh^{105}—100%)	$4{,}08\cdot10^{-2}$	$6.94\cdot10^{-4}$	$1.7\cdot10^{-2}$	Cd^{113}	$7{,}29\cdot10^{-3}$	$5.67\cdot10^{-5}$	$0.78\cdot10^{-2}$
Pd	$2{,}49\cdot10^{-2}$	$2.76\cdot10^{-4}$	$1{,}1\cdot10^{-2}$	Cd^{114}	$6{,}29\cdot10^{-3}$	$7.19\cdot10^{-6}$	$1.14\cdot10^{-3}$
Ag	$4{,}07\cdot10^{-2}$	$1.1\cdot10^{-3}$	$2.7\cdot10^{-2}$	Cd^{116}	$3.36\cdot10^{-3}$	$1{,}7\cdot10^{-6}$	$0{,}51\cdot10^{-3}$
Cd	$2.07\cdot10^{-2}$	$9{,}74\cdot10^{-4}$	$4.7\cdot10^{-2}$	Sn^{114}	$1{,}77\cdot10^{-2}$	$3{,}21\cdot10^{-4}$	$1{,}81\cdot10^{-2}$
In	$1.91\cdot10^{-2}$	$7.18\cdot10^{-5}$	$3{,}8\cdot10^{-3}$	Sn^{120}	$5{,}74\cdot10^{-3}$	$4.78\cdot10^{-6}$	$0{,}33\cdot10^{-3}$
Sn	$1{,}09\cdot10^{-2}$	$2.21\cdot10^{-5}$	$2{,}1\cdot10^{-3}$	Sn^{122}	$4.58\cdot10^{-3}$	$4{,}0\cdot10^{-7}$	$0{,}78\cdot10^{-4}$
$(I + Cs)_{av}$	$1{,}39\cdot10^{-2}$	$1.5\cdot10^{-5}$	$1.1\cdot10^{-3}$	Sn^{124}	$3{,}29\cdot10^{-3}$	$6{,}38\cdot10^{-8}$	$1{,}94\cdot10^{-5}$
Ba	$8.62\cdot10^{-3}$	$7{,}5\cdot10^{-6}$	$0{,}91\cdot10^{-3}$				
Ce	$8.82\cdot10^{-3}$	$6{,}8\cdot10^{-6}$	$0{,}78\cdot10^{-3}$				

Note. Asterisks indicate those nuclei in which the photoproton evaporation process plays a dominant role.

The cross section for compound nucleus formation was calculated in a quantum-mechanically exact fashion by Weisskopf and Blatt [123] and by Shapiro [124] for a nucleus with uniform density and sharp boundaries. It is well known, however, that the nucleus has a diffuse boundary which should increase barrier penetrability. In more recent calculations, $\sigma_c(\varepsilon)$ calculated from the optical model is often used. Unfortunately, data resulting from these calculations have not been published. As Evans [72] and Scott [125] have shown, however, the cross section for compound nucleus formation by charged particles in the diffuse boundary case can be approximated by the cross section for a nucleus with uniform density if the nuclear radius is increased (r_0 should be made equal to approximately 1.65 F).

The level density of excited nuclei which was used in evaporation theory was obtained on the basis of a model having two noninteracting proton and neutron Fermi gases. Considered most acceptable from the modern theoretical point of view is the dependence of level density on excitation energy proposed by Bethe, which in simplified form is

$$\rho(E^*) \sim \frac{\exp 2\,[a'(E^* - \Delta)]^{1/2}}{(E^* - \Delta)^2}, \tag{4.8}$$

where E^* is the nuclear excitation energy; Δ is the energy gap near the Fermi surface resulting from nuclear pair correlations ($\Delta = 0$ in nuclei with odd N and Z; $\Delta = \delta_p$ or $\Delta = \delta_n \approx 1$ MeV in nuclei with even Z or N, respectively; $\Delta = \delta_p + \delta_n \approx 2$ MeV in nuclei with even Z and N; here, δ_p and δ_n are the pairing energies of the last proton or neutron, respectively); a' is a constant depending on the atomic weight A and determined by the density of single-particle states close to the Fermi surface.

The calculation of F by means of tabulated data for $\sigma_c(\varepsilon)$ and level densities in the form (4.8) must be performed by numerical integration, which is extremely inconvenient if it is necessary to make a large number of calculations.

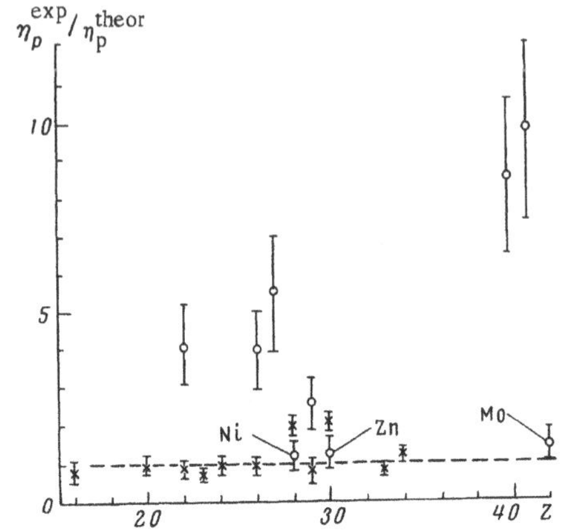

Fig. 31. A comparison of the ratio $\eta_p^{exp}/\eta_p^{theor}$ for the (γ, p) reaction at $E_{\gamma\ max}$=24 MeV from this work (circles) and for the (n, p) reaction with 14-MeV neutrons from Allan's data [75] (crosses).

In order to represent the function F in integrated form, the often-used dependence of level density in the final nucleus on excitation energy E* in the form

$$\rho_R(E^*) \sim \exp 2\,[a\,(E^* - \Delta)], \tag{4.9}$$

was employed, and the cross section for compound nucleus formation in inverse reactions was approximated, in accordance with [126], by the formulas

$$\frac{\sigma_p(\varepsilon)}{\pi R^2}\begin{cases}(1+c)\left(1-\frac{kV_{coul}}{\varepsilon}\right) & \text{for } \varepsilon \geqslant kV_{coul};\\ 0 & \text{for } \varepsilon < kV_{coul};\end{cases} \tag{4.10}$$

$$\frac{\sigma_n(\varepsilon)}{\pi R^2} = \alpha\left(1+\frac{\beta}{\varepsilon}\right), \tag{4.11}$$

where $V_{coul} = Ze^2/R$ is the Coulomb barrier for protons, and the constants k, c, α, and β are selected in such a way as to give the best agreement of the values obtained from formulas (4.10) and (4.11) with the results of exact calculations [123, 124].

After substituting (4.9)-(4.11) into (4.7), we have

$$F_p = \frac{2M_p}{\hbar^2}\pi R^2(1+c)\int\limits_{kV_{coul}}^{\varepsilon_p^{max}} \varepsilon\left(1-\frac{kV_{coul}}{\varepsilon}\right)\exp 2\,[a\,(\varepsilon_p^{max}-\varepsilon)]^{1/2}; \tag{4.12}$$

$$F_n = \frac{2M_n}{\hbar^2}\int\limits_0^{\varepsilon_n^{max}} \varepsilon\sigma_c(\varepsilon)\,\rho_R(\varepsilon_n^{max}-\varepsilon)\,d\varepsilon. \tag{4.13}$$

As a result of the integration of expressions (4.12) and (4.13) and subsequent algebraic transformations, analytic expressions for the functions $E_p(E)$ and $F_n(E)$, which are very convenient for numerical computations, were obtained:

$$F_p(E_\gamma) = \frac{2M_p}{\hbar^2}\pi R^2 \frac{1+c}{8a^2}\varphi_1(X_p); \tag{4.14}$$

$$F_n(E_\gamma) = \frac{2M_n}{\hbar^2}\pi R^2 \frac{\alpha}{8a^2}\{\varphi_1(X_n) + 4a\beta\varphi_2(X_n)\}, \tag{4.15}$$

where

$$\varphi_1(X) = 2(X - 3X^{1/2} + 3)\exp X^{1/2} + X - 6; \tag{4.16}$$

$$\varphi_2(X) = (X^{1/2} - 1)\exp X^{1/2} + 1; \tag{4.17}$$

$$X_p = 4a(\varepsilon_p^{\max} - kV_{\text{coul}}); \tag{4.18}$$

$$X_n = 4a\varepsilon_n^{\max}. \tag{4.19}$$

Furthermore, the maximum proton and neutron energies are

$$\varepsilon_p^{\max} = E_\gamma - B_p - \Delta(p); \tag{4.20}$$

$$\varepsilon_n^{\max} = E_\gamma - B_n - \Delta(n). \tag{4.21}$$

Here, $\Delta(p)$ and $\Delta(n)$ equal the sum of the pairing energies of the last even protons and neutrons $\Delta = \delta_p + \delta_n$ (in nuclei which remain after the emission of a proton or neutron, respectively), if one takes into consideration the gap Δ in the level densities of the Fermi gas close to the ground state.

The quantity r_0 was assumed to be 1.7 F for the calculation of F_p (to allow for the effect of the diffuseness of the nuclear boundary [72, 125]) and 1.5 F for the calculation of F_n.

A value of the level density parameter $a = A/10$, where A is the atomic weight, was used, with which Allan obtained agreement of the theoretical cross sections for (n, p) reactions of 14-MeV neutrons with experimentally measured cross sections [75]. It is known at the present time that the parameter a is not a smooth function and undergoes fluctuations depending on A [69, 70]. However, the approximation that was made does not introduce a large error. The point is, this parameter is important for predicting the shape of the spectrum of evaporated particles, but it has little effect on the ratio of the total proton and neutron widths F_p/F_n and, therefore, on the quantity $\eta_p = F_p/(F_p + F_n)$. This is verified by evaluations of the quantity F_p/F_n made by Dostrovsky, et al. [126] and by Allan [75] for $a = A/10$, $A/20$, and $A/40$.

However, the quantity F_p/F_n is extremely sensitive to the difference

$$[B_n + \Delta(n)] - [B_p + \Delta(n)] = B_n - B_p + \Delta(n) - \Delta(p). \tag{4.22}$$

For that reason, the most precise experimental binding energies B_p and B_n, taken from [127-133], were used in the calculations. If such data were lacking, B_n calculated by Cameron [134] from the mass formula was used, and the quantity B_p was determined from β-decay data:

$$B_p = B_n + Q_{\beta^-} - (m_H - m_n) \tag{4.23}$$

in the case of an unstable residual nucleus as the result of a (γ, p) reaction;

$$B_p = B_n - Q_{EC} - (m_H - m_n) \tag{4.24}$$

in the case of an unstable nucleus following a (γ, n) reaction.

The hydrogen-neutron mass difference $m_H - m_N$ was assumed to be 0.783 MeV in accordance with [135]. Values for Q_β and Q_{EC} were taken from [136, 137]. The quantities δ_p and δ_n were taken from Cameron's paper [138].

In those cases where targets with a natural mixture of isotopes were used for the measurement of proton yields, calculations were made for each isotope and averaging over isotopic composition was performed.

Results of the calculations of η_p^{theor} and a comparison of η_p^{theor} and η_p^{exp} are given in Table 6.

The following is evident from the table:

1. Both η_p^{theor} and η_p^{exp} decrease with increasing Z (because of the rise in the Coulomb barrier) and with increasing neutron number N for isotopes of the same element (effect of the rise in proton binding energy B_p), and also display an identical direction of variation, which is associated with the evenness of the proton number in neighboring nuclei. However, the observed variation of the quantity η_p as a function of Z, A, and the evenness of Z is less strongly evident than expected on the basis of statistical theory.

2. The ratio $\eta_p^{theor}/\eta_p^{exp}$, as a rule, is less (often considerably less) than unity, i.e., in contrast to the result of Halpern and Weinstock [76], the observed photoproton yield exceeds that expected on the basis of statistical theory even in the case of nuclei with Z = 20-40.

Evaporation plays a dominant role only in the case of Ni, Zn, and Mo. All those elements contain a considerable quantity of the light, even-even isotopes Ni^{58}, Zn^{64}, and Mo^{92} where conditions for proton evaporation are extremely favorable: 1) the Coulomb barrier is relatively low; 2) the proton binding energy B_p is several MeV lower than the neutron binding energy; 3) the effect associated with different evenness of the final nuclei which remain as the result of proton and neutron emission does not depress proton evaporation (both final nuclei have an odd number of nucleons). The last effect mentioned plays an important role. One can see that in the case of the odd-even nuclei (Co^{59}, Cu^{63}, and Nb^{93}) neighboring on Ni^{58}, Ni^{64}, and Mo^{92}, which have approximately the same ratio of proton and neutron binding energies, the theoretical and experimental data are considerably different.

The reason for this is the following. In first approximation, the probability for the evaporation of any particle is proportional to the level density in the nucleus which remains after emission of the particle from the compound nucleus. As the result of the emission of a proton or neutron from a compound nucleus with odd Z and even N, there remains an even-even or odd-odd final nucleus, respectively. Other conditions being equal, it is known that the level density of even-even nuclei is less than the level density of odd-odd nuclei. Therefore, conditions are unfavorable for the evaporation of protons from nuclei with odd Z and even N.

In this connection, it is of interest to note the recent establishment of the fact that the even-odd types of nuclear excited level densities are not described by the Weisskopf relation [48]

$$\rho_{\text{odd-odd}}(E^*) = 2\rho_{\substack{\text{even-odd}\\ \text{odd-even}}}(E^*) = 4\rho_{\text{even-even}}(E^*), \tag{4.25}$$

but arise from an energy gap close to the nuclear ground state produced by pair correlations, i.e.,

$$\rho(E^*) = \rho(E^* - \Delta).$$

It has been pointed out [71] that relation (4.25) gives a ratio of level densities for even-odd and odd-odd nuclei which is several times too high when compared with experimental values. It then follows that the earlier estimates (without consideration of pair correlations) certainly gave enhanced yields of evaporation photoprotons from nuclei with odd Z and even N, and that the agreement between theory and experiment for such nuclei (Cu and Co, for example [77, 78]) was accidental.

In conclusion, we point out that the divergence of experiment from statistical theory by a factor greater than 10 conclusively demonstrates that the reaction mechanism is something other than evaporation. However, there may be completely reasonable doubt that it is necessary to attach great importance to disagreement by a factor of 3-5 in the values of η^{theor} and η^{exp}. In our opinion, such a disagreement is not associated with experimental errors or with inexact theoretical estimates. In the case of copper, for example, this disagreement is confirmed by the presence of a very large anisotropic component in the angular distribution of the photoprotons [14]. Furthermore, in the case of Mo^{92} where the proton yield is described by statistical theory, one can see that its spectrum, measured in [72], is also determined mainly by the evaporation spectrum. In the case of Nb^{93} where $\eta_p^{theor} < \eta_p^{exp}$, the observed spectrum (our data) is markedly different from an evaporation spectrum (see Fig. 32d).

An indication that the theoretical parameters we used are completely acceptable is furnished by the agreement between η_p^{theor} and η_p^{exp} obtained by Allen [75] for the same parameters in the case of the reaction with 14-MeV neutrons (Fig. 31).

2. Photoproton Spectra

Comparison with evaporation spectra. Measured photoproton spectra and spectra calculated on the basis of the evaporation model are shown in Fig. 32 for a number of nuclei. Among these nuclei are those for which the measured yield Y_p is close to that predicted by statistical theory Y_p^{evap} (Ni^{58}, Zn, and Mo^{92}), and those with neighboring atomic weights for which $Y_p > Y_p^{evap}$ (Ni^{60}, Cu (natural mixture), Cu^{65}, and Nb^{93}).

The evaporation spectra were calculated from the formula

$$F(\varepsilon_p) = \varepsilon_p \sigma_c(\varepsilon_p) \int_0^{E_\gamma \max} \frac{\sigma_\gamma(E_\gamma)\,\varphi(E_\gamma E_{\gamma \max})\,\rho_R\,[\varepsilon_p^{\max}(E_\gamma) - \varepsilon]\,dE_\gamma}{F_p(E_\gamma) + F(E_\gamma)}, \tag{4.26}$$

where the quantities have the same meaning and numerical values as in the calculation of photoproton yields (see previous section). An exception is the value for the cross section for compound nucleus formation $\sigma_c(\varepsilon)$ for which the exact values from Shapiro [124] for $r_0 = 1.7$ F were taken in place of the approximation (4.10) which strongly distorts the shape of the proton spectra in the low-energy region.

Pair correlations were not taken into account because, as has been shown [126], they have practically no effect on the shape of the spectrum (Fig. 32d). The computed spectra were normalized to the maximum of the observed spectra.

The comparison shows that the photoproton spectra from Ni^{58}, Zn, and Mo^{92}, for which the yield is close to that predicted by statistical theory ($\eta_p^{theor} \approx \eta_p^{exp}$), basically agree with the evaporation spectrum. In the high-energy proton region, however, there is an excess over the evaporation spectrum amounting to about 20%. Thus it is impossible to assign the entire observed proton yield to evaporation even in these nuclei.

Furthermore, along with spectra clearly in disagreement with statistical theory (Nb^{93}, Cu^{65}) for the case $\eta_p^{exp} > \eta_p^{theor}$, there are spectra very much like the evaporation spectrum (Cu, Ni^{60}). Consequently, the agreement of an observed spectrum predicted by statistical theory cannot be considered as a conclusive argument in favor of the evaporation mechanism.

Consideration of Spectral Shape from the Shell Model Point of View. In principle, the shell model can quantitatively predict the spectra of photonucleons emitted in the decay of a particle-hole configuration. A comparison of the observed spectra of photonucleons with those predicted is of the greatest interest for establishing the nature of the mechanism other than evaporation which is mainly responsible for the photoproton yields, as mentioned above. The agreement of theory and experiment would be evidence

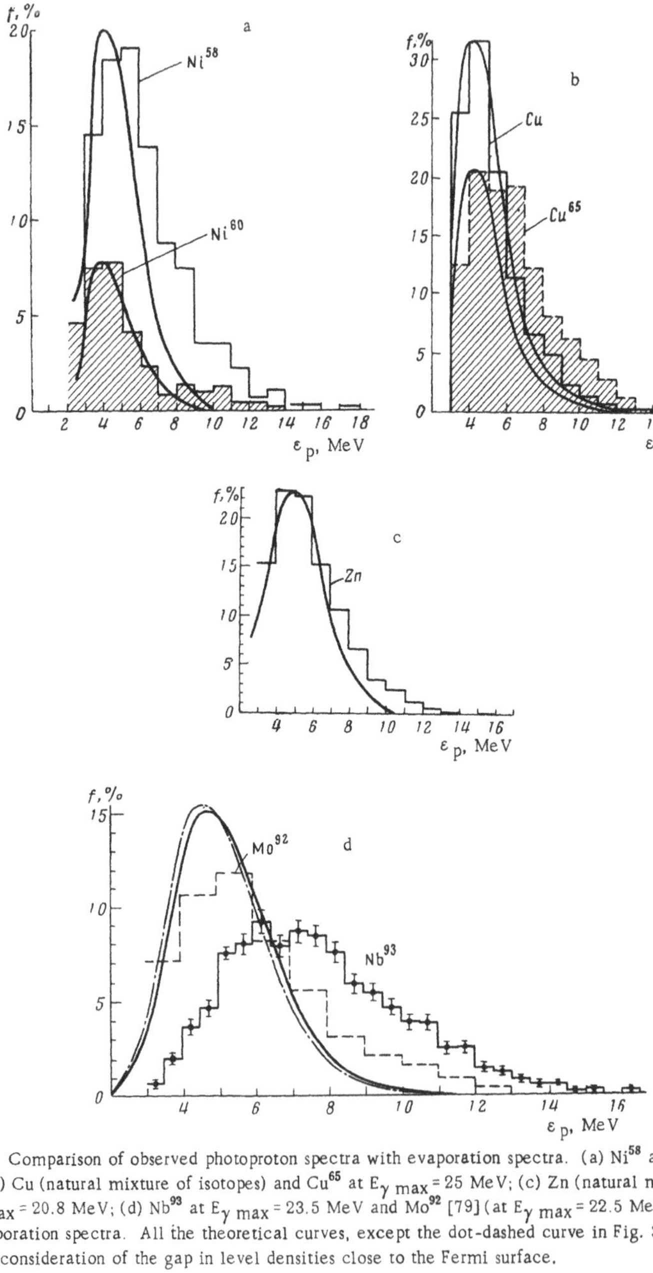

Fig. 32. Comparison of observed photoproton spectra with evaporation spectra. (a) Ni^{58} and Ni^{60} at $E_{\gamma\ max} = 24$ MeV; (b) Cu (natural mixture of isotopes) and Cu^{65} at $E_{\gamma\ max} = 25$ MeV; (c) Zn (natural mixture of isotopes) at $E_{\gamma\ max} = 20.8$ MeV; (d) Nb^{93} at $E_{\gamma\ max} = 23.5$ MeV and Mo^{92} [79] (at $E_{\gamma\ max} = 22.5$ MeV). The smooth curves are evaporation spectra. All the theoretical curves, except the dot-dashed curve in Fig. 32d, were obtained without consideration of the gap in level densities close to the Fermi surface.

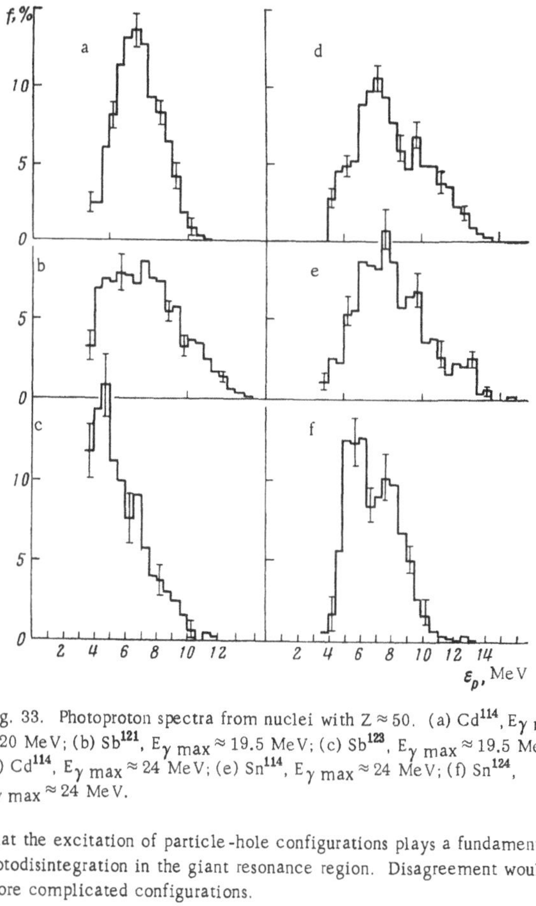

Fig. 33. Photoproton spectra from nuclei with $Z \approx 50$. (a) Cd^{114}, $E_{\gamma\,max} \approx 20$ MeV; (b) Sb^{121}, $E_{\gamma\,max} \approx 19.5$ MeV; (c) Sb^{123}, $E_{\gamma\,max} \approx 19.5$ MeV; (d) Cd^{114}, $E_{\gamma\,max} \approx 24$ MeV; (e) Sn^{114}, $E_{\gamma\,max} \approx 24$ MeV; (f) Sn^{124}, $E_{\gamma\,max} \approx 24$ MeV.

in favor of the idea that the excitation of particle-hole configurations plays a fundamental role in the cross section for nuclear photodisintegration in the giant resonance region. Disagreement would be a point in favor of the excitation of more complicated configurations.

Unfortunately, shell model calculations are extremely laborious and require a knowledge of the binding energy of nucleons in inner shells which participate in dipole excitation of the nucleus. These calculations have been performed only for nuclei with double filled shells, Ca^{40} [40] and Zr^{90} [41]. Agreement between theoretical and observed spectra was not obtained in the latter of the two papers mentioned. It should be pointed out, however, that the comparison was not altogether proper; the calculations were performed for the isotope Zr^{90} while the measurements were made with the natural mixture of a large number of isotopes.

Photoproton spectra from nuclei with $Z \approx 50$ measured by us are shown in Fig. 33. These spectra are quite varied; their shape depends on Z and changes both with changes in $E_{\gamma\,max}$ and with the transition from isotope to isotope of given Z. This indicates the extreme sensitivity of the spectra to actual nuclear structure.

Following is a qualitative discussion of the photoproton spectra we measured from various isotopes of the same element and of the changes in spectrum shape as a function of energy from the point of view of the shell (particle-hole) model.

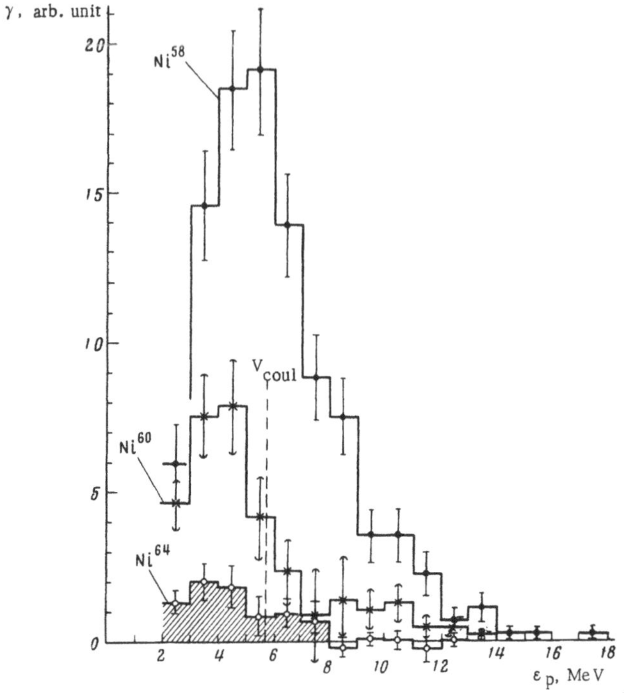

Fig. 34. Comparison of photoproton spectra from the nickel isotopes Ni^{58}, Ni^{60}, and Ni^{64} observed at $E_{\gamma\ max} = 24$ MeV.

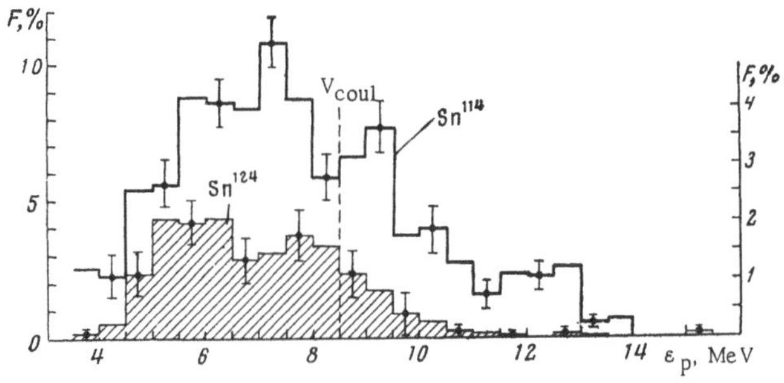

Fig. 35. Comparison of photoproton spectra from the tin isotopes Sn^{114} (left-hand scale) and Sn^{124} (right-hand scale) at $E_{\gamma\ max} = 23.5$ MeV.

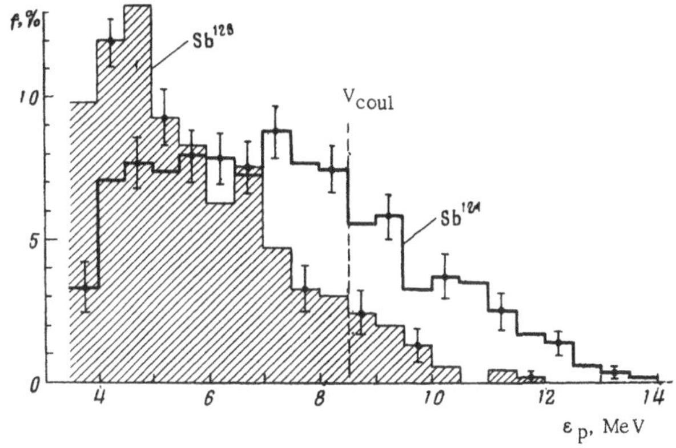

Fig. 36. Comparison of photoproton spectra from the antimony isotopes Sb^{121} and Sb^{123} at $E_{\gamma\ max} = 19.5$ MeV.

1. Photoproton Spectra from Various Isotopes of a Given Element. Photoproton yields from vari^us isotopes of a given element drop in proportion to the increase in mass number A (see Fig. 30). There is no doubt but that such a variation in yields is associated with an increase in proton binding energy as the number of neutrons in the nucleus increases. This circumstance, in particular, leads to a sharp increase in competition from (γ, n) reactions during evaporation. However, it was shown above that statistical theory fails to explain the observed variations in photoproton yields as a function of A (and the yields themselves, in the majority of cases). It is most likely that these variations are associated with various structural features of the dipole levels in individual isotopes. One can expect that this will be reflected in the spectral shape and angular distribution of the photoprotons.

The spectra we obtained from isotopes of nickel (Ni^{58}, Ni^{60}, and Ni^{64}) and of tin (Sn^{114} and Sn^{124}) at $E_{\gamma\ max} \approx 24$ MeV, and from antimony (Sb^{121} and Sb^{123}) at $E_{\gamma\ max} \approx 19.5$ MeV, are shown in Figs. 34-36. The total number of protons from the lightest isotope was assumed to be 100. The areas bounded by the histograms are proportional to the yields. In the case of tin, for which it was impossible to make corrections for the contributions from other isotopes, the spectra shown correspond to spectra from targets having the compositions given in Table 1. As the table shows, the enrichment of the Sn^{124} isotope was close to 100%. In the target enriched in Sn^{114}, there was a considerable admixture of other light tin isotopes (Sn^{116}, Sn^{117}, and Sn^{118}), but the admixture of heavy isotopes was insignificant.

TABLE 7. Comparison of the Shapes of Photoproton Spectra from Various Isotopes of a Given Element

Isotope	Thres. B_p, MeV	V_{coul}, MeV	Energy at the maximum of the proton spectrum, MeV	Fraction of protons (%) with energy $\varepsilon_p < V_{coul}$
Ni^{58}	7.86	5.94	5	58
Ni^{60}	9.54	5.88	4	76
Ni^{64}	14.41	5.75	4	90
Sn^{114}	8.4	8.58	7.5	67
Sn^{124}	12.0	8.34	6	83
Sb^{121}	5.81	8.59	7	72
Sb^{123}	6.65	8.54	5	93

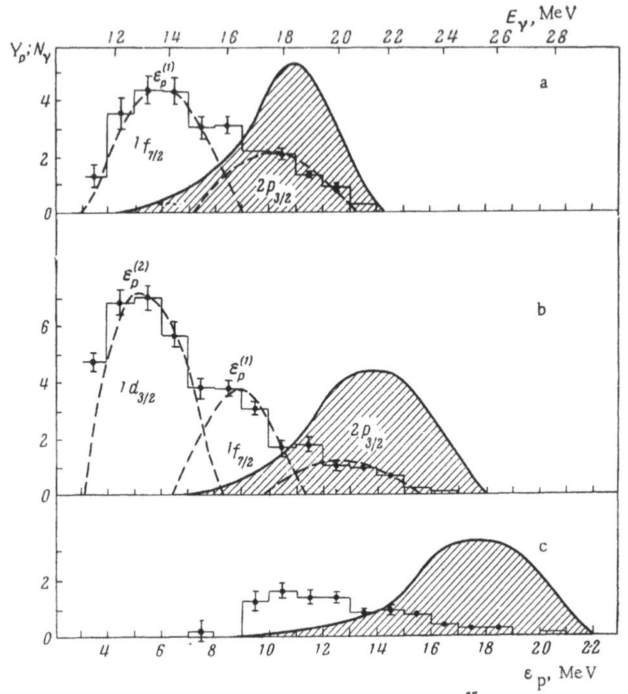

Fig. 37. Difference spectra for photoprotons from Cu^{65} at varying upper limits $E_{\gamma\ max}$. (a) $E_{\gamma\ max}$ = 20.3-18 MeV; (b) 24.5-20.3 MeV; (c) 28.5-24.5 MeV. γ-Ray spectra are shaded.

One can see that the observed spectra are different for the different isotopes. Furthermore, a general rule for all the nuclei considered is worthy of note: the photoproton spectra become softer with increasing isotopic weight. The shift in the direction of lower energies is so substantial that the energy ε_p of the greater portion of the emitted protons becomes less than the height of the Coulomb barrier V_{coul} in thecase of the heaviest isotopes (Table 7).

The rule is qualitatively explainable from the point of view of the shell (particle-hole) model. According to this model, the energy of the dipole states E_d is practically identical for neighboring nuclei. The energy of the protons emitted as the result of the decay of a particle-hole configuration is equal to the difference of the energy of the absorbed γ-ray E_d and the binding energy of the protons B in the shells which are excited by dipole absorption. The binding energy of the protons increases with increasing number of neutrons in nickel and tin isotopes; consequently, the energy with which the protons can be emitted for a given value of γ-ray energy must decrease.* Softening of the spectra leads to a reduction in proton penetration through the barrier (or even to their falling into the negative energy region). Thus the reduction in photoproton yield occurring with increasing A for a given Z finds a reasonable explanation in the fact that the photoproton spectra are shifted towards lower energies.

*It is interesting to note that the spectrum of photoprotons from Ni^{58}, where the evaporation contribution is evidently large, is harder than the spectra for Ni^{60} and Ni^{64}, where evaporation is certainly slight.

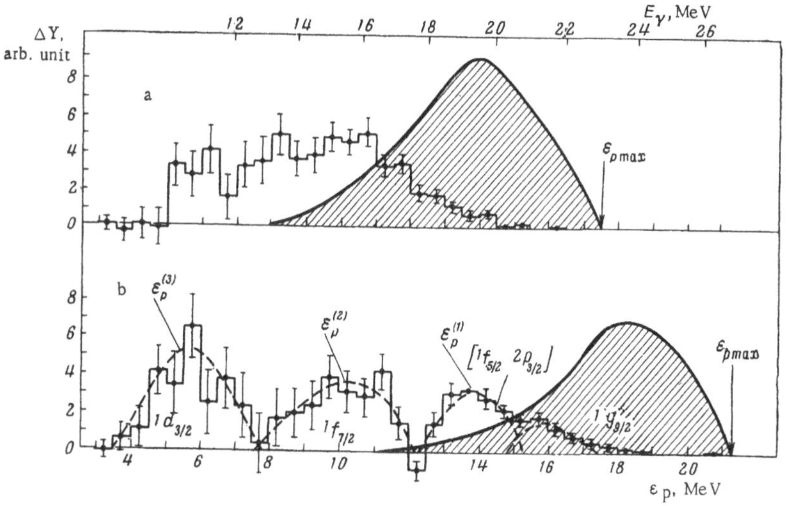

Fig. 38. Difference spectra for photoprotons from Nb^{93}. (a) $E_{\gamma\,max}$ = 23.5-19.5 MeV; (b) 27.5-23.5 MeV.

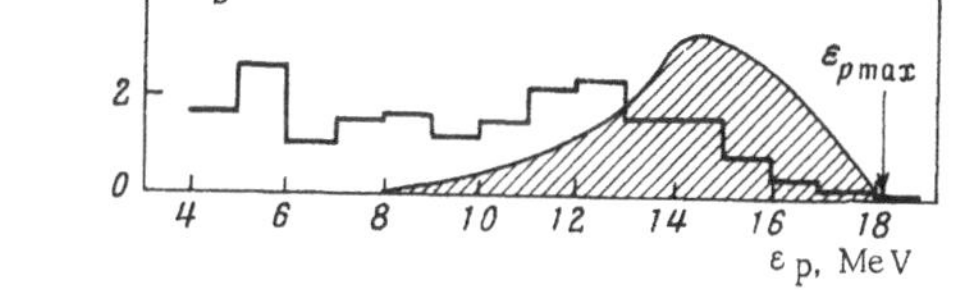

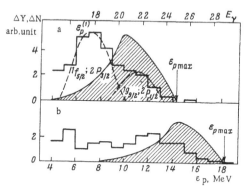

Fig. 39. Difference spectra for photoprotons from Cd^{114}.
(a) $E_{\gamma\,max}$ = 24.5-20.3 MeV; (b) 28.5-24.5 MeV.

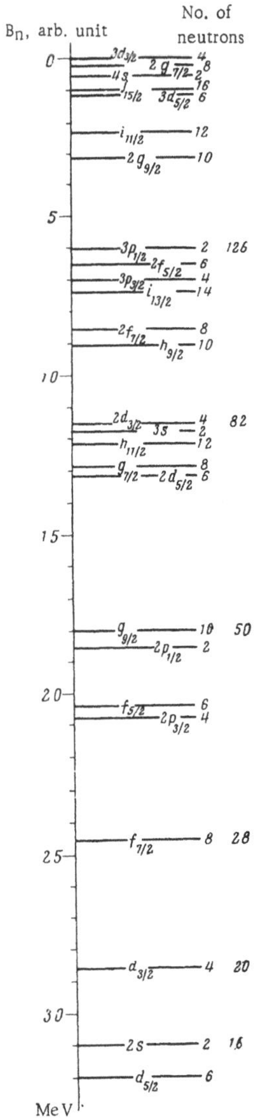

Fig. 40. Semiempirical scheme for the sequence of nuclear levels (taken from [139]).

In the case of tin, still another interesting fact is to be noted. The angular distribution of photoprotons for Sn^{114} (see Table 4) is considerably more isotropic than in the case of Sn^{124}. The degree of anisotropy b/a for all detected protons with energies $\varepsilon_p \geq 3.5$ MeV is 0.10 ± 0.17 for Sn^{114} and 0.61 ± 0.28 for Sn^{124}. This difference is even greater for fast protons ($\varepsilon_p \geq 7$-8 MeV); b/a is about 0.5 for Sn^{114} and 2.25 Sn^{124}. In tin, the shell structure for 50 protons is completed. The highest level is the $1g_{9/2}$, which is followed by the $2p_{1/2}$, and $2p_{3/2}$ levels. The large anisotropy of the angular distributions in the case of Sn^{124} is evidence of the fact that the relative role of the protons emitted with small momenta l is significantly greater in the case of Sn^{124}. This conclusion is in complete accord with the results of measurements of the yields of the ground (spin $+\,^9/_2$) and isomeric (spin $-\,^1/_2$) states of In produced by (γ, p) reactions in various tin isotopes [108, 137]. In these measurements, it was found that in the case of Sn^{124} (γ, p)In^{123} reactions the dominant yield was In* (spin $-\,^1/_2$), which can be considered as tin with a hole in the $2p_{1/2}$ shell, while the yield of In (spin $^9/_2$), which can be represented as tin with a hole in the $1g_{9/2}$ shell, was a few percent at most. In the case of the lighter isotopes (Sn^{118}, ^{120}, ^{122}), the relative yield of In (spin $^9/_2$) increased. The ratio of the yields of In (spin $^9/_2$) and In* (spin $-\,^1/_2$) was 0.07 for the Sn^{124}(γ, p) reaction and 0.65 for the Sn^{118}(γ, p) reaction

It is clear from Table 7 that the difference of the binding energies B_p in the isotopes Sb^{121} and Sb^{123} is small. Consequently, the explanation given above involving the softening of spectra with increasing neutron number N is unsuitable. However, the difference in spectra can be explained qualitatively in terms of the decay of particle-hole configurations in the case of Sb also. Spectra from antimony isotopes were measured at a relatively low energy $E_{\gamma\,max}$ = 19.5 MeV.* At this energy, one should expect that the 51-st "valence" proton found in the Sb nucleus outside the Z = 50 shell would play an important role. Its role should be particularly noticeable in the high-energy proton region. The difference in spectra observed in this region is evidently associated with the fact that the "valence" proton is in the $2d_{5/2}$ state in the Sb^{122} nucleus and in the $1g_{7/2}$ state in the Sb^{123} nucleus. The reduction in penetration, which is associated with a higher value of the centrifugal barrier for the "valence" proton in the Sb^{123} nucleus qualitatively explains the considerably smaller fraction of protons with energies $\varepsilon_p > 8$-9 MeV in the case of this nucleus in comparison with that observed for the Sb^{121} isotope.

2. Changes in Spectral Shape as a Function of $E_{\gamma\,max}$. It was observed [14] that a sharp increase in the fraction of fast protons with energies $\varepsilon_p \geq 10$ MeV occurred in the case of copper with a change in the upper limit of the bremsstrahlung spectrum $E_{\gamma\,max}$ from 24 to 28 MeV. At the same time, there was a change in the shape of the angular distributions from $f(\theta) \approx \sin^2\theta$ to $f(\theta) = a + b\sin^2\theta$ where b/a was 0.1-0.2 at most.

* This is connected with the considerable admixture of CH_2 in the irradiated target. The (γ, p) threshold for C^{12} is about 16 MeV. With increasing $E_{\gamma\,max}$, the C^{12} background becomes considerably greater than the photoproton yield from the antimony being investigated.

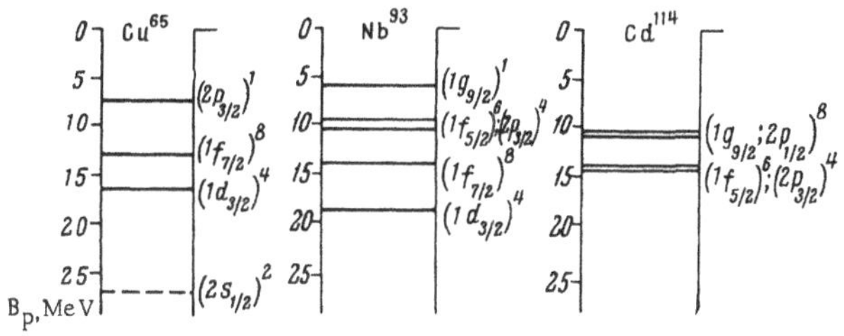

Fig. 41. Diagram of the higher proton levels in Cu^{65}, Nb^{93}, and Cd^{114} obtained from an analysis of photoproton difference spectra.

These results were interpreted in the following manner. For $E_{\gamma\ max}$ = 24 MeV, the yield of fast protons, which is produced by the "tail" of the bremsstrahlung spectrum, is associated with photoeffects in the uppermost $2p_{3/2}$ level where a single proton is located. For $E_{\gamma\ max}$ = 28 MeV, particles from the deeper $1f_{7/2}$ level, where there are eight particles, are also included in the fast protons. This leads to an abrupt change in the shape of the angular distribution. Knowing the energy $E_{\gamma\ max}$ for which this occurs, one can estimate the spacing between the $2p_{3/2}$ and $1f_{7/2}$ levels, which turns out to be 5-6 MeV.

As already noted, more definite data with respect to the binding energy of protons in the deeper levels should be obtained by means of monochromatic γ-rays. However, by working with bremsstrahlung, one can also obtain a particle spectrum produced by γ-rays with energies in a comparatively narrow range if one measures with sufficient accuracy spectra for two not very different values of the upper limit $E_{\gamma\ max}$, normalizes them in the proper manner, and takes their difference.

Measurements of spectra and angular distributions were made by us for several values of the energy $E_{\gamma\ max}$ with the nuclei Cu^{65}, Nb^{93}, and Cd^{114}. The spectral differences $\Delta Y(\varepsilon_p)$, normalized to unit solid angle and the same ionization in the thick-walled chamber, are shown in Figs. 37-39. The quantity $\Delta Y(\varepsilon_p)$ corresponds to the increment in the number of protons with energies in the range from $\varepsilon_p = -0.5$ to $\varepsilon_p = +0.5$ for a change in the upper limit of the bremsstrahlung spectrum from $E^{(1)}_{\gamma\ max}$ to $E^{(2)}_{\gamma\ max}$. This quantity was determined from the formula

$$\Delta Y(\varepsilon_p) = [Y_p f(\varepsilon_p)]_{E^{(2)}_{\gamma\ max}} - [Y_p f(\varepsilon_p)]_{E^{(2)}_{\gamma\ max}}, \qquad (4.27)$$

where Y_p is the measured photoproton yield (values for the yields are given in Table 5); and $f(\varepsilon_p)$ is the fraction of protons with energies in the range from $\varepsilon_p = -0.5$ MeV to $\varepsilon_p = +0.5$ MeV (obtained from the spectra in Figs. 13, 15, and 16).

Also shown in Figs. 37-39 are the differences of the corresponding bremsstrahlung spectra:

$$\Delta N_\gamma(E_\gamma) = N_\gamma(E_\gamma E^{(2)}_{\gamma\ max}) - N_\gamma(E_\gamma, E^{(1)}_{\gamma\ max}), \qquad (4.28)$$

i.e., the γ-ray spectra which produce the photoproton difference spectra $\Delta Y(\varepsilon_p)$. The energy scale of the γ-rays is shifted with respect to the energy scale of the protons by an amount equal to the binding energy B_p of the least bound proton. On such a representation, it is evident that the protons emitted from the highest level (they leave the final nucleus in the ground state) must produce a spectrum which overlaps the γ-ray spectrum. Protons emitted as the result of direct, or quasidirect, photoeffect in deeper levels, or as the result of the decay of multiparticle configurations, will give a softer spectrum.

In Figs. 37-39, one can notice the following:

1. The shape of the photoproton spectra for all three nuclei changes significantly as a function of γ-ray energy (this, incidentally, indicates once again that the emission of protons cannot be described by a statistical

theory in which the spectral shape is determined by the level density of the final nucleus $\rho_R(E_R^*)$ and is slightly dependent on the excitation of the compound nucleus $E_C^* = E_\gamma$).

2. The formation of excited states in the final nucleus as the result of proton emission plays an important role (all the spectra contain a soft component).

3. The contribution from the least bound protons increases with increasing γ-ray energy in the observed spectra for each of the nuclei studied.

4. At γ-ray energies $E_\gamma = 20$-22 MeV (Figs. 37b, 38a, and 39a), which correspond to the maximum in the cross section for the (γ, p) reaction in medium nuclei,* the role of the photoprotons from the highest level is unimportant in the case of Cu^{65} and Nb^{93} and significant in the case of Cd^{114}. The latter fact can be explained from the point of view of the shell model; in Cu^{65} and Nb^{93} nuclei, the highest level contains a single proton in the $2p_{3/2}$ and $1g_{9/2}$ states, respectively, above the completed shells at Z = 28 and Z = 40. In Cd^{114} (Z = 48), the filling of the shell at Z = 50 is almost complete, and there are eight particles in the highest level.

5. In the case of monochromatic γ-rays, as already pointed out, the spectra of photoprotons which are emitted as the result of a "direct" (resonance or nonresonance) photoeffect can be a source of information about proton hole states in the nucleus or about proton binding energies in the deeper shells. Maxima should be observed in such spectra at energies $\varepsilon_p^{(i)} = E_\gamma - B_i$. Furthermore, the difference $E_\gamma - \varepsilon_p^{(i)} = B_p^{(i)}$ should remain constant with changes in γ-ray energy.

An attempt is made below to determine the energies $B_p^{(i)}$ by attributing the $\Delta Y(\varepsilon_p)$ spectra to the processes mentioned on the basis of the empirical scheme for filling of nuclear levels given in [139].†

(A) Cu^{65} (Z = 29). According to the scheme shown in Fig. 40, the upper proton levels $(2p_{3/2})^1$, $(1f_{7/2})^8$, $(1d_{3/2})^4$, $(2s_{1/2})^2$, and $(1f_{5/2})^6$ can participate in dipole absorption of γ-rays by the copper nucleus. As the result of the emission of an excited proton into the continuum, it is possible to product the $(1f_{7/2})^{-1}$, $(1d_{3/2})^{-1}$, $(2s_{1/2})^{-1}$, and $(1f_{5/2})^{-1}$ hole states. It should be pointed out that any of the states mentioned can only be produced in a direct, nonresonance photoeffect. In the case of a resonance process, the protons in the $(2s_{1/2})^2$ and $(1d_{5/2})^6$ levels are transferred by dipole absorption into the higher unfilled $(2p_{3/2})^1$ level and appear, most likely, in bound states of the type $(2s_{1/2})^{-1}$, ..., $(2p_{3/2})^2$, and $(1d_{5/2})^{-1}$, ..., $(2p_{3/2})^2$, respectively (it is reasonable to assume that pairing of protons in the $2p_{3/2}$ level leads to an increase in their binding energy in this level rather than to a decrease in comparison with the energy in the ground state of the copper nucleus).

We now consider the difference spectra for photoprotons from Cu^{65}, which are shown in Fig. 37. No structure is observed in these spectra (although structure is noticeable in the original spectra shown in Fig. 13). However, the asymmetric shape of the proton spectra for $E_\gamma = 17$-20 MeV (Fig. 37a) allows one to consider it as made up of two peaks. Attributing them to a photoeffect in the upper $(2p_{3/2})^1$ level and in the next lower $(1f_{7/2})^8$ level, we obtain for the proton binding energy in the $(1f_{7/2})^8$ level of the Cu^{65} nucleus a value

$$B_p(1f_{7/2}) = E_\gamma^{av} - \varepsilon_p^{(1)} = 13 \text{ MeV}.$$

At $E_\gamma = 20$-23 MeV in Fig. 37b, there are shoulders at proton energies of 7 and 10 MeV which exceed more than four statistical errors. This permits one to consider the particular spectrum is made up of three peaks. Attributing them to photoeffects in the $(2p_{3/2})^1$, $(1f_{7/2})^8$, and $(1d_{3/2})^4$ levels, we obtain

$$B_p(1f_{7/2}) = E_\gamma^{av} - \varepsilon_p^{(1)} = 13.5 \text{ MeV},$$

---

*The position of the maximum of the (γ, p) cross section, which is shifted several MeV to the right with respect to the maximum of the giant resonance and which is independent of A, is located at an energy $E_\gamma = 20$-22 MeV [107].

†One can expect the appearance of separate maxima in the $\Delta Y(\varepsilon_p)$ spectra with sufficiently good resolution, i.e., under conditions where the spacing between hole levels is greater than the sum of the γ-ray width and the proton energy resolution.

which is in good agreement with the data from the spectrum in Fig. 37a considered above. In a similar manner, we obtain for the binding energy of the protons in the $(1d_{3/2})^4$ level:

$$B_p(1d_{3/2}) = E_\gamma^{av} - \varepsilon_p^{(2)} = 16.5 \text{ MeV}.$$

The data from the spectrum in Fig. 37a, if it is treated as the result of nuclear photoeffects in the $2p_{3/2}$, $1f_{7/2}$, and $1d_{3/2}$ levels, does not contradict the data from Figs. 38a and 38b. An interesting feature of Fig. 38c is the absence from the spectrum of protons with energies in the range $\varepsilon_p = 3$-10 MeV. This situation apparently indicates that protons from the $(2s_{1/2})^2$ level drop into a bound state in the unfilled $2p_{3/2}$ level as the result of a resonance photoeffect, as has already been pointed out, and that the direct nonresonance process for emission of protons with energies $\varepsilon_p > 3$ MeV from the $(2s_{1/2})^2$ level is impossible on the basis of energy considerations. If the latter is correct, the binding energy of protons in the $(2s_{1/2})^2$ level of the Cu^{65} nucleus is not less than 25 MeV

$$B_p(2s_{1/2}) \gtrsim E_{\gamma\,\max} - \varepsilon_{p\,\min} = 25 \text{ MeV}.$$

(B) Nb^{93} (Z = 41). According to Fig. 40, the arrangement of proton levels in the ground state of the Nb^{93} nucleus has the form $(1g_{9/2})^1$, $(1f_{5/2})^6$, $(2p_{3/2})^4$, $(1f_{7/2})^8$, $(1d_{3/2})$, etc. As the result of dipole photoeffect, it is possible to produce the hole states $(1f_{5/2})^{-1}$, $(2p_{3/2})^{-1}$, $(1f_{7/2})^{-1}$, and $(1d_{3/2})^{-1}$.

The spectrum in Fig. 38a does not allow one to reach any sort of conclusion with respect to the proton binding energy in the deeper levels although it is clearly evident that the protons ejected from those levels play an important role.

Three peaks were resolved in the spectrum of Fig. 38b at proton energies ε_p^i of 6, 10, and 14 MeV. According to the level scheme shown in Fig. 40, the first of them can be attributed to photoeffect in the $1d_{3/2}$ level, the second to photoeffect in the $1f_{7/2}$ level, and the third is apparently produced by the two neighboring $1f_{5/2}$ and $2p_{3/2}$ levels. In a manner similar to that used above, we obtain for the proton binding energies of these levels in the Nb^{93} nucleus

$$B_p(1f_{5/2};\ 2p_{3/2})_{av} = E_{\gamma\,av} - \varepsilon_p^{(1)} = 10 \text{ MeV};$$

$$B_p(1f_{7/2}) = 14 \text{ MeV};$$

$$B_p(1d_{3/2}) = 18.5 \text{ MeV}.$$

(C) Cd^{114} (Z = 48). The level scheme of the Cd^{114} nucleus is the following: $[(1g_{9/2})(2p_{1/2})]^8$, $[(1f_{5/2})^6 (2p_{3/2})^4]$, $(1f_{7/2})^8$, $(1d_{3/2})^4$, etc.

Closely spaced levels (according to the scheme of Fig. 40) are enclosed in the square brackets. Evidently, precisely those levels determine the proton spectrum shown in Fig. 39a. On the basis of Fig. 39a, one can conclude that the group of $1f_{5/2}$, $2p_{3/2}$ levels is located 4 MeV lower than the least bound group of $1g_{9/2}$, $2p_{1/2}$ levels, i.e., the average binding energy for the first group of levels is

$$B_p[(1f_{5/2};\ 2p_{3/2})]_{av} = 14 \text{ MeV}.$$

The slower protons in the spectrum of Fig. 39b, in all probability, are produced in transitions from the still deeper $1f_{7/2}$ shell and others.

A composite diagram of the location of proton levels in Cu^{65}, Nb^{93}, and Cd^{114} nuclei, based on an analysis of the difference spectra $\Delta Y(\varepsilon_p)$, is shown in Fig. 41.

Our data do not contradict the level scheme shown in Fig. 40; the spacing between proton levels as determined by the analysis just made is 3-5 MeV.

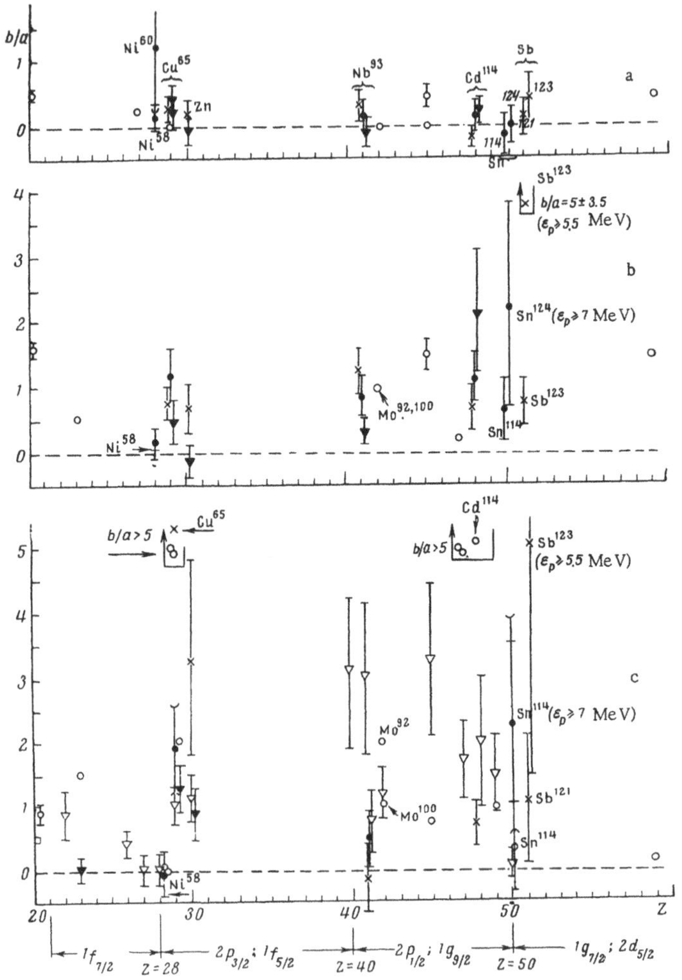

Fig. 42. Anisotropy of photoproton angular distributions (ratio b/a for a distribution of the type $f(\theta) = a + b\sin^2(\theta)$. Shown below is a diagram of completed proton shells. (a) Slow protons (ε_p = 3-6 MeV); (b) medium-energy protons (ε_p = 6-9 MeV); (c) fast protons ($\varepsilon_p \geq 9$ MeV); (×) data of the present work at E_γ max ≈ 20 MeV; (●) at E_γ max = 24 MeV; (▼) at E_γ max ≈ 28 MeV; (▽) data from [140]; (○) data from [14, 15, 78, 79, 81, 111, 141-143].

It should be emphasized, however, that the preceding discussion is of quite a qualitative nature. Consequently, it is impossible to consider as convincing any conclusion that the proton spectra actually confirm the particle-hole model of the giant resonance and that they can serve as a source of information about proton binding energies in deeper nuclear levels. More specifically analyzed data on photonucleon spectra produced by monochromatic γ-rays are required.

3. Angular Distribution of the Photoprotons

As can be seen from Figs. 21-29, the observed angular distributions are quite varied in shape, and depend on γ-ray energy, on the selected proton energy range, on Z, and on A for a given Z. As a rule, they can be approximated by the relation

$$f(\theta) = a + b\sin^2\theta, \tag{4.29}$$

which indicates the dipole nature of γ-ray absorption in the processes we investigated ($50 \leqslant A \leqslant 125$; $E_{\gamma\,max} \leq 28$ MeV). A shift of the maximum toward the forward direction, which is evidence of interference with quadrupole absorption of quanta, was observed only in the case of Ni isotopes ($E_{\gamma\,max}$=24 MeV), Cu^{65} (24.5 and 28.5 MeV), and Nb^{93} (28.5 MeV, the fastest protons). The results obtained agree with the data of Shevchenko and Yur'ev [111], who showed that the absorption is essentially quadrupole in the γ-ray energy region beyond the giant resonance in the case of the specified range of atomic weights A.

Because of the accuracy in the determination of the quantity b/a from the observed distributions is low, and because the data were obtained with a continuous γ-ray spectrum, it was decided to point out some qualitative regularities in the observed angular distributions without going into any detailed discussion of them.

Values of the ratio b/a found by ourselves and other authors for protons of various energy groups are shown in Fig. 42. One can note that the distribution of the slowest protons, as a rule, reveals an anisotropy. However, the value of b/a does not exceed 0.5, and is about 0.2 on the average. Protons of medium energy are more anisotropic, and the ratio b/a for them is 0.5-0.6 on the average.

The greatest anisotropy is displayed by the fastest protons. An interesting characteristic is the appearance of values $b/a > 2$-3. For the nuclei investigated, where an s-state is certainly not among the higher levels, such an anisotropy is impossible according to the models of Courant [2] and Wilkinson [28] if one does not take into account the interference of $l \to l+1$ and $l \to l-1$ transitions.*

Eichler and Weidenmuller [144] thoroughly discussed the angular distributions we obtained for fast protons from Cu and Zn where the highest level is $2p_{3/2}$. They established that the degree of anisotropy increases as the result of interference of p→s and p→d transitions in comparison with that which is observed for each of the transitions individually. The distribution takes the form $f(\theta) = \sin^2\theta$ when the relative phases for each of these transitions becomes 90°, i.e., the resonance energies of the specified transitions coincide.

Therefore, the observed large anisotropy of fast protons in the case of nuclei with $29 \leq Z \leq 40$ is evidently associated with the contribution from photoeffects in the $2p_{3/2}$ level, and in the case of nuclei with $Z \approx 50$, it is produced by the $2p_{1/2}$ level (see the level scheme in Fig. 40). This is confirmed by the fact that p-state protons are absent in the levels which can make a contribution to dipole absorption in nuclei with $Z < 28$, and the anisotropy of the angular distributions is small [$(b/a) < 1$].

*In these models, the value of the ratio b/a is determined by the relations

$$b/a = \frac{l+2}{2l} \quad \text{for} \quad l \to l+1 \quad \text{transitions}$$
$$b/a = \frac{l-1}{2(l+1)} \quad \text{for} \quad l \to l-2 \quad \text{transitions.} \tag{4.30}$$

The existence of angular distributions close to $\sin^2\theta$ in nuclei for which the photoeffect in p shells can play a role is evidence that the resonance energies of the p→d and p→s transitions coincide. This can evidently be considered as direct indication that configurations are mixed in the dipole states.

Conclusions

From the study of the (γ, p) reaction in medium-weight nuclei, one can arrive at the following conclusions:

1. As a rule, yields of photoprotons from medium-weight nuclei ($20 \leq Z \leq 60$) for excitations in the giant resonance region are considerably greater than predicted by the evaporation model.

2. The spectra and angular distribution parameters for photoprotons from nuclei where evaporation is small are quite varied and reveal dependence on γ-ray energy, on atomic number, and on the number of neutrons N in the case of isotopes with a given Z. This indicates the cross section for photoproton emission is sensitive to nuclear shell structure.

3. As a rule, a large fraction of slow protons is observed in the spectra even in cases where there is a considerable excess (by factors of 10-100) of the observed yield over that predicted by statistical theory. Sometimes the spectra are quite like evaporation spectra (Cu, Ni^{60}). This shows that a resemblance of the observed spectrum to an evaporation spectrum is insufficient for confirmation that the evaporation mechanism actually occurs; this is something that is often done.

4. An interesting feature observed for practically the first time in spectra of photoprotons from isotopes of the same element (Ni, Sn, Nb) is that they become softer with increasing isotopic weight.* Evidently, exactly this fact is the reason for the observed reduction in photoproton yield with increase in A for a given Z [107, 108].

5. As a rule, the angular distributions can be approximated by the relation $f(\theta) = a + b\sin^2\theta$, which is evidence of the dipole nature of γ-ray absorption in the processes we investigated ($50 \leq A \leq 125$; $E_{\gamma\,max} = 28$ MeV).

Comparison with the calculations of Eichler and Weidenmuller [144] shows that the value of b/a for angular distributions of fast protons from nuclei where there is a contribution from photoeffect in p-levels (Zn, Cu, Cd, and Sn) can only be explained on the assumption that the resonance energies of p→s and p→d transitions coincide. This fact can obviously be considered as direct confirmation of considerable mixing of particle-hole configurations in nuclear dipole states.

From the results we obtained, and from the data of other authors presented in the review, one can arrive at a number of conclusions which are of interest with respect to clarification of the mechanisms for photonuclear processes and which are useful for clarifying the as yet obscure causes for the width of the giant resonance.

A. That statistical theory (at least in the form in which it is successfully used to describe nucleon-nucleus interactions) poorly explains the set of data for photonuclear reactions in the region of the dipole giant resonance can obviously be regarded as reliably established.

In the case of nuclei with $A > 40$, in which the giant resonance is certainly located in a region where a large number of compound nucleus levels overlap, the contribution from processes other than evaporation is not 5-10% of the γ-ray absorption cross section, as is often assumed, but is at least comparable to evaporation.

The fact that photonuclear reactions are more poorly described by statistical theory than nucleon interactions at the same excitations may be explained in two ways: either by a large contribution from "direct processes" corresponding to the immediate ejection of nucleons from the nucleus in γ-ray interactions, or by the fact that the initial ("input") nucleon configuration produced by the absorption of γ-rays has a smaller probability of forming a long-lived compound nucleus as a result of interactions with the other nucleons (with

*The only previous data of this kind are the quite inaccurate results of Butler and Almy [79], who measured the spectra of photoprotons from targets enriched in the isotopes Mo^{92} and Mo^{100} for $E_{\gamma\,max} = 22.5$ MeV.

equilibrium established among the numerous degrees of freedom of intranuclear motion and with the decay properties described by the evaporation model) in comparison with the "input" configuration produced in the absorption of nucleons.

Cross sections considerably less than those observed experimentally were obtained in calculations of the probability for the direct photoeffect made by Courant [2] and other investigators [61, 62]. Therefore, the deviation from statistical theory is more likely the result of the second cause.

One can point out two causes which may be responsible, at equal excitation energies, for the lower probability of long-lived compound nucleus formation (with "thermal" equilibrium) in the case of γ-ray absorption:

(1) According to modern shell theory, the giant resonance is made up of states which are a mixture of particle-hole configurations. At the same time, an "input" configuration of two particles above the Fermi surface and a hole is formed by the absorption of a nucleon. It is evident that the probability of emission of an excited nucleon into the continuum without interaction with the other nucleons is considerably greater in the case of a particle-hole "input" configuration than for a two particle-hole configuration because all the excitation energy is concentrated in a single nucleon in the first case. In the second case, this energy is distributed between two particles and the probability that the excited nucleons will be found in bound states is large.

(2) The forbidding of internucleon collisions associated with the conversion of angular momentum is considerably stronger in the case of γ-ray interactions. The point is that dipole quanta can only excite those compound nucleus states in which the angular momentum differs from the spin of the target nucleus by $\Delta J = \pm 1$. Nucleons with energy ε can be absorbed with a set of momenta from 0 to $l_{max} \approx R/\lambda \approx 0.3\ A^{1/3}\varepsilon^{1/2}$ (here, ε is expressed in MeV). Thus in the case of medium-weight nuclei and a nucleon energy $\varepsilon = 10$-15 MeV, which leads to excitation in the giant resonance region, one obtains a value $l_{max} = 4$-5; consequently, the nucleon will excite any states with ΔJ from 0 to ± 5.

It is, therefore, completely possible that out of the tremendous number of compound nucleus levels in the giant resonance region which are excited by nucleon absorption, so few levels are excited by γ-ray dipole absorption that the condition $\Gamma_C \gg D_c$, necessary for the application of statistical theory, is not fulfilled. It is also completely possible that in photonuclear processes there fully appears the first clearly observed mechanism intermediate between quasidirect breakdown of an "input" configuration and evaporation.

B. An explanation of the nature of the mechanism other than evaporation observed in photonuclear reactions is of considerable interest. In particular, to understand the causes underlying the width of the giant resonance $\Gamma(\sigma_\gamma)$ (see Chaper I) in doubly magic nuclei, it would be important to establish whether this mechanism is the result of the decay of dipole shell configurations of the particle-hole type (quasidirect effect) or whether it was produced by other processes of an intermediate nature (between direct photoeffect and evaporation), for example, by the decay of a two particle-hole pair configuration which was initially excited by γ-ray absorption (Balashov and Chernov [58]) or which was produced as the result of a single collision of an initially excited nucleon with other nucleons in the nucleus (Danos and Greiner [60]).

In our opinion, the observed pronounced dependence of spectra and angular distributions of photoprotons on E, Z, and N, which indicates extreme sensitivity to nuclear shell structure, is more likely evidence in favor of the quasidirect effect.

Qualitative consideration of the photoproton spectra from various isotopes of Ni, Sn, and Sb, and also of the difference spectra for Cu^{65}, Nb^{93}, and Cd^{114}, demonstrated that these spectra can be reasonably interpreted in the framework of the particle-hole model, and can serve as a source of information about nucleon binding energies in deeper nuclear levels. It should be pointed out, however, that it is impossible to consider these conclusions as final because the presence of a large number of slow particles in the photoproton spectra can be associated not only with the emission of protons from internal nuclear shells, but also with the breakdown of two particle-hole pair configurations.

For a decisive choice of the model for a mechanism other than evaporation, it is necessary to measure experimentally the photonucleon spectra for monochromatic γ-ray interactions and to compare them with exact shell-model calculations. It would also be very interesting to obtain some kind of theoretical prediction with regard to the shape of the photonucleon spectrum which one ought to expect from the decay of two particle-hole pair configurations.

C. If one considers that the deviation from the evaporation model results from the quasidirect photoeffect, the comparability of the contribution from a process other than evaporation to the contribution from evaporation, in terms of the shell model of the giant resonance, mean the following. The portion of the decay width of the dipole configurations corresponding to direct emission of excited nucleons into the continuum, Γ_{part}, is comparable to the broadening of these states $\Delta\Gamma$ because of their decay through the excitation of more complex configurations (if one considers, as is usually done, that in the latter case a large number of compound nucleus levels are excited, the decay of which occurs according to the laws of statistical theory).

As already noted $\Gamma_{part} \approx 200$-$300$ keV in Zr^{90} [41], and there is no reason to suppose that this quantity is significantly greater for other nuclei in the region of A under consideration (except for the lightest nuclei). Thus the sum $\Gamma_{part} + \Delta\Gamma$ makes a small contribution to the width $\Gamma(\sigma_\gamma)$ of the giant resonance in doubly magic nuclei (and to the width of each of the peaks in deformed nuclei), which is approximately 4 MeV. Consequently, the broadening of the giant resonance not associated with static deformation of the nucleus results from the energy spread ΔE_d of the dipole particle-hole states rather than from their final lifetimes. Therefore, the concept of the existence of a Brown-Bolsterli state [34], which exhausts almost the entire dipole sum, is incorrect.

In conclusion, the author considers it a great pleasure to express her deep appreciation to B. S. Ratner, with whom this work was jointly begun, for constant support and consideration; to V. V. Balashov and S. F. Semenko for valuable discussions; to N. V. Lin'kova, V. G. Seryapin, R. Sh. Amirov, and V. V. Akindinov for participating in various stages of the investigation; to R. D. Rozhdesvenska and N. V. Ponomareva for emulsion scanning and for assistance in the analysis of the results; to N. I. Isotov and V. P. Lyubimov for assistance in experimental design; to R. A. Latypova and V. P. Fomina, members of the FIAN computing group, who wrote programs for analyzing the experimental data and for calculating solid angles; to the staff members who prepared the targets and did the isotope enrichment, and to synchrotron operating group.

Appendix. Evaluation of the Magnitude of the Equivalent Range ΔR_{equ}

The energy loss per unit length for particles is determined from the Bethe-Block formula, which can be represented in the nonrelativistic case by

$$-\frac{d\varepsilon}{dR} = \frac{4\pi e^4 z^2}{m} \cdot \frac{NB}{v^2} = \frac{2\pi z^2 e^4}{\varepsilon}\left(\frac{M}{m}\right) NB, \tag{1}$$

where e and m are the absolute values of the electronic charge and mass; z, M, v, and ε are the atomic number, mass, velocity, and energy of the incident particle; N is the number of atoms per cm^3 of the stopping material; B is a dimensionless quantity which depends on the nature of the stopping material (the so-called stopping power) and which is given by

$$B = Z \ln \frac{2mv^2}{I} = Z \ln\left(4\,\frac{m}{M} \cdot \frac{\varepsilon}{I}\right), \tag{2}$$

where Z is the atomic number of the medium; and I is the average excitation potential of its atoms.

It is assumed that I is independent of particle velocity and is equal to I_0 (I_0 is an experimentally determined constant and is equal to the Rydberg constant (13.5 eV) in order of magnitude). The data with respect to the value of I_0 are contradictory (for example, see the review presented in [145]). In addition, it is not clear how valid the assumption about the velocity-independent of I_0 is. In this connection, Lindhard and Scharff [146] proposed the use of another relation for the computation of stopping power which was based on the Thomas-Fermi model:

$$B = ZL(x), \tag{3}$$

where $L(x)$ is an experimentally determined universal function depending neither on atomic number Z of the medium nor on particle velocity v; x is a dimensionless parameter given by

$$x = \left(\frac{v}{u_0}\right)^2 Z^{-1}, \tag{4}$$

where $u_0 = e^2/\hbar = 0.218 \times 10^9$ cm/sec.

It turned out that the empirical dependence of $L(x)$ could be expressed in analytic form

$$L(x) = 1.36\, x^{1/2} - 0.016\, x^{3/2} = 1.36\, x^{1/2}(1 - 0.012\, x). \tag{5}$$

An estimate shows that the value of x does not exceed 10 in the range of materials and proton energies ($\varepsilon_p < 20$ MeV) of interest to us. Consequently, one can neglect the second term in the parentheses in formula (5) and obtain a simple and convenient relation for computing the stopping power:

$$B = Z^{-1} \cdot 1.36 \frac{v}{u_0} Z_{\mathrm{av}}^{1/2} = 1.36 \frac{v}{u_0} Z^{1/2} = 1.92 \frac{\varepsilon^{1/2} Z^{1/2} M^{-1/2}}{u_0}. \tag{6}$$

Substituting (6) into (1), we obtain the expression

$$-\frac{d\varepsilon}{dR} = \frac{2\pi z^2 e^4 M^{1/2}}{m u_0} \cdot 1.92 N \varepsilon^{-1/2} Z^{1/2}, \tag{7}$$

which, after substitution of the numerical values appropriate to the proton ($z = 1$), takes the form

$$-\frac{d\varepsilon}{dR} = CNZ^{1/2}\varepsilon^{-1/2}\ (\mathrm{MeV/cm}), \tag{8}$$

where $C = 2.01 \times 10^{-21}$ $\mathrm{MeV}^{3/2}\text{-cm}^2$.

The energy loss in a sufficiently thin layer ΔR (cm) is

$$|\Delta\varepsilon| = CNZ^{1/2}\varepsilon^{-1/2}\, \Delta R\ \ (\mathrm{MeV}). \tag{9}$$

Hence it follows at once that the thickness of an emulsion layer equivalent in energy loss to a target half-thickness $t/2$ (cm) is given by

$$\Delta R_{\mathrm{equ}} = \frac{N_M Z_M^{1/2}}{\sum_i N_{i\,\mathrm{em}} Z_{i\,\mathrm{em}}^{1/2}} \cdot \frac{t}{2}, \tag{10}$$

where Z_M and N_M are the atomic number and the number of atoms per cm³ of the target; and $Z_{i\,\mathrm{em}}$ and $N_{i\,\mathrm{em}}$ are the atomic number and the number of atoms per cm³ of each of the components of the emulsion.

Dividing and multiplying (10) by the total number of atoms per cm³ in the emulsion $N_{\mathrm{em}} = \sum_i N_{i\,\mathrm{em}}$, we obtain

$$\Delta R_{\mathrm{equ}} = \frac{N_M Z_M^{1/2}}{N_{\mathrm{em}} (Z_{\mathrm{em}}^{\mathrm{av}})^{1/2}} \cdot \frac{t}{2}, \tag{11}$$

where $N_{em} = 7.88 \times 10^{22}$ atoms/cm^3; $Z_{em}^{av} = 13.17$, according to [92]; $N_M = N_A\ \rho_M/A_M = 6.02 \times 10^{23}\ \rho_M/A_M$ atoms/cm^3, ρ_M (g/cm^3) is the density and A_M the atomic weight of the target.

After substitution of numerical values for the NIKFI Ya-2 emulsion with a composition like that of the Ilford C-2 emulsion, we obtain

$$\Delta R_{equ} = 2.09\ Z_M^{1/2} \frac{\rho_M}{A_M} \cdot \frac{t}{2}\ (\text{cm}), \qquad (12)$$

if the target half-thickness t/2 is expressed in centimeters, or

$$\Delta R_{equ} = 20.9\ \frac{Z_M^{1/2}}{A_M} \cdot \frac{t}{2}\ (\text{m}\mu), \qquad (13)$$

if t/2 is expressed in milligrams per square centimeter. The constants in formulas (12) and (13) are 6% greater in the case of the T-3 emulsion, which has a lower density.

LITERATURE CITED

1. O. Hirzel and H. Waffler, Helv. Phys. Acta, 20:373 (1947).
2. E. D. Courant, Phys. Rev., 82:703 (1951).
3. P. S. Jensen, Naturwiss., 35:190 (1948).
4. J. S. Levinger and H. A. Bethe, Phys. Rev., 78:115 (1950).
5. S. C. Fultz, R. L. Bramblett, et al., Phys. Rev., 128:2345 (1962).
6. R. M. Osokina and B. S. Ratner, Zh. Eks. i Teor. Fiz., 32:20 (1957).
7. V. V. Akindinov, R. Sh. Amirov, R. M. Osokina, and B. S. Ratner, In: Proceedings 1st All-Union Conference on Nuclear Reactions at Low and Medium Energies (November, 1957). Izd. Akad. Nauk SSSR, Moscow (1958), p. 403.
8. R. M. Osokina and B. S. Ratner, Physica, 22:1147A (1956).
9. N. V. Lin'kova, R. M. Osokina, B. S. Ratner, R. Sh. Amirov, and V. V. Akindinov, Zh. Eks. i Teor. Fiz., 38:780 (1960).
10. R. M. Osokina, In: Proceedings 2nd All-Union Conference on Nuclear Reactions at Low and Medium Energies (July, 1960). Izd. Akad. Nauk SSSR, Moscow (1962), p. 498.
11. R. M. Osokina and V. G. Seryapin, In: Proceedings 2nd All-Union Conference on Nuclear Reactions at Low and Medium Energies (July, 1960). Izd. Akad. Nauk SSSR, Moscow (1962), p. 504.
12. R. M. Osokina, Zh. Eks. i Teor. Fiz., 44:444 (1963).
13. R. M. Osokina, Proc. of the Conference on Direct. Interactions and Nuclear Reaction Mechanisms, Padua, 1962: Gordon Publ., New York (1963), p. 297.
14. E. M. Leikin, R. M. Osokina, and B. S. Ratner, Dokl. Akad. Nauk SSSR, 102:245 (1955).
15. E. M. Leikin, R. M. Osokina, and B. S. Ratner, Dokl. Akad. Nauk SSSR, 102:493 (1955).
16. E. M. Leikin, R. M. Osokina, and B. S. Ratner, Nuovo Cimento, Vol. 3, Suppl. 1, p. 105 (1956).
17. B. I. Goryachev, Atomic Energy Review, Internat. Atom. Energy Agency, Vienna, 2(3) 71 (1964).
18. J. H. Carver and D. C. Peaslee, Phys. Rev., 120:2155 (1960).
19. J. Levinger, Nuclear Photodisintegration. Oxford Univ. Press (1960).
20. D. H. Wilkinson, Ann. Rev. Nucl. Sci., 9:1 (1959).
21. E. G. Fuller and E. Hayward, Phys. Rev., 101:692 (1956).
22. H. Tyren and T. A. J. Maris, Nucl. Phys., 6:82 (1958); 6:446 (1958); 7:24 (1958).
23. K. Okamoto, Progr. Theoret. Phys., 15:75L (1956).
24. O. V. Bogdankevich, B. I. Goryachev, and V. A. Zapevalov, Zh. Eks. i Teor. Fiz., 42:1502 (1962).
25. A. B. Migdal, Zh. Eks. i Teor. Fiz., 15:81 (1945).
26. M. Goldhaber and E. Teller, Phys. Rev., 74:1046 (1948).
27. H. Steinwedel and J. H. D. Jensen, Z. Naturforsch. a., 5:413 (1950).

28. D. H. Wilkinson, Physica, 22:1039 (1956); Proc. Glasgow Conf. Nucl. and Meson Phys. (1955), p. 161.
29. M. Danos, Bull. Amer. Phys. Soc., 1:135A (1956).
30. A. M. Lane, et al., Phys. Rev., 13:281 (1958).
31. D. M. Brink, Nucl. Phys., 4:215 (1957).
32. B. L. Cohen, Phys. Rev., 130:227 (1963).
33. J. P. Elliott and B. H. Flowers, Proc. Phys. Soc., A242:57 (1957).
34. G. Brown and M. Bolsterli, Phys. Rev. Letters, 3:472 (1959).
35. G. Brown, L. Castillejo, and J. A. Evans, Nucl. Phys., 22:1 (1960).
36. G. Brown, J. A. Evans, and D. J. Thounless, Nucl. Phys., 24:1 (1961).
37. V. G. Neudachin, V. G. Shevchenko, and N. P. Yudin, Proc. Internat. Conf. on Nucl. Structure, Canada, 1960. North Holland Publ. Co., (1960), p. 732.
38. V. V. Balashov, V. G. Shevchenko, and N. P. Yudin, Proceedings 2nd All-Union Conference on Nuclear Reactions. Moscow (1960), p. 435.
39. V. V. Balashov, V. G. Shevchenko, and N. P. Yudin, Zh. Eks. i Teor. Fiz., 42:275 (1961).
40. V. V. Balashov, V. G. Shevchenko, and N. P. Yudin, Nucl. Phys., 27:323 (1961).
41. B. S. Ishkhanov, K. V. Shitikova, and V. A. Yur'ev, Izv. Akad. Nauk SSSR, Ser. Fiz., 29:216 (1965).
42. A. B. Migdal, A. A. Lushnikov, and D. F. Zaretski, Nucl. Phys., 66:193 (1965).
43. A. A. Lushnikov and D. F. Zatetski, Nucl. Phys., 66:35 (1965).
44. S. G. Nilsson, J. Sawicki, and N. R. Glendenning, Nucl. Phys., 33:239 (1962).
45. B. Mottelson and S. G. Nilsson, Nucl. Phys., 13:281 (1958).
46. M. Fabry de la Rinella, Nucl. Phys., 27:561 (1961).
47. E. V. Inopin, Zh. Eks. i Teor. Fiz., 38:992 (1960).
48. S. F. Semenko, Nucl. Phys., 37:486 (1962).
49. S. F. Semenko, Zh. Eks. i Teor. Fiz., 43:2188 (1962).
50. M. Danos and W. Greiner, Phys. Rev., 134B:284 (1964).
51. S. F. Semenko, Phys. Letters, 13:157 (1964).
52. A. K. Kerman and Ho Kim Quang, Phys. Rev., 135B:883 (1964).
53. K. Wuldermuth and H. Wittern, Z. Naturforsch. a., 12:39 (1957).
54. M. Danos, Bull. Amer. Phys. Soc., 2:354A (1957).
55. U. L. Businaro and S. Gallone, Nuovo Cimento, 1:1285 (1955).
56. S. Fujii and S. Takagi, Progr. Theoret. Phys., 14:402 (1955).
57. B. S. Dolbilkin, V. I. Korin, et al., Phys. Letters, 17:49 (1965).
58. V. V. Balashov and B. M. Chernov, Zh. Eks. i Teor. Fiz., 43:227 (1962).
59. M. V. Michailovic and M. Rosina, Nucl. Phys., 40:252 (1963).
60. M. Danos and W. Greiner, Phys. Rev., 138B:876 (1965).
61. Yu. V. Orlov, Zh. Eks. i Teor. Fiz., 37:1834 (1959).
62. S. Fujii and O. Sugimoto, Nuovo Cimento, 12:513 (1959).
63. V. Weisskopf, Phys. Rev., 52:295 (1937).
64. V. Weisskopf, Statistical Theory of Nuclear Reactions. [Russian translation], Moscow, IL (1952).
65. L. Wolfenstein, Phys. Rev., 82:629 (1951).
66. V. V. Daragan, Proceedings 1st All-Union Conference on Nuclear Reactions at Low and Medium Energies. (November, 1957). Izd. Akad. Nauk SSSR, Moscow (1962), p. 476.
67. T. Ericson, Advances Phys., 9:425 (1960).
68. T. Ericson, Ann. Phys., 23:390 (1963).
69. T. D. Newton, Canad. J. Phys., 34:804 (1956).
70. D. W. Lang, Nucl. Phys., 26:434 (1961).
71. Y. P. Varshni, Nuovo Cimento, 22:145 (1961).
72. J. A. Evans, Proc. Phys. Soc., A73:33 (1959).
73. D. Bodanskii, Direct Processes in Nuclear Reactions, In: Selected Papers from the Conference at Padua, September 3-8, 1962, p. 77 of Russian translation published by Atomizdat (1965).
74. U. Facchini, Direct Processes in Nuclear Reactions, In: Selected Papers from the Conference at Padua, September 3-8, 1962, p. 89 of Russian translation published by Atomizdat (1965).

75. D. L. Allan, Nucl. Phys., 24:274 (1961).
76. J. Weinstock and E. V. Halpern, Phys. Rev., 94:1651 (1954).
77. P. R. Byerly and W. E. Stephens, Phys. Rev., 83:54 (1951).
78. M. E. Toms and W. E. Stephens, Phys. Rev., 95:1209 (1954).
79. W. A. Butler and G. M. Almy, Phys. Rev., 91:58 (1953).
80. G. N. Zatsepina, L. E. Lazareva, and A. N. Pospelov, Zh. Eks. i Teor. Fiz., 32:27 (1957).
81. M. E. Toms and W. E. Stephens, Phys. Rev., 92:362 (1953); 98:626 (1955).
82. R. Sagane, Phys. Rev., 85:926 (1952).
83. S. N. Goshal, Phys. Rev., 80:939 (1950).
84. F. Ferrero, et al., Nuovo Cimento, 10:423 (1959).
85. V. Emma, et al., Nuovo Cimento, 22:135 (1961).
86. E. S. Anashkina, Zh. Eks. i Teor. Fiz., 43:1197 (1962).
87. W. Bertozzi, F. R. Paolini, and C. P. Sargent, Phys. Rev., 110:790 (1958).
88. G. N. Zatsepina, V. V. Igonin, et al., Proceedings 2nd All-Union Conference on Nuclear Reactions at Low and Medium Energies. (July, 1960). Izd. Akad. Nauk SSSR (1962), p. 479.
89. M. Galloud and G. Haenny, Sci. Inds. Phot., 23(2):221 (1952).
90. M. Galloud and C. Haenny, Mem. Soc. Vand. Sci. Nat., 10:271 (1952).
91. A. M. Goldstein and C. H. Sherman, Rev. Sci. Instr., 23:267 (1952).
92. A. Bonetti, S. Dilworth, S. R. Pelk, and L. Scarci, Nuclear Emulsions [Russian translation], Fizmatgiz (1961).
93. A. D. Bondar', A. S. Emlyaninov, et al., PTÉ, 3:134 (1960).
94. F. A. El Bedewi, Proc. Phys. Soc., A64:581 (1951).
95. A. J. Rotblat, Nature, 167:550 (1951).
96. B. J. Catala and W. M. Gibson, Nature, 167:551 (1950).
97. H. Bradner, et al., Phys. Rev., 77:462 (1950).
98. W. M. Gibson, D. J. Prowse, and A. I. Rotblat, Nature, 173:1181 (1954).
99. E. S. Anashkina, PTÉ, 4:148 (1961).
100. M. E. Toms and J. McElhinney, Phys. Rev., 111:561 (1958).
101. F. Heinrich, H. Wäffler, and M. Wallter, Helv. Phys. Acta, 29:3 (1956).
102. P. Erdos, et al., Helv. Phys. Acta, 30:639 (1957).
103. B. Forkman, Arkiv Fys., 11:265 (1956).
104. J. J. Grant, H. A. Medicus, and P. Demers, Photographic Corpusc. II – Les Presses Univ. de Montreal (1959), p. 255.
105. B. Forkman, Nucl. Phys., 23:269 (1961).
106. L. Katz and A. G. W. Cameron, Canad. J. Phys., 29:518 (1961).
107. Go Tsi-di and B. S. Ratner, Dokl. Akad. Nauk SSSR, 125:761 (1958).
108. H. Yuta and H. Morinaga, Nucl. Phys., 16:119 (1960).
109. R. B. Taylor, Nucl. Phys., 19:453 (1960).
110. J. H. Carver and W. Turchinetz, Proc. Phys., 73A:585 (1959).
111. V. G. Shevchenko and B. A. Yur'ev, Nucl. Phys., 37:495 (1962).
112. G. A. Price and D. W. Kerst, Phys. Rev., 77:806 (1950).
113. R. Montalbetti, L. Katz, and J. Goldemberg, Phys. Rev., 91:659 (1953).
114. R. Nathans and J. Halpern, Phys. Rev., 93:437 (1950).
115. L. J. Schiff, Phys. Rev., 83:252 (1951).
116. P. F. Yergin and B. P. Fabricand, Phys. Rev., 104:1334 (1956).
117. L. Katz and J. Chidley, Proceedings 2nd All-Union Conference on Nuclear Reactions at Low and Medium Energies. (July, 1960). Izd. Akad. Nauk SSSR (1962), p. 371.
118. B. I. Gabrilov and L. E. Lazareva, Zh. Eks. i Teor. Fiz., 30:855 (1956).
119. Go Tsi-di, B. S. Ratner, and B. S. Sergeev, Zh. Eks. i Teor. Fiz., 40:85 (1961).
120. E. G. Fuller and M. S. Weiss, Phys. Rev., 112:560 (1958).
121. L. Katz and A. Cameron, Canad. J. Phys., 29:518 (1951).

122. J. Miller, C. Schuhl, and C. Tzara, Nucl. Phys., 32:236 (1962).
123. J. Blatt and V. Weisskopf, Theoretical Nuclear Physics. Wiley, New York (1952).
124. M. M. Shapiro, Phys. Rev., 90:171 (1953).
125. J. M. C. Scott, Phil. Mag., Ser. 7, 45:441 (1954).
126. J. Dostrovsky and Z. Fraenkel, Phys. Rev., 116:683 (1959).
127. K. S. Quisenberry, et al., Phys. Rev., 104:461 (1956).
128. W. E. Ogle and R. E. England, Phys. Rev., 78:63 (1950).
129. R. Sher, J. Halpern, and W. E. Stephens, Phys. Rev., 81:1541 (1951).
130. R. Sher, J. Halpern, and A. K. Mann, Phys. Rev., 84:387 (1951).
131. K. N. Geller, J. Halpern, and E. G.Muirhead, Phys. Rev., 118:1302 (1960).
132. N. Mutsuro and J. Ohnuki, Phys. Soc. Japan, 14:1469 (1959).
133. B. G. Chidley, L. Katz, and S. Kowalsky, Canad. J. Phys., 36:407 (1958).
134. A. G. N. Cameron, AECL Report CRP-690, Chalk River, Ontario (1957).
135. D. Menzel, Ed., Fundamental Formulas of Physics. Dover, New York (1959).
136. D. Stominger, I. M. Hollander, and G. T. Seaborg, Rev. Mod. Phys., 30:585 (1958).
137. J. P. Hummel, Phys. Rev., 123:950 (1961).
138. A. G. W. Cameron, Canad. J. Phys., 36:1040 (1958).
139. R. G. Baker and K. G. McNeill, Canad. J. Phys., 39:1158 (1961).
140. O. M. M. Mitchell and K. G. McNeill, Canad. J. Phys., 41:871 (1963).
141. B. S. Diven and G. M. Almy, Phys. Rev., 80:407 (1950).
142. S. A. E. Johansson and B. Forkman, Nucl. Phys., 36:141 (1962).
143. A. K. Mann, J. Halpern, and M. Rothman, Phys. Rev., 87:146 (1952).
144. J. Eichler and H. A. Weidenmuller, Z. Physik, 152:261 (1958).
145. S. V. Starodubtsev and A. M. Romanov, Penetration of Charged Particles through Matter. Izd. Akad. Nauk UzSSR, Tashkent (1962).
146. I. Lindhard and M. Scharff, Dan. Mat.-Fys. Medd., 27:15 (1953).

Printed in the United States
By Bookmasters